DIGITAL RADIO SYSTEM DESIGN

DIGITAL RADIO SYSTEM DESIGN

Grigorios Kalivas

University of Patras, Greece

A John Wiley and Sons, Ltd, Publication

This edition first published 2009

© 2009 John Wiley & Sons Ltd.,

Registered office

John Wiley & Sons Ltd, The Atrium, Southern Gate, Chichester, West Sussex, PO19 8SQ, United Kingdom

For details of our global editorial offices, for customer services and for information about how to apply for permission to reuse the copyright material in this book please see our website at www.wiley.com.

Library of Congress Cataloging-in-Publication Data

Kalivas, Grigorios.
    Digital radio system design / Grigorios Kalivas.
      p. cm.
    Includes bibliographical references and index.
    ISBN 978-0-470-84709-1 (cloth)
      1. Radio—Transmitter-receivers—Design and construction. 2. Digital communications—Equipment and supplies—Design and construction. 3. Radio circuits—Design and construction. 4. Signal processing—Digital techniques. 5. Wireless communication systems—Equipment and supplies—Design and construction. I. Title.
    TK6553.K262 2009
    621.384′131—dc22
                                               2009015936

A catalogue record for this book is available from the British Library.

ISBN 9780470847091 (H/B)

Set in 10/12 Times Roman by Macmillan Typesetting

Printed in Singapore by Markono Print Media Pte Ltd

To Stella, Maria and Dimitra and to the memory of my father

Contents

Preface

Radio communications is a field touching upon various scientific and engineering disciplines. From cellular radio, wireless networking and broadband indoor and outdoor radio to electronic surveillance, deep space communications and electronic warfare. All these applications are based on radio electronic systems designed to meet a variety of requirements concerning reliable communication of information such as voice, data and multimedia. Furthermore, the continuous demand for quality of communication and increased efficiency imposes the use of digital modulation techniques in radio transmission systems and has made it the dominant approach in system design. Consequently, the complete system consists of a radio transmitter and receiver (front-end) and a digital modulator and demodulator (modem).

This book aims to introduce the reader to the basic principles of radio systems by elaborating on the design of front-end subsystems and circuits as well as digital transmitter and receiver sections.

To be able to handle the complete transceiver, the electronics engineer must be familiar with diverse electrical engineering fields like digital communications and RF electronics. The main feature of this book is that it tries to accomplish such a demanding task by introducing the reader to both digital modem principles and RF front-end subsystem and circuit design. Furthermore, for effective system design it is necessary to understand concepts and factors that mainly characterize and impact radio transmission and reception such as the radio channel, noise and distortion. Although the book tackles such diverse fields, it treats them in sufficient depth to allow the designer to have a solid understanding and make use of related issues for design purposes.

Recent advancements in digital processing technology made the application of advanced schemes (like turbo coding) and transmission techniques like diversity, orthogonal frequency division multiplexing and spread spectrum very attractive to apply in modern receiver systems.

Apart from understanding the areas of digital communications and radio electronics, the designer must also be able to evaluate the impact of the characteristics and limitations of the specific radio circuits and subsystems on the overall RF front-end system performance. In addition, the designer must match a link budget analysis to specific digital modulation/transmission techniques and RF front-end performance while at the same time taking into account aspects that interrelate the performance of the digital modem with the characteristics of the RF front-end. Such aspects include implementation losses imposed by transmitter–receiver nonidealities (like phase noise, power amplifier nonlinearities, quadrature mixer imbalances) and the requirements and restrictions on receiver synchronization subsystems.

This book is intended for engineers working on radio system design who must account for every factor in system and circuit design to produce a detailed high-level design of the required system. For this reason, the designer must have an overall and in-depth understanding of a variety of concepts from radio channel characteristics and digital modem principles to silicon technology and RF circuit configuration for low noise and low distortion design. In addition, the book is well suited for graduate students who study transmitter/receiver system design as it presents much information involving the complete transceiver chain in adequate depth that can be very useful to connect the diverse fields of digital communications and RF electronics in a unified system concept.

To complete this book several people have helped in various ways. First of all I am indebted to my colleagues Dimitrios Toumpakaris and Konstantinos Efstathiou for reading in detail parts of the manuscript and providing me with valuable suggestions which helped me improve it on various levels.

Further valuable help came from my graduate and ex-graduate students Athanasios Doukas, Christos Thomos and Dr Fotis Plessas, who helped me greatly with the figures. Special thanks belong to Christos Thomos, who has helped me substantially during the last crucial months on many levels (proof-reading, figure corrections, table of contents, index preparation etc.).

1

Radio Communications: System Concepts, Propagation and Noise

A critical point for the development of radio communications and related applications was the invention of the 'super-heterodyne' receiver by Armstrong in 1917. This system was used to receive and demodulate radio signals by down-converting them in a lower intermediate frequency (IF). The demodulator followed the IF amplification and filtering stages and was used to extract the transmitted voice signal from a weak signal impaired by additive noise. The super-heterodyne receiver was quickly improved to demodulate satisfactorily very weak signals buried in noise (high sensitivity) and, at the same time, to be able to distinguish the useful signals from others residing in neighbouring frequencies (good selectivity). These two properties made possible the development of low-cost radio transceivers for a variety of applications. AM and FM radio were among the first popular applications of radio communications. In a few decades packet radios and networks targeting military communications gained increasing interest. Satellite and deep-space communications gave the opportunity to develop very sophisticated radio equipment during the 1960s and 1970s. In the early 1990s, cellular communications and wireless networking motivated a very rapid development of low-cost, low-power radios which initiated the enormous growth of wireless communications.

The biggest development effort was the cellular telephone network. Since the early 1960s there had been a considerable research effort by the AT&T Bell Laboratories to develop a cellular communication system. By the end of the 1970s the system had been tested in the field and at the beginning of the 1980s the first commercial cellular systems appeared. The increasing demand for higher capacity, low cost, performance and efficiency led to the second generation of cellular communication systems in the 1990s. To fulfill the need for high-quality bandwidth-demanding applications like data transmission, Internet, web browsing and video transmission, 2.5G and 3G systems appeared 10 years later.

Along with digital cellular systems, wireless networking and wireless local area networks (WLAN) technology emerged. The need to achieve improved performance in a harsh propagation environment like the radio channel led to improved transmission technologies like spread spectrum and orthogonal frequency division multiplexing (OFDM). These technologies were

Digital Radio System Design   Grigorios Kalivas
© 2009 John Wiley & Sons, Ltd

put to practice in 3G systems like wideband code-division multiple access (WCDMA) as well as in high-speed WLAN like IEEE 802.11a/b/g.

Different types of digital radio system have been developed during the last decade that are finding application in wireless personal area networks (WPANs). These are Bluetooth and Zigbee, which are used to realize wireless connectivity of personal devices and home appliances like cellular devices and PCs. Additionally, they are also suitable for implementing wireless sensor networks (WSNs) that organize in an ad-hoc fashion. In all these, the emphasis is mainly on short ranges, low transmission rates and low power consumption.

Finally, satellite systems are being constantly developed to deliver high-quality digital video and audio to subscribers all over the world.

The aims of this chapter are twofold. The first is to introduce the variety of digital radio systems and their applications along with fundamental concepts and challenges of the basic radio transceiver blocks (the radio frequency, RF, front-end and baseband parts). The second is to introduce the reader to the technical background necessary to address the main objective of the book, which is the design of RF and baseband transmitters and receivers. For this purpose we present the basic concepts of linear systems, stochastic processes, radio propagation and channel models. Along with these we present in some detail the basic limitations of radio electronic systems and circuits, noise and nonlinearities. Finally, we introduce one of the most frequently used blocks of radio systems, the phase-locked loop (PLL), which finds applications in a variety of subsystems in a transmitter/receiver chain, such as the local oscillator, the carrier recovery and synchronization, and coherent detection.

1.1 Digital Radio Systems and Wireless Applications

The existence of a large number of wireless systems for multiple applications considerably complicates the allocation of frequency bands to specific standards and applications across the electromagnetic spectrum. In addition, a number of radio systems (WLAN, WPAN, etc.) operating in unlicensed portions of the spectrum demand careful assignment of frequency bands and permitted levels of transmitted power in order to minimize interference and permit the coexistence of more than one radio system in overlapping or neighbouring frequency bands in the same geographical area.

Below we present briefly most of the existing radio communication systems, giving some information on the architectures, frequency bands, main characteristics and applications of each one of them.

1.1.1 Cellular Radio Systems

A cellular system is organized in hexagonal cells in order to provide sufficient radio coverage to mobile users moving across the cell. A base station (BS) is usually placed at the centre of the cell for that purpose. Depending on the environment (rural or urban), the areas of the cells differ. Base stations are interconnected through a high-speed wired communications infrastructure. Mobile users can have an uninterrupted session while moving through different cells. This is achieved by the MTSOs acting as network controllers of allocated radio resources (physical channels and bandwidth) to mobile users through the BS. In addition, MTSOs are responsible for routing all calls associated with mobile users in their area.

Second-generation (2G) mobile communications employed digital technology to reduce cost and increase performance. Global system for mobile communications (GSM) is a very successful 2G system that was developed and deployed in Europe. It employs Gaussian minimum shift keying (MSK) modulation, which is a form of continuous-phase phase shift keying (PSK). The access technique is based on time-division multiple access (TDMA) combined with slow frequency hopping (FH). The channel bandwidth is 200 kHz to allow for voice and data transmission.

IS-95 (Interim standard-95) is a popular digital cellular standard deployed in the USA using CDMA access technology and binary phase-shift keying (BPSK) modulation with 1.25 MHz channel bandwidth. In addition, IS-136 (North American Digital Cellular, NADC) is another standard deployed in North America. It utilizes 30 kHz channels and TDMA access technology.

2.5G cellular communication emerged from 2G because of the need for higher transmission rates to support Internet applications, e-mail and web browsing. General Packet Radio Service (GPRS) and Enhanced Data Rates for GSM Evolution (EGDE) are the two standards designed as upgrades to 2G GSM. GPRS is designed to implement packet-oriented communication and can perform network sharing for multiple users, assigning time slots and radio channels [Rappaport02]. In doing so, GPRS can support data transmission of 21.4 kb/s for each of the eight GSM time slots. One user can use all of the time slots to achieve a gross bit rate of $21.4 \times 8 = 171.2$ kb/s.

EDGE is another upgrade of the GSM standard. It is superior to GPRS in that it can operate using nine different formats in air interface [Rappaport02]. This allows the system to choose the type and quality of error control. EDGE uses 8-PSK modulation and can achieve a maximum throughput of 547.2 kb/s when all eight time slots are assigned to a single user and no redundancy is reserved for error protection. 3G cellular systems are envisaged to offer high-speed wireless connectivity to implement fast Internet access, Voice-over-Internet Protocol, interactive web connections and high-quality, real-time data transfer (for example music).

UMTS (Universal Mobile Telecommunications System) is an air interface specified in the late 1990s by ETSI (European Telecommunications Standards Institute) and employs WCDMA, considered one of the more advanced radio access technologies. Because of the nature of CDMA, the radio channel resources are not divided, but they are shared by all users. For that reason, CDMA is superior to TDMA in terms of capacity. Furthermore, each user employs a unique spreading code which is multiplied by the useful signal in order to distinguish the users and prevent interference among them. WCDMA has 5 MHz radio channels carrying data rates up to 2 Mb/s. Each 5 MHz channel can offer up to 350 voice channels [Rappaport02].

1.1.2 Short- and Medium-range Wireless Systems

The common characteristic of these systems is the range of operation, which is on the order of 100 m for indoor coverage and 150–250 m for outdoor communications. These systems are mostly consumer products and therefore the main objectives are low prices and low energy consumption.

1.1.2.1 Wireless Local Area Networks

Wireless LANs were designed to provide high-data-rate, high-performance wireless connectivity within a short range in the form of a network controlled by a number of central points (called access points or base stations). Access points are used to implement communication between two users by serving as up-link receivers and down-link transmitters. The geographical area of operation is usually confined to a few square kilometres. For example, a WLAN can be deployed in a university campus, a hospital or an airport.

The second and third generation WLANs proved to be the most successful technologies. IEEE 802.11b (second generation) operates in the 2.4 GHz ISM (Industral, Scientific and Medical) band within a spectrum of 80 MHz. It uses direct sequence spread spectrum (DSSS) transmission technology with gross bit rates of 1, 2, 5 and 11 Mb/s. The 11 Mb/s data rate was adopted in late 1998 and modulates data by using complementary code keying (CCK) to increase the previous transmission rates. The network can be formulated as a centralized network using a number of access points. However, it can also accommodate peer-to-peer connections.

The IEEE 802.11a standard was developed as the third-generation WLAN and was designed to provide even higher bit rates (up to 54 Mb/s). It uses OFDM transmission technology and operates in the 5 GHz ISM band. In the USA, the Federal Communications Commission (FCC) allocated two bands each 100 MHz wide (5.15–5.25 and 5.25–5.35 GHz), and a third one at 5.725–5.825 GHz for operation of 802.11a. In Europe, HIPERLAN 2 was specified as the standard for 2G WLAN. Its physical layer is very similar to that of IEEE 802.11a. However, it uses TDMA for radio access instead of the CSMA/CA used in 802.11a.

The next step was to introduce the 802.11g, which mostly consisted of a physical layer specification at 2.4 GHz with data rates matching those of 802.11a (up to 54 Mb/s). To achieve that, OFDM transmission was set as a compulsory requirement. 802.11g is backward-compatible to 802.11b and has an extended coverage range compared with 802.11a. To cope with issues of quality of service, 802.11e was introduced, which specifies advanced MAC techniques to achieve this.

1.1.2.2 WPANs and WSNs

In contrast to wireless LANs, WPAN standardization efforts focused primarily on lower transmission rates with shorter coverage and emphasis on low power consumption. Bluetooth (IEEE 802.15.1), ZigBee (IEEE 802.15.4) and UWB (IEEE 802.15.3) represent standards designed for personal area networking. Bluetooth is an open standard designed for wireless data transfer for devices located a few metres apart. Consequently, the dominant application is the wireless interconnection of personal devices like cellular phones, PCs and their peripherals. Bluetooth operates in the 2.4 GHz ISM band and supports data and voice traffic with data rates of 780 kb/s. It uses FH as an access technique. It hops in a pseudorandom fashion, changing frequency carrier 1600 times per second (1600 hops/s). It can hop to 80 different frequency carriers located 1 MHz apart. Bluetooth devices are organized in groups of two to eight devices (one of which is a master) constituting a piconet. Each device of a piconet has an identity (device address) that must be known to all members of the piconet. The standard specifies two modes of operation: asynchronous connectionless (ACL) in one channel (used for data transfer at 723 kb/s) and synchronous connection-oriented (SCO) for voice communication (employing three channels at 64 kb/s each).

A scaled-down version of Bluetooth is ZigBee, operating on the same ISM band. Moreover, the 868/900 MHz band is used for ZigBee in Europe and North America. It supports transmission rates of up to 250 kb/s covering a range of 30 m.

During the last decade, WSNs have emerged as a new field for applications of low-power radio technology. In WSN, radio modules are interconnected, formulating ad-hoc networks.

WSN find many applications in the commercial, military and security sectors. Such applications concern home and factory automation, monitoring, surveillance, etc. In this case, emphasis is given to implementing a complete stack for ad hoc networking. An important feature in such networks is multihop routing, according to which information travels through the network by using intermediate nodes between the transmitter and the receiver to facilitate reliable communication. Both Bluetooth and ZigBee platforms are suitable for WSN implementation [Zhang05], [Wheeler07] as they combine low-power operation with network formation capability.

1.1.2.3 Cordless Telephony

Cordless telephony was developed to satisfy the needs for wireless connectivity to the public telephone network (PTN). It consists of one or more base stations communicating with one or more wireless handsets. The base stations are connected to the PTN through wireline and are able to provide coverage of approximately 100 m in their communication with the handsets. CT-2 is a second-generation cordless phone system developed in the 1990s with extended range of operation beyond the home or office premises.

On the other hand, DECT (Digital European Cordless Telecommunications) was developed such that it can support local mobility in an office building through a private branch exchange (PBX) system. In this way, hand-off is supported between the different areas covered by the base stations. The DECT standard operates in the 1900 MHz frequency band. Personal handyphone system (PHS) is a more advanced cordless phone system developed in Japan which can support both voice and data transmission.

1.1.2.4 Ultra-wideband Communications

A few years ago, a spectrum of 7.5 GHz (3.1–10.6 GHz) was given for operation of ultra-wideband (UWB) radio systems. The FCC permitted very low transmitted power, because the wide area of operation of UWB would produce interference to most commercial and even military wireless systems. There are two technology directions for UWB development. Pulsed ultra-wideband systems (P-UWB) convey information by transmitting very short pulses (of duration in the order of 1 ns). On the other hand, multiband-OFDM UWB (MB-OFDM) transmits information using the OFDM transmission technique.

P-UWB uses BPSK, pulse position modulation (PPM) and amplitude-shift keying (ASK) modulation and it needs a RAKE receiver (a special type of receiver used in Spread Spectrum systems) to combine energy from multipath in order to achieve satisfactory performance. For very high bit rates (on the order of 500 Mb/s) sophisticated RAKE receivers must be employed, increasing the complexity of the system. On the other hand, MB-UWB uses OFDM technology to eliminate intersymbol interference (ISI) created by high transmission rates and the frequency selectivity of the radio channel.

Ultra-wideband technology can cover a variety of applications ranging from low-bit-rate, low-power sensor networks to very high transmission rate (over 100 Mb/s) systems designed to wirelessly interconnect home appliances (TV, PCs and consumer electronic appliances). The low bit rate systems are suitable for WSN applications.

P-UWB is supported by the UWB Forum, which has more than 200 members and focuses on applications related to wireless video transfer within the home (multimedia, set-top boxes,

DVD players). MB-UWB is supported by WiMedia Alliance, also with more than 200 members. WiMedia targets applications related to consumer electronics networking (PCs TV, cellular phones). UWB Forum will offer operation at maximum data rates of 1.35 Gb/s covering distances of 3 m [Geer06]. On the other hand, WiMedia Alliance will provide 480 Mb/s at distances of 10 m.

1.1.3 Broadband Wireless Access

Broadband wireless can deliver high-data-rate wireless access (on the order of hundreds of Mb/s) to fixed access points which in turn distribute it in a local premises. Business and residential premises are served by a backbone switch connected at the fixed access point and receive broadband services in the form of local area networking and video broadcasting.

LMDS (local multipoint distribution system) and MMDS (multichannel multipoint distribution services) are two systems deployed in the USA operating in the 28 and 2.65 GHz bands. LMDS occupies 1300 MHz bandwidth in three different bands around 28, 29 and 321 GHz and aims to provide high-speed data services, whereas MMDS mostly provides telecommunications services [Goldsmith05] (hundreds of digital television channels and digital telephony). HIPERACCESS is the European standard corresponding to MMDS.

On the other hand, 802.16 standard is being developed to specify fixed and mobile broadband wireless access with high data rates and range of a few kilometres. It is specified to offer 40 Mb/s for fixed and 15 Mb/s for mobile users. Known as WiMAX, it aims to deliver multiple services in long ranges by providing communication robustness, quality of service (QoS) and high capacity, serving as the 'last mile' wireless communications. In that capacity, it can complement WLAN and cellular access. In the physical layer it is specified to operate in bands within the 2–11 GHz frequency range and uses OFDM transmission technology combined with adaptive modulation. In addition, it can integrate multiple antenna and smart antenna techniques.

1.1.4 Satellite Communications

Satellite systems are mostly used to implement broadcasting services with emphasis on high-quality digital video and audio applications (DVB, DAB). The Digital Video Broadcasting (DVB) project specified the first DVB-satellite standard (DVB-S) in 1994 and developed the second-generation standard (DVB-S2) for broadband services in 2003. DVB-S3 is specified to deliver high-quality video operating in the 10.7–12.75 GHz band. The high data rates specified by the standard can accommodate up to eight standard TV channels per transponder. In addition to standard TV, DVB-S provides HDTV services and is specified for high-speed Internet services over satellite.

In addition to DVB, new-generation broadband satellite communications have been developed to support high-data-rate applications and multimedia in the framework of fourth-generation mobile communication systems [Ibnkahla04].

Direct-to-Home (DTH) satellite systems are used in North America and constitute two branches: the Broadcasting Satellite Service (BSS) and the Fixed Satellite Service (FSS). BSS operates at 17.3–17.8 GHz (uplink) and 12.2–12.7 GHz (downlink), whereas the bands for FSS are 14–14.5 and 10.7–11.2 GHz, respectively.

Finally, GPS (global positioning satellite) is an ever increasing market for providing localization services (location finding, navigation) and operates using DSSS in the 1500 MHz band.

1.2 Physical Layer of Digital Radio Systems

Radio receivers consist of an RF front-end, a possible IF stage and the baseband platform which is responsible for the detection of the received signal after its conversion from analogue to digital through an A/D converter. Similarly, on the transmitter side, the information signal is digitally modulated and up-converted to a radio-frequency band for subsequent transmission.

In the next section we use the term 'radio platform' to loosely identify all the RF and analogue sections of the transmitter and the receiver.

1.2.1 Radio Platform

Considering the radio receiver, the main architectures are the super-heterodyne (SHR) and the direct conversion receiver (DCR). These architectures are examined in detail in Chapter 3, but here we give some distinguishing characteristics as well as their main advantages and disadvantages in the context of some popular applications of radio system design. Figure 1.1 illustrates the general structure of a radio transceiver. The SHR architecture involves a mixing stage just after the low-noise amplifier (LNA) at the receiver or prior to the transmitting medium-power and high-power amplifiers (HPA). Following this stage, there is quadrature mixing bringing the received signal down to the baseband. Following mixers, there is variable gain amplification and filtering to increase the dynamic range (DR) and at the same time improve selectivity.

When the local oscillator (LO) frequency is set equal to the RF input frequency, the received signal is translated directly down to the baseband. The receiver designed following this approach is called Direct conversion Receiver or zero-IF receiver. Such an architecture eliminates the IF and the corresponding IF stage at the receiver, resulting in less hardware but, as we will see in Chapter 3, it introduces several shortcomings that can be eliminated with careful design.

Comparing the two architectures, SHR is advantageous when a very high dynamic range is required (as for example in GSM). In this case, by using more than one mixing stage, amplifiers with variable gain are inserted between stages to increase DR. At the same time, filtering inserted between two mixing stages becomes narrower, resulting in better selectivity [Schreir02].

Furthermore, super-heterodyne can be advantageous compared with DCR when large in-band blocking signals have to be eliminated. In DCR, direct conversion (DC) offset would change between bursts, requiring its dynamic control [Tolson99].

Regarding amplitude and phase imbalances of the two branches, In-phase (I-phase) and Q-phase considerably reduce the image rejection in SHR. In applications where there can be no limit to the power of the neighbouring channels (like the ISM band), it is necessary to have an image rejection (IR) on the order of 60 dB. SHR can cope with the problem by suitable choice of IF frequencies [Copani05]. At the same time, more than one down-converting stage relaxes the corresponding IR requirements. On the other hand, there is no image band in DCR and hence no problem associated with it. However, in DCR, imbalances at the I–Q

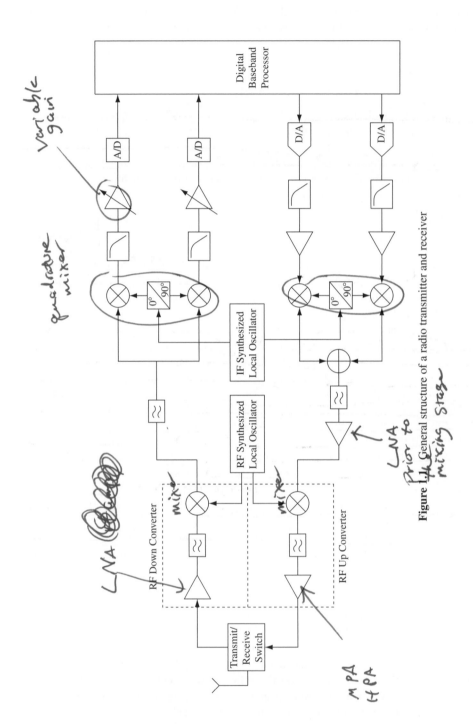

Figure 1.4 General structure of a radio transmitter and receiver

mixer create problems from the self-image and slightly deteriorate the receiver signal-to-noise ratio (SNR) [Razavi97]. This becomes more profound in high-order modulation constellations (64-QAM, 256-QAM, etc.)

On the other hand, DCR is preferred when implementation cost and high integration are the most important factors. For example, 3G terminals and multimode transceivers frequently employ the direct conversion architecture. DC offset and $1/f$ noise close to the carrier are the most frequent deficiencies of homodyne receivers, as presented in detail in Chapter 3. Furthermore, second-order nonlinearities can also create a problem at DC. However, digital and analogue processing techniques can be used to eliminate these problems.

Considering all the above and from modern transceiver design experience, SHR is favoured in GSM, satellite and millimetre wave receivers, etc. On the other hand, DCR is favoured in 3G terminals, Bluetooth and wideband systems like WCDMA, 802.11a/b/g, 802.16 and UWB.

1.2.2 Baseband Platform

The advent of digital signal processors (DSP) and field-programmable gate arrays (FPGAs), dramatically facilitated the design and implementation of very sophisticated digital demodulators and detectors for narrowband and wideband wireless systems. 2G cellular radio uses GMSK, a special form of continuous-phase frequency-shift keying (CPFSK). Gaussian minimum-shift keying (GMSK) modem (modulator–demodulator) implementation can be fully digital and can be based on simple processing blocks like accumulators, correlators and look-up tables (LUTs) [Wu00], [Zervas01]. FIR (Finite Impulse Response) filters are always used to implement various forms of matched filters. Coherent demodulation in modulations with memory could use more complex sequential receivers implementing the Viterbi algorithm.

3G cellular radios and modern WLAN transceivers employ advanced transmission techniques using either spread spectrum or OFDM to increase performance. Spread spectrum entails multiplication of the information sequence by a high-bit-rate pseudorandom noise (PN) sequence operating at speeds which are multiples of the information rate. The multiple bandwidth of the PN sequence spreads information and narrowband interference to a band with a width equal to that of the PN sequence. Suitable synchronization at the receiver restores information at its original narrow bandwidth, but interference remains spread due to lack of synchronization. Consequently, passing the received signal plus spread interference through a narrow band filter corresponding to the information bandwidth reduces interference considerably. In a similar fashion, this technique provides multipath diversity at the receiver, permitting the collection and subsequent constructive combining of the main and the reflected signal components arriving at the receiver. This corresponds to the RAKE receiver principle, resembling a garden rake that is used to collect leaves. As an example, RAKE receivers were used to cope with moderate delay spread and moderate bit rates (60 ns at the rate of 11 Mb/s [VanNee99]. To face large delay spreads at higher transmission rates, the RAKE receiver was combined with equalization. On the other hand, OFDM divides the transmission bandwidth into many subchannels, each one occupying a narrow bandwidth. In this way, owing to the increase in symbol duration, the effect of dispersion in time of the reflected signal on the receiver is minimized. The effect of ISI is completely eliminated by inserting a guard band in the resulting composite OFDM symbol. Fast Fourier transform (FFT) is an efficient way to produce (in the digital domain) the required subcarriers over which the information will be embedded. In practice, OFDM is used in third-generation WLANs, WiMAX and DVB to eliminate ISI.

From the above discussion it is understood that, in modern 3G and WLAN radios, advanced digital processing is required to implement the modem functions which incorporate transmission techniques like spread spectrum and OFDM. This can be performed using DSPs [Jo04], FPGAs [Chugh05], application-specific integrated circuits (ASICs) or a combination of them all [Jo04].

1.2.3 Implementation Challenges

Many challenges to the design and development of digital radio systems come from the necessity to utilize the latest process technologies (like deep submicron complementary metal-oxide semiconductor, CMOS, processes) in order to save on chip area and power consumption. Another equally important factor has to do with the necessity to develop multistandard and multimode radios capable of implementing two or more standards (or more than one mode of the same standard) in one system. For example, very frequently a single radio includes GSM/GPRS and Bluetooth. In this case, the focus is on reconfigurable radio systems targeting small, low-power-consumption solutions.

Regarding the radio front-end and related to the advances in process technology, some technical challenges include:

- reduction of the supply voltage while dynamic range is kept high [Muhammad05];
- elimination of problems associated with integration-efficient architectures like the direct conversion receiver; such problems include DC offset, $1/f$ noise and second order nonlinearities;
- low-phase-noise local oscillators to accommodate for broadband and multistandard system applications;
- wideband passive and active components (filters and low-noise amplifiers) just after the antenna to accommodate for multistandard and multimode systems as well as for emerging ultrawideband receivers;

For all the above RF front-end-related issues a common target is to minimize energy dissipation.

Regarding the baseband section of the receiver, reconfigurability poses considerable challenges as it requires implementation of multiple computationally intensive functions (like FFT, spreading, despreading and synchronization and decoding) in order to:

- perform hardware/software partition that results in the best possible use of platform resources;
- define the architecture based on the nature of processing; for example, parallel and computationally intensive processing vs algorithmic symbol-level processing [Hawwar06];
- implement the multiple functionalities of the physical layer, which can include several kinds of physical channels (like dedicated channels or synchronization channels), power control and parameter monitoring by measurement (e.g. BER, SNR, signal-to-interference ratio, SIR).

The common aspect of all the above baseband-related problems is to design the digital platform such that partition of the functionalities in DSP, FPGAs and ASICs is implemented in the most efficient way.

1.3 Linear Systems and Random Processes

1.3.1 Linear Systems and Expansion of Signals in Orthogonal Basis Functions

A periodic signal $s(t)$ of bandwidth B_S can be fully reproduced by N samples per period T, spaced $1/(2B_S)$ seconds apart (Nyquist's theorem). Hence, $s(t)$ can be represented by a vector of dimension $N = 2B_S T$. Consequently, most of the properties of vector spaces are true for time waveforms like $s(t)$. Hence, we can define the inner product of two signals $s(t)$ and $y(t)$ in an interval $[c_1, c_2]$ as:

$$\langle s(t), y(t) \rangle = \int_{c_1}^{c_2} s(t) y^*(t) dt \tag{1.1}$$

Using this, a group of signals $\psi_n(t)$ is defined as orthonormal basis if the following is satisfied:

$$\langle \psi_n(t), \psi_m(t) \rangle = \delta_{mn} = \begin{cases} 1, & m = n \\ 0, & m \neq n \end{cases} \tag{1.2}$$

This is used to expand all signals $\{s_m(t), \ m = 1, 2, \ldots M\}$ in terms of the functions $\psi_n(t)$. In this case, $\psi_n(t)$ is defined as a complete basis for $\{s_m(t), \ m = 1, 2, \ldots M\}$ and we have:

$$s_m(t) = \sum_k s_{mk} \psi_k(t), \quad s_{mk} = \int_0^T s_m(t) \psi_k(t) dt \tag{1.3}$$

This kind of expansion is of great importance in digital communications because the group $\{s_m(t), \ m = 1, 2, \ldots M\}$ represents all possible transmitted waveforms in a transmission system.

Furthermore, if $y_m(t)$ is the output of a linear system with input $s_m(t)$ which performs operation $H[\cdot]$, then we have:

$$y_m(t) = H[s_m(t)] = \sum_k s_{mk} H[\psi_k(t)] \tag{1.4}$$

The above expression provides an easy way to find the response of a system by determining the response of it, when the basis functions $\psi_n(t)$ are used as inputs.

In our case the system is the composite transmitter–receiver system with an overall impulse response $h(t)$ constituting, in most cases, the cascading three filters, the transmitter filter, the channel response and the receiver filter. Hence the received signal will be expressed as the following convolution:

$$y_m(t) = s_m(t) * h(t) \tag{1.5}$$

For example, as shown in Section 2.1, in an ideal system where the transmitted signal is only corrupted by noise we have:

$$r(t) = s_m(t) + n(t), \quad r_k = \int_0^T r(t) \psi_k(t) dt = s_{mk} + n_k \tag{1.6}$$

Based on the orthonormal expansion

$$s_m(t) = \sum_k s_{mk} \psi_k(t)$$

of signal $s_m(t)$ as presented above, it can be shown [Proakis02] that the power content of a periodic signal can be determined by the summation of the power of its constituent harmonics. This is known as the Parseval relation and is mathematically expressed as follows:

$$\frac{1}{T_0}\int_{t_C}^{t_C+T_0}|s_m(t)|^2\mathrm{d}t = \sum_{k=-\infty}^{+\infty}|s_{mk}|^2 \tag{1.7}$$

1.3.2 Random Processes

Figure 1.2 shows an example for a random process $X(t)$ consisting of sample functions $x_{Ei}(t)$. Since, as explained above, the random process at a specific time instant t_C corresponds to a random variable, the mean (or expectation function) and autocorrelation function can be defined as follows:

$$E\{X(t_C)\} = m_X(t_C) = \int_{-\infty}^{\infty} x p_{X(t_C)}(x)\mathrm{d}x \tag{1.8}$$

$$R_{XX}(t_1, t_2) = E[X(t_1)X(t_2)] = \int_{-\infty}^{+\infty}\int_{-\infty}^{+\infty} x_1 x_2 p_{X(t_1)X(t_2)}(x_1, x_2)\mathrm{d}x_1\mathrm{d}x_2 \tag{1.9}$$

Wide-sense stationary (WSS) is a process for which the mean is independent of t and its autocorrelation is a function of the time difference $t_1 - t_2 = \tau$ and not of the specific values of t_1 and t_2 $[R_X(t_1 - t_2) = R_X(\tau)]$.

A random process is stationary if its statistical properties do not depend on time. Stationarity is a stronger property compared with wide-sense stationarity.

Two important properties of the autocorrelation function of stationary processes are:

(1) $R_X(-\tau) = R_X(\tau)$, which means that it is an even function;
(2) $R_X(\tau)$ has a maximum absolute value at $\tau = 0$, i.e. $|R_X(\tau)| \le R_X(0)$.

Ergodicity is a very useful concept in random signal analysis. A stationary process is ergodic if, for all outcomes E_i and for all functions $f(x)$, the statistical averages are equal to time averages:

$$E\{f[X(t)]\} = \lim_{T\to\infty}\frac{1}{T}\int_{-T/2}^{T/2} f[x_{Ei}(t)]\mathrm{d}t \tag{1.10}$$

1.3.2.1 Power Spectral Density of Random Processes

It is not possible to define a Fourier transform for random signals. Thus, the concept of power spectral density (PSD) is introduced for random processes. To do that the following steps are taken:

(1) Truncate the sample functions of a random process to be nonzero for $t < T$:

$$x_{Ei}(t; T) = \begin{cases} x_{Ei}(t), & 0 \le t \le T \\ 0, & \text{otherwise} \end{cases} \tag{1.11}$$

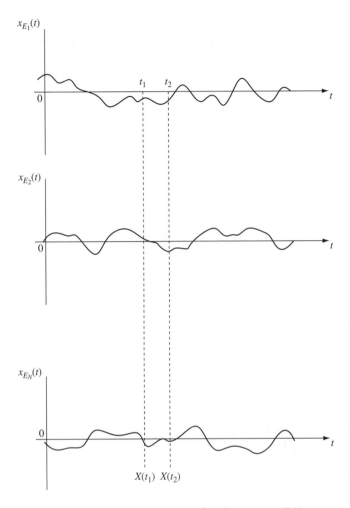

Figure 1.2 Sample functions of random process $X(t)$

(2) Determine $|X_{Ti}(f)|^2$ from the Fourier transform $X_{Ti}(f)$ of the truncated random process $x_{Ei}(t; T)$. The power spectral density $S_{x_{Ei}}(f)$ for $x_{Ei}(t; T)$ is calculated by averaging over a large period of time T:

$$S_{x_{Ei}}(f) = \lim_{T \to \infty} \frac{|X_{Ti}(f)|^2}{T} \tag{1.12}$$

(3) Calculate the average $E\{|X_{Ti}(f)|^2\}$ over all sample functions $x_{Ei}(t; T)$ [Proakis02]:

$$S_X(f) = E_i \left\{ \lim_{T \to \infty} \frac{|X_{Ti}(f)|^2}{T} \right\} = \lim_{T \to \infty} \frac{E\{|X_{Ti}(f)|^2\}}{T} \tag{1.13}$$

The above procedure converts the power-type signals to energy-type signals by setting them to zero for $t > T$. In this way, power spectral density for random processes defined as above corresponds directly to that of deterministic signals [Proakis02].

In practical terms, $S_X(f)$ represents the average power that would be measured at frequency f in a bandwidth of 1 Hz.

Extending the definitions of energy and power of deterministic signals to random processes, we have for each sample function $x_{Ei}(t)$:

$$E_i = \int x_{Ei}^2(t)\mathrm{d}t, \quad P_i = \lim_{T \to \infty} \frac{1}{T} \int x_{Ei}^2(t)\mathrm{d}t \tag{1.14}$$

Since these quantities are random variables the energy and power of the random process $X(t)$ corresponding to sample functions $x_{Ei}(t)$ are defined as:

$$E_X = E\left\{ \int X^2(t)\mathrm{d}t \right\} = \int R_X(t, t)\mathrm{d}t \tag{1.15}$$

$$P_X = E\left[\lim_{T \to \infty} \frac{1}{T} \int_{-T/2}^{T/2} X^2(t)\mathrm{d}t \right] = \frac{1}{T} \int_{-T/2}^{T/2} R_X(t, t)\mathrm{d}t \tag{1.16}$$

For stationary processes, the energy and power are:

$$P_X = R_X(0)$$

$$E_X = \int_{-\infty}^{+\infty} R_X(0)\mathrm{d}t \tag{1.17}$$

1.3.2.2 Random Processes Through Linear Systems

If $Y(t)$ is the output of a linear system with input the stationary random process $X(t)$ and impulse response $h(t)$, the following relations are true for the means and correlation (crosscorrelation and autocorrelation) functions:

$$m_Y = m_X \int_{-\infty}^{+\infty} h(t)\mathrm{d}t \tag{1.18}$$

$$R_{XY}(\tau) = R_X(\tau) * h(-\tau) \tag{1.19}$$

$$R_Y(\tau) = R_X(\tau) * h(\tau) * h(-\tau) \tag{1.20}$$

Furthermore, translation of these expressions in the frequency domain [Equations (1.21)–(1.23)] provides powerful tools to determine spectral densities along the receiver chain in the presence of noise.

$$m_Y = m_X H(0) \tag{1.21}$$

$$S_Y(f) = S_X |H(f)|^2 \tag{1.22}$$

$$S_{YX}(f) = S_X(f)H^*(f) \tag{1.23}$$

1.3.2.3 Wiener–Khinchin Theorem and Applications

The power spectral density of a random process $X(t)$ is given as the following Fourier transform:

$$S_X(f) = F\left[\lim_{T \to \infty} \frac{1}{T} \int_{-T/2}^{T/2} R_X(t + \tau, t) dt\right] \qquad (1.24)$$

provided that the integral within the brackets takes finite values.

If $X(t)$ is stationary then its PSD is the Fourier transform of the autocorrelation function:

$$S_X(f) = \mathcal{F}[R_X(\tau)] \qquad (1.25)$$

An important consequence of the Wiener–Khinchin is that the total power of the random process is equal to the integral of the power spectral density:

$$P_X = E\left[\lim_{T \to \infty} \frac{1}{T} \int_{-T/2}^{T/2} X^2(t) dt\right] = \int_{-\infty}^{\infty} S_X(f) df \qquad (1.26)$$

Another useful outcome is that, when the random process is stationary and ergodic, its power spectral density is equal to the PSD of each sample function of the process $x_{Ei}(t)$:

$$S_X(f) = S_{x_{Ei}}(f) \qquad (1.27)$$

1.3.3 White Gaussian Noise and Equivalent Noise Bandwidth

White noise is a random process $N(t)$ having a constant power spectral density over all frequencies. Such a process does not exist but it was experimentally shown that thermal noise can approximate $N(t)$ well in reasonably wide bandwidth and has a PSD of value $kT/2$ [Proakis02]. Because the PSD of white noise has an infinite bandwidth, the autocorrelation function is a delta function:

$$R_N(\tau) = \frac{N_0}{2} \delta(\tau) \qquad (1.28)$$

where $N_0 = kT$ for white random process. The above formula shows that the random variables associated with white noise are uncorrelated [because $R_N(\tau) = 0$ for $\tau \neq 0$]. If white noise is also a Gaussian process, then the resulting random variables are independent. In practical terms the noise used for the analysis of digital communications is considered white Gaussian, stationary and ergodic process with zero mean. Usually, this noise is additive and is called additive white Gaussian noise (AWGN).

If the above white noise passes through an ideal bandpass filter of bandwidth B, the resulting random process is a bandpass white noise process. Its spectral density and autocorrelation function are expressed as follows:

$$S_{BN}(f) = \begin{cases} N_0, & |f| \leq B/2 \\ 0, & |f| \geq B/2 \end{cases}, \quad R_{BN}(\tau) = N_0 \frac{\sin(\pi B \tau)}{\pi \tau} \qquad (1.29)$$

Noise equivalent bandwidth of a specific system refers to the bandwidth of an ideal reference filter that will produce the same noise power at its output with the given system. More specifically, let the same white noise with PSD equal to $N_0/2$ pass through a filter F_r with a given frequency response $|H_F(f)|$ and a fictitious ideal (rectangular) filter, as shown in Figure 1.3. We can define the constant magnitude of the ideal filter equal to the magnitude $|H(f)|_{\text{ref}}$ of

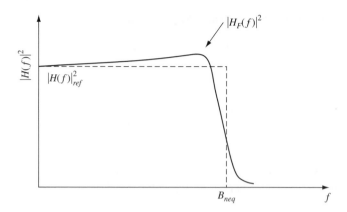

Figure 1.3 Equivalence between frequency response $|H_F(f)|^2$ and ideal brick-wall filter with bandwidth B_{neq}

F_r at a reference frequency f_{ref}, which in most cases represents the frequency of maximum magnitude or the 3 dB frequency.

In this case, noise equivalent bandwidth is the bandwidth of the ideal brick-wall filter, which will give the same noise power at its output as filter F_r. The output noise power of the given filter and the rectangular filter is:

$$P_N = N_0 \int_0^\infty |H(f)|^2 df, \quad P_{Nr} = N_0 |H(f)|_{f_{ref}}^2 B_{neq} \tag{1.30}$$

To express B_{neq} we equate the two noise powers. Thus, we get:

$$B_{neq} = \frac{\int_0^\infty |H(f)|^2 df}{|H(f)|_{f_{ref}}^2} \tag{1.31}$$

1.3.4 Deterministic and Random Signals of Bandpass Nature

In communications, a high-frequency carrier is used to translate the information signal into higher frequency suitable for transmission. For purposes of analysis and evaluation of the performance of receivers, it is important to formulate such signals and investigate their properties.

A bandpass signal is defined as one for which the frequency spectrum $X(f)$ is nonzero within a bandwidth W around a high frequency carrier f_C:

$$X(f) = \begin{cases} \text{nonzero,} & |f - f_C| \leq W \\ \text{zero,} & |f - f_C| \geq W \end{cases} \tag{1.32}$$

It is customary to express high frequency modulated signals as:

$$x(t) = A(t) \cos[2\pi f_C t + \theta(t)] = \text{Re}[A(t) \exp(j2\pi f_C t) \exp(j\theta(t))] \tag{1.33}$$

$A(t)$ and $\theta(t)$ correspond to amplitude and phase which can contain information. Since the carrier f_C does not contain information, we seek expressions for the transmitted signal in

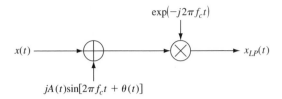

Figure 1.4 The lowpass complex envelope signal $x_{LP}(t)$ produced from $x(t)$

which dependence on f_C is eliminated. Inevitably, this signal will be of lowpass (or baseband) nature and can be expressed in two ways:

$$x_{LP}(t) = A(t) \exp\left(j\theta(t)\right), \quad x_{LP}(t) = x_I(t) + jx_Q(t) \tag{1.34}$$

To obtain the lowpass complex envelope signal $x_{LP}(t)$ from $x(t)$, the term $jA(t) \sin\left[2\pi f_C t + \theta(t)\right]$ must be added to the passband signal $x(t)$ and the carrier must be removed by multiplying by $\exp\left(-2\pi f_C t\right)$. This is depicted in Figure 1.4.

Consequently, $x_{LP}(t)$ can be expressed as [Proakis02]:

$$x_{LP}(t) = [x(t) + \bar{x}(t)] \exp\left(-j2\pi f_C t\right) \tag{1.35}$$

where $\bar{x}(t)$ is defined as the Hilbert transform of $x(t)$ and is analytically expressed in time and frequency domains as:

$$\bar{x}(t) = \frac{1}{\pi t} * x(t) \tag{1.36}$$

From the above we realize that the Hilbert transform is a simple filter which shifts by $-\pi/2$ the phase of the positive frequencies and by $+\pi/2$ the phase of the negative frequencies. It is straightforward to show that the relation of the bandpass signal $x(t)$ and quadrature lowpass components $x_I(t), x_Q(t)$ is:

$$x(t) = x_I(t) \cos\left(2\pi f_C t\right) - x_Q(t) \sin\left(2\pi f_C t\right) \tag{1.37a}$$

$$\bar{x}(t) = x_I(t) \sin\left(2\pi f_C t\right) + x_Q(t) \cos\left(2\pi f_C t\right) \tag{1.37b}$$

The envelope and phase of the passband signal are:

$$A(t) = \sqrt{x_I^2(t) + x_Q^2(t)}, \quad \theta(t) = \tan^{-1}\left[\frac{x_Q(t)}{x_I(t)}\right] \tag{1.38}$$

Considering random processes, we can define that a random process $X_N(t)$ is bandpass if its power spectral density is confined around the centre frequency f_C:

$$X_N(t) \text{ is a bandpass process if}: S_{X_N}(f) = 0 \text{ for } |f - f_C| \geq W, \quad W < f_C \tag{1.39}$$

It is easy to show that the random process along with its sample functions $x_{Ei}(t)$ can be expressed in a similar way as for deterministic signals in terms of two new processes $X_{nI}(t)$ and $X_{nQ}(t)$, which constitute the in-phase and quadrature components:

$$X_N(t) = A(t) \cos\left[2\pi f_C t + \theta(t)\right] = X_{nI}(t) \cos\left(2\pi f_C t\right) - X_{nQ}(t) \sin\left(2\pi f_C t\right) \tag{1.40}$$

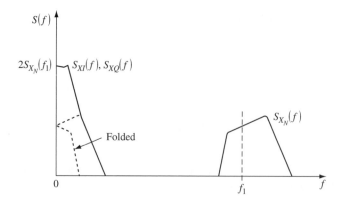

Figure 1.5 Lowpass nature of the PSDs $S_{XI}(f)$, $S_{XQ}(f)$ of quadrature components $x_I(t)$, $x_Q(t)$

If $X_N(t)$ is a stationary bandpass process of zero mean, processes $X_{nI}(t)$ and $X_{nQ}(t)$ are also zero mean [Proakis02].

Considering autocorrelation functions $R_{nI}(\tau)$, $R_{nQ}(\tau)$ of $X_{nI}(t)$ and $X_{nQ}(t)$, it can be shown that:

$$R_{nI}(\tau) = R_{nQ}(\tau) \tag{1.41}$$

The spectra $S_{XI}(f)$ and $S_{XQ}(f)$ of processes $X_{nI}(t)$ and $X_{nQ}(t)$ become zero for $|f| \geq W$ and consequently they are lowpass processes. Furthermore, their power spectral densities can be calculated and are given as [Proakis02]:

$$S_{XI}(f) = S_{XQ}(f) = \frac{1}{2}[S_{X_N}(f - f_C) + S_{X_N}(f + f_C)] \tag{1.42}$$

Figure 1.5 gives the resulting lowpass spectrum of $X_{nI}(t)$ and $X_{nQ}(t)$. Similarly, as for deterministic signals, the envelope and phase processes $A(t)$ and $\theta(t)$ are defined as:

$$X_{LP}(t) = A(t) \exp[j\theta(t)], \quad A(t) = \sqrt{X_{nI}^2(t) + X_{nQ}^2(t)}, \quad \theta(t) = \tan^{-1}\left[\frac{X_{nQ}(t)}{X_{nI}(t)}\right] \tag{1.43}$$

where $X_{LP}(t)$ is the equivalent lowpass process for $X_N(t)$, which now can be expressed as:

$$X_N(t) = A(t) \cos[2\pi f_C t + \theta(t)] \tag{1.44}$$

The amplitude p.d.f. follows the Rayleigh distribution with mean $\overline{A}$ and variance $\overline{A^2}$ [Gardner05]:

$$E[A] = \overline{A} = \sigma_n \sqrt{\pi/2}, \quad E[A^2] = \overline{A^2} = 2\sigma_n^2 \tag{1.45}$$

Regarding the phase, if we assume that it takes values in the interval $[-\pi, \pi]$, $\theta(t)$ follows a uniform distribution with p.d.f. $p(\theta) = 1/(2\pi)$ within the specified interval. Furthermore, its mean value is equal to zero and its variance is $\overline{\theta^2} = \pi^2/3$.

1.4 Radio Channel Characterization

Transmission of high frequency signals through the radio channel experiences distortion and losses due to reflection, absorption, diffraction and scattering. One or more of these mechanisms is activated depending on the transceiver position. Specifically, in outdoor environments important factors are the transmitter–receiver (Tx–Rx) distance, mobility of the transmitter or the receiver, the formation of the landscape, the density and the size of the buildings. For indoor environments, apart from the Tx–Rx distance and mobility, important factors are the floor plan, the type of partitions between different rooms and the size and type of objects filling the space.

A three-stage model is frequently used in the literature to describe the impact of the radio channel [Pra98], [Rappaport02], [Proakis02], [Goldsmith05]:

- large-scale path loss;
- medium-scale shadowing;
- small-scale multipath fading.

Large-scale attenuation (or path loss) is associated with loss of the received power due to the distance between the transmitter and the receiver and is mainly affected by absorption, reflection, refraction and diffraction.

Shadowing or shadow fading is mainly due to the presence of obstacles blocking the line-of-sight (LOS) between the transmitter and the receiver. The main mechanisms involved in shadowing are reflection and scattering of the radio signal.

Small-scale multipath fading is associated with multiple reflected copies of the transmitted signal due to scattering from various objects arriving at the receiver at different time instants. In this case, the vector summation of all these copies with different amplitude and phase results in fading, which can be as deep as a few tens of decibels. Successive fades can have distances smaller than $\lambda/2$ in a diagram presenting received signal power vs. distance. In addition, the difference in time between the first and the last arriving copy of the received signal is the time spread of the delay of the time of arrival at the receiver. This is called *delay spread of the channel* for the particular Tx–Rx setting. Figure 1.6 depicts the above three attenuation and fading mechanisms.

1.4.1 Large-scale Path Loss

The ratio between the transmitted power P_T and the locally-averaged receiver signal power P_{Rav} is defined as the path loss of the channel:

$$P_L = \frac{P_T}{P_{Rav}} \tag{1.46}$$

The receiver signal power is averaged within a small area (with a radius of approximately 10 wavelengths) around the receiver in order to eliminate random power variations due to shadow fading and multipath fading.

The free-space path loss for a distance d between transmitter and receiver, operating at a frequency $f = c/\lambda$, is given by [Proakis02]:

$$L_S = \left(\frac{4\pi d}{\lambda}\right)^2 \tag{1.47}$$

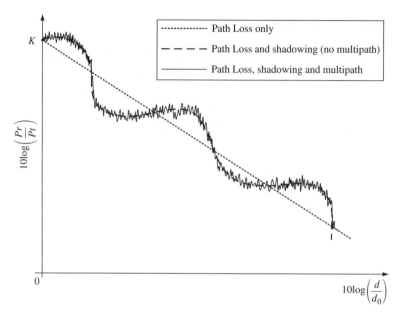

Figure 1.6 The three mechanisms contributing to propagation losses (reprinted from A. Goldsmith, 'Wireless Communications', copyright © 2005 by Cambridge Academic Press)

whereas the power at the input of the receiver for antenna gains of the transmitter and the receiver G_T, G_R, respectively, is:

$$P_{\text{Rav}} = \frac{P_T G_T G_R}{(4\pi d/\lambda)^2} \tag{1.48}$$

With these in mind, the free-space path loss is given as:

$$P_L(\text{dB}) = -10 \log_{10} \left[\frac{G_T G_R \lambda^2}{(4\pi d)^2} \right] \tag{1.49}$$

However, in most radio systems the environment within which communication between the transmitter and the receiver takes place is filled with obstacles which give rise to phenomena like reflection and refraction, as mentioned above. Consequently, the free-space path loss formula cannot be used to accurately estimate the path losses. For this reason, empirical path loss models can be used to calculate path loss in macrocellular, microcellular and picocellular environments. The most important of these models are the Okumura model and the Hata model [Rappaport02], [Goldsmith05], which are based on attenuation measurements recorded in specific environments as a function of distance.

The Okumura model refers to large urban macrocells and can be used for distances of 1–100 km and for frequency ranges of 150–1500 MHz. The Okumura path-loss formula is associated with the free-space path loss and also depends on a mean attenuation factor $A_M(f_C d)$ and gain factors $G_T(h_T)$, $G_R(h_R)$ and G_{ENV} related to base station antenna, mobile antenna and type of environment respectively [Okumura68], [Rappaport02], [Goldsmith05]:

$$P_L(d) = L_F(f_C, d) + A_M(f_C, d) - G_T(h_T) - G_R(h_R) - G_{\text{ENV}} \tag{1.50}$$

$$G_T(h_T) = 20 \log_{10}(h_T/200), \text{ for } 30\,\text{m} < h_T < 1000\,\text{m} \tag{1.51}$$

$$G_R(h_R) = \begin{cases} 10 \log_{10}(h_R/3) & h_R \leq 3\,\text{m} \\ 20 \log_{10}(h_R/3) & 3\,\text{m} < h_R < 10\,\text{m} \end{cases} \tag{1.52}$$

The Hata model [Hata80] is a closed form expression for path loss based on the Okumura data and is valid for the same frequency range (150–1500 MHz):

$$P_{Lu}(d) = 69.55 + 26.16 \log_{10}(f_C) - 13.82 \log_{10}(h_T) - C(h_R)$$
$$+ [44.9 - 6.55 \log_{10}(h_T)] \log_{10}(d) \text{ dB} \tag{1.53}$$

h_T and h_R represent the base station and mobile antenna heights as previously whereas $C(h_R)$ is a correction factor associated with the antenna height of the mobile and depends on the cell radius. For example, for small or medium size cities it is [Goldsmith05]:

$$C(h_R) = [1.1 \log_{10} f_C - 0.7]h_R - [1.56 \log_{10}(f_C) - 0.8] \text{ dB} \tag{1.54}$$

There is a relation associating the suburban and rural models to the urban one. For example, the suburban path-loss model is:

$$P_{L,\text{sub}}(d) = P_{L,u}(d) - 2(\log_{10}(f_C/28))^2 - 5.4 \tag{1.55}$$

COST 231 [Euro-COST 231-1991] is an extension of the Hata model for specific ranges of antenna heights and for frequencies between 1.5 and 2.0 GHz.

An empirical model for path loss in a microcellular environment (outdoor and indoor) is the so-called 'piecewise linear' model [Goldsmith05]. It can consist of N linear sections (segments) of different slopes on a path loss (in decibels) vs. the logarithm of normalized distance $[\log_{10}(d/d_0)]$ diagram.

The most frequently used is the dual-slope model, giving the following expression for the received power [Goldsmith05]:

$$P_R(d) = \begin{cases} P_T + K - 10\gamma_1 \log_{10}(d/d_0) \text{ dB} & d_0 \leq d \leq d_B \\ P_T + K - 10\gamma_1 \log_{10}(d_B/d_0) - 10\gamma_2 \log_{10}(d/d_C) \text{ dB} & d > d_B \end{cases} \tag{1.56}$$

where K is an attenuation factor depending on channel attenuation and antenna patterns. K is usually less than 1, corresponding to negative values in decibels; d_0 is a reference distance marking the beginning of the antenna far field; d_B is a breakpoint beyond which the diagram of P_R vs the logarithm of normalized distance changes slope; and γ_1 and γ_2 represent the two different slopes (for distances up to d_B and beyond d_B) of the P_R vs $\log_{10}(d/d_0)$ diagram.

For system design and estimation of the coverage area, it is frequently very useful to employ a simplified model using a single path loss exponent γ covering the whole range of transmitter–receiver distances. Hence the corresponding formula is:

$$P_R(d) = P_T + K - 10 \log_{10}(d/d_0) \text{ dBm}, \quad d > d_0 \tag{1.57}$$

The above model is valid for both indoor environments ($d_0 = 1$–10 m) and outdoor environments ($d_0 = 10$–100 m). In general, the path loss exponent is between 1.6 and 6 in most applications depending on the environment, the type of obstructions and nature of the walls in indoor communication. For example, in an indoor environment $1.6 \leq \gamma \leq 3.5$, when transmitter and receiver are located on the same floor [Rappaport02].

Finally, a more detailed model for indoor propagation can be produced by taking into account specific attenuation factors for each obstacle that the signal finds in its way from the transmitter to the receiver. Hence, the above formulas can be augmented as follows:

$$P_R(d) = P_T - P_L(d) - \sum_{i=1}^{N} AF_i \text{ dB} \tag{1.58}$$

where $P_L(d)$ is the losses using a path loss model and AF_i is the attenuation factor of the ith obstacle. For example, if the obstacle is a concrete wall, AF is equal to 13 dB.

1.4.2 Shadow Fading

As mentioned in the beginning of this section, shadow fading is mainly due to the presence of objects between the transmitter and the receiver. The nature, size and location of the objects are factors that determine the amount of attenuation due to shadowing. Hence, the randomness due to shadow fading stems from the size and location of the objects and not from the distance between the transmitter and the receiver. The ratio of the transmitted to the received power $\psi = P_T/P_R$ is a random variable with log–normal distribution [Goldsmith05]:

$$p(\psi) = \frac{10/\ln 10}{\psi\sqrt{2\pi}\sigma_{\psi dB}} \exp\left[-\frac{(10\log_{10}\psi - m_{\psi dB})^2}{2\sigma_{\psi dB}^2}\right] \tag{1.59}$$

where $m_{\psi dB}$ and $\sigma_{\psi dB}$ are the mean and variance (both in decibels) of the random variable $\Psi = 10\log_{10}\psi$. The mean $m_{\psi dB}$ represents the empirical or analytical path loss, as calculated in the 'large-scale path-loss' subsection above.

1.4.3 Multipath Fading in Wideband Radio Channels

1.4.3.1 Input–Output Models for Multipath Channels

The objective of this section is to obtain simple expressions for the impulse response of a radio channel dominated by multipath. For this purpose, we assume a transmitting antenna, a receiving antenna mounted on a vehicle and four solid obstructions (reflectors) causing reflected versions of the transmitted signal to be received by the vehicle antenna, as illustrated in Figure 1.7. We examine two cases: one with static vehicle and one with moving vehicle.

Taking into account that we have discrete reflectors (scatterers), we represent by a_n, x_n, τ_n and φ_n, the attenuation factor, the length of the path, the corresponding delay and the phase change due to the nth arriving version of the transmitted signal (also called the nth path), respectively.

Let the transmitted signal be:

$$s(t) = \text{Re}\left\{s_L(t)e^{j2\pi f_c t}\right\} = \text{Re}\left\{|s_L(t)|e^{j(2\pi f_c t + \varphi_{s_L}(t))}\right\} = |s_L(t)|\cos\left(2\pi f_c t + \varphi_{s_L}(t)\right) \tag{1.60}$$

where $s_L(t)$ represents the baseband equivalent received signal.

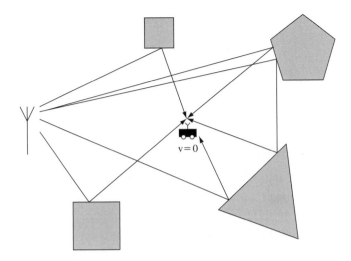

Figure 1.7 Signal components produced by reflections on scatterers arrive at the mobile antenna

Static Transmitter and Reflectors (Scatterers), Static Vehicle
The received bandpass and baseband signals are given as:

$$r(t) = \mathrm{Re}\left\{r_{\mathrm{L}}(t)e^{j2\pi f_c t}\right\} = \mathrm{Re}\left\{\left[\sum_n a_n e^{-j2\pi f_c \tau_n} s_{\mathrm{L}}(t - \tau_n)\right]e^{j2\pi f_c t}\right\} \tag{1.61}$$

with

$$a_n = \overline{a_n}e^{j\varphi_n} \tag{1.62}$$

The attenuation coefficient a_n is complex so as to include the magnitude of the attenuation factor $\overline{a_n}$ and the effect of the change of phase φ_n due to reflections. τ_n is related to x_n by $\tau_n = x_n/c = x_n/(\lambda f_C)$. Consequently, the lowpass channel impulse response is given by:

$$c(\tau; t) = \sum_n a_n e^{-j2\pi f_C \tau_n}\delta(\tau - t_n) \tag{1.63}$$

Static Transmitter and Reflectors, Moving Vehicle
In this case we have to briefly present the impact of the Doppler effect. For a vehicle moving in the horizontal direction, Doppler is associated with the small difference Δx in the distance that the transmitter signal must cover in the two different positions of the vehicle. As shown in Figure 1.8, when the vehicle is located at point X, it receives the signal from point S in an angle θ.

After time t, the vehicle has moved to point Y where we assume that the angle of arrival is still θ (a valid assumption if the transmitter S is far away from the vehicle). Hence, the distance difference Δx is:

$$\Delta x = -vt\cos\theta \tag{1.64}$$

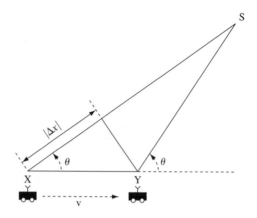

Figure 1.8 Generation of Doppler effect due to a moving vehicle

This change results in a phase change and consequently in a change of instantaneous frequency which is:

$$v_d = \frac{f_C}{c} v \cos \theta \tag{1.65}$$

whereas the maximum frequency change (for $\theta = 0$) is:

$$v_D = \frac{f_C}{c} v \tag{1.66}$$

After some elementary calculations one can show that the received equivalent baseband signal is the summmation of n different paths:

$$r_L(t) = \sum_n a_n e^{-j2\pi x_n/\lambda} e^{j2\pi \frac{v}{\lambda} \cos \theta_n t} s_L\left(t - \frac{x_n}{c} + \frac{v \cos \theta_n t}{c}\right)$$

$$= \sum_n a_n e^{-j2\pi x_n/\lambda} e^{j2\pi f_{Dn} t} s_L\left(t - \frac{x_n}{c} + \frac{v \cos \theta_n t}{c}\right) \tag{1.67}$$

Disregarding $v \cos \theta_n t/c$ because it is very small, Equation (1.67) gives:

$$r_L(t) = \sum_n a_n e^{j2\pi v_D \cos \theta_n t} s_L(t - \tau_n) \tag{1.68}$$

Consequently:

$$c(\tau; t) = \sum_n a_n e^{j2\pi v_D \cos \theta_n t} \delta(t - \tau_n) \tag{1.69}$$

The next step is to assume that there is a 'continuous' set of scatterers instead of discrete scatterers located in the surrounding area. In that case, summations are replaced with integrals and it can be shown that the received signals (passband and baseband) and impulse responses are [Proakis02], [Goldsmith05]:

$$r(t) = \int_{-\infty}^{+\infty} \alpha(\tau; t) s(t - \tau) d\tau \tag{1.70}$$

$$r_L(t) = \int_{-\infty}^{+\infty} \alpha(\tau;t)e^{-j2\pi f_c \tau} s_L(t - \tau)d\tau \qquad (1.71)$$

where $\alpha(\tau;t)$ represents the attenuation at delay equal to τ at time instant t.

The lowpass channel impulse response in this case is:

$$c(\tau;t) = \alpha(\tau;t)e^{-j2\pi f_c \tau} \qquad (1.72)$$

The input–output relations between $r_L(t)$ and $s_L(t)$ are given [Fleury00], [Goldsmith05]:

$$r_L(t) = \int_{-\infty}^{+\infty} c(\tau;t)s_L(t - \tau)d\tau \qquad (1.73)$$

$$r_L(t) = \int_{-\infty}^{+\infty} C(f;t)S_L(f)e^{-j2\pi ft}df \qquad (1.74)$$

where $C(f;t)$ represents the Fourier transform of $c(\tau;t)$ with respect to variable τ. It is called the time-variant transfer function and is given by:

$$C(f;t) = \int_{-\infty}^{+\infty} c(\tau;t)e^{-j2\pi f \tau}d\tau \qquad (1.75)$$

which, for discrete impulse response, becomes:

$$C(f;t) = \sum_n \alpha_n(t)e^{-j2\pi f_c \tau_n(t)}e^{-j2\pi f \tau_n(t)} \qquad (1.76)$$

In addition, $S_L(f)$ represents the power spectrum of $s_L(t)$.

Another expression for $r_L(t)$ is [Fleury00]:

$$r_L(t) = \int_{-\infty}^{+\infty} \int_{-\infty}^{+\infty} e^{j2\pi \nu t} s_L(t - \tau)h(\tau,\nu)d\tau d\nu \qquad (1.77)$$

where $h(\tau,\nu)$ is called the delay-Doppler spread function expressed as:

$$h(\tau,\nu) = \sum_n \alpha_n \delta(\nu - \nu_n)\delta(\tau - \tau_n) \qquad (1.78)$$

and consequently $h(\tau,\nu)$ is the Fourier transform of $c(\tau;t)$ with respect to variable t:

$$h(\tau,\nu) = \int_{-\infty}^{+\infty} c(\tau;t)e^{-j2\pi \nu t}dt \qquad (1.79)$$

Furthermore, $h(\tau,\nu)$ is the two-dimensional Fourier transform of $C(f;t)$:

$$h(\tau,\nu) = \int_{-\infty}^{+\infty} \int_{-\infty}^{+\infty} C(f;t)e^{j2\pi f \tau}e^{-j2\pi \nu t}dtdf \qquad (1.80)$$

1.4.3.2 Spectral Densities and Autocorrelation Functions

It is now necessary to produce quantities which can be used to determine the distribution of power with respect to time delay τ and Doppler frequency ν. These quantities are associated with autocorrelation functions of the impulse response and the frequency response of the radio channel.

With respect to the distribution of power the cross-power delay spectrum $\varphi_c(\tau; \Delta t)$ is needed, which is given by (for wide-sense stationary uncorrelated scattering, WSSUS):

$$\varphi_c(\tau_1; \Delta t)\delta(\tau_1 - \tau_2) = R_c(\tau_1, \tau_2; t_1, t_2) \equiv \frac{1}{2}E\big[c^*(\tau_1, t_1)c(\tau_2, t_2)\big] \qquad (1.81)$$

where $\Delta t = t_2 - t_1$.

$\varphi_c(\tau_1; \Delta t)$ gives the average power of the output of the channel as a function of τ and Δt. For $\Delta t = 0$ the resulting autocorrelation $\varphi_c(\tau; 0) \equiv \varphi_c(\tau)$ is called the power delay profile (PDP) and illustrates how the power at the radio channel output is distributed in the delay τ domain.

Furthermore, we define the frequency–time correlation function $\varphi_C(\Delta f; \Delta t)$:

$$\varphi_C(\Delta f; \Delta t) = \frac{1}{2}E\big[C^*(f; t)C(f + \Delta f; t + \Delta t)\big]$$

$$\equiv \int_{-\infty}^{\infty} \varphi_c(\tau_1; \Delta t)e^{j2\pi\tau_1(f_2 - f_1)}d\tau_1 = R_C(f_1, f_2; t_1, t_2) \qquad (1.82)$$

At this point, it is important to introduce the delay-Doppler power spectrum, or scattering function $S(\tau; \nu)$ which can be shown to be [Fleury00]:

$$S(\tau; \nu) = \int_{-\infty}^{\infty} \varphi_c(\tau; \Delta t)e^{-j2\pi\nu\Delta t}d(\Delta t) = \int_{-\infty}^{\infty}\int_{-\infty}^{\infty} \varphi_C(\Delta f; \Delta t)e^{-j2\pi\nu\Delta t}e^{j2\pi\tau\Delta f}d(\Delta t)d(\Delta f) \qquad (1.83)$$

Also, its relation to the autocorrelation of $h(\tau, \nu)$ can be shown to be [Fleury00]:

$$R_h(\tau_1, \tau_2; \nu_1, \nu_2) = S(\tau_1; \nu_1)\delta(\nu_2 - \nu_1)\delta(\tau_2 - \tau_1) \qquad (1.84)$$

The importance of the scattering function lies in the fact that it reveals the way the average power at the receiver is distributed in two domains, the delay domain and the Doppler frequency domain.

In addition the double integral of $S(\tau; \nu)$ can be expressed as:

$$\int_{-\infty}^{+\infty}\int_{-\infty}^{+\infty} S(\tau; \nu)d\tau d\nu = \varphi_C(0; 0) = \frac{1}{2}E\big[|C(f; t)|^2\big] \qquad (1.85)$$

which implies that the bandpass output power is the same regardless of the fact that the input to the channel may be a narrowband or a wideband signal.

Finally, the Doppler cross-power spectrum $S_C(\Delta f; \Delta \nu)$ is the Fourier transform of $\varphi_C(\Delta f; \Delta t)$:

$$S_C(\Delta f; \nu) = \int_{-\infty}^{+\infty} \varphi_C(\Delta f; \Delta t)e^{-j2\pi\nu\Delta t}d\tau = \int_{-\infty}^{+\infty} S(\tau; \nu)e^{-j2\pi\tau\Delta f}d\tau \qquad (1.86)$$

Letting $\Delta f = 0$, we define $S_C(0; \Delta \nu) \equiv S_C(\nu)$ as the Doppler power spectrum which takes the form:

$$S_C(\nu) = \int_{-\infty}^{+\infty} \varphi_C(\Delta t)e^{-j2\pi\nu\Delta t}d\tau \qquad (1.87)$$

It is important to note that $S_C(\nu)$ depicts the power distribution in Doppler frequency.

In addition we have:

$$\varphi_c(\tau) = \int_{-\infty}^{+\infty} S(\tau; v)dv \tag{1.88}$$

$$S_C(v) = \int_{-\infty}^{+\infty} S(\tau; v)d\tau \tag{1.89}$$

In conclusion, all three functions $S(\tau; v)$, $S_C(v)$ and $\varphi_c(\tau)$ represent power spectral densities and will produce narrowband or wideband power after integration. Furthermore, function $\varphi_C(\Delta f; \Delta t)$ characterizes channel selectivity in frequency and time. By eliminating one of the variables, we obtain channel selectivity with respect to the other. Hence, $\varphi_C(\Delta f)$ gives the channel selectivity with respect to frequency and $\varphi_C(\Delta t)$ with respect to time.

For discrete scatterer positions, the power delay profile is given by:

$$\varphi_c(\tau) = \sum_n \frac{1}{2}E\left[|a_n|^2\right]\delta(\tau - \tau_n) \tag{1.90}$$

Furthermore, the Doppler power spectrum and the corresponding autocorrelation function are given by:

$$S_C(v) = \frac{\sigma^2}{\pi v_D} \frac{1}{\sqrt{1 - \left(\frac{v}{v_D}\right)^2}}, \quad \text{for } v \in (-v_D, v_D) \tag{1.91}$$

$$\varphi_C(\Delta t) = \sigma^2 J_0(2\pi v_D \Delta t) \tag{1.92}$$

Figure 1.9 shows the Doppler power spectrum for omnidirectional receiving antenna and uniformly distributed scatterers.

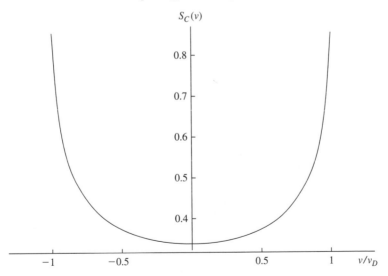

Figure 1.9 shows the Doppler power spectrum for omnidirectional receiving antenna and uniformly distributed scatterers.

Figure 1.9   U-Shaped Doppler power spectrum for uniform distributed scatterers

1.4.3.3 Parameters of Multipath Radio Channels

Dispersion in Time Delay and Frequency Selectivity
By choosing the peak values in a continuous power delay profile, we obtain the corresponding discrete PDP. Furthermore, it is convenient to set the time of arrival of the first ray equal to zero ($\tau_1 = 0$).

Figure 1.10 illustrates the above notions. $\varphi_c(\tau)$ is used to depict time dispersion because it gives the way power is distributed as a function of time delay. For equivalence between the continuous and the corresponding discrete PDP, the following must hold:

$$\sigma_{a_n}^2 = \int_{\tau_n}^{\tau_{n+1}} \varphi_c(\tau)d\tau, \quad \text{for } n = 1, \ldots, N-1$$

$$\sigma_{a_N}^2 = \int_{\tau_N}^{+\infty} \varphi_c(\tau)d\tau, \quad \text{for } n = N \tag{1.93}$$

The mean delay for continuous and discrete power delay profiles, is given respectively by:

$$m_\tau = \frac{\int_0^\infty \tau\varphi_c(\tau)d\tau}{\int_0^\infty \varphi_c(\tau)d\tau} \tag{1.94}$$

$$m_\tau = \frac{\sum_n \tau_n \int_{\tau_n}^{\tau_{n+1}} \varphi_c(\tau)d\tau}{\sum_n \int_{\tau_n}^{\tau_{n+1}} \varphi_c(\tau)d\tau} = \frac{\sum_n \sigma_{a_n}^2 \tau_n}{\sum_n \sigma_{a_n}^2} \tag{1.95}$$

where the denominator is necessary for PDP normalization. If the PDP power is normalized to one, the denominators are equal to one and can be eliminated.

The rms delay spread for both continuous and discrete profiles is given by:

$$\sigma_\tau \equiv \sqrt{\frac{\int_0^{+\infty} (\tau - m_\tau)^2 \varphi_c(\tau)d\tau}{\int_0^\infty \varphi_c(\tau)d\tau}} \tag{1.96}$$

$$\sigma_\tau = \sqrt{\frac{\sum_n \sigma_{a_n}^2 \tau_n^2}{\sum_n \sigma_{a_n}^2} - m_\tau^2} \tag{1.97}$$

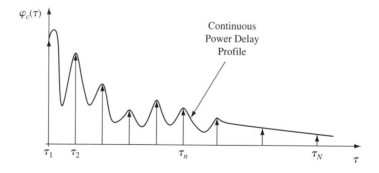

Figure 1.10 Transformation of continuous power delay profile into a discrete one

Mean excess delay and rms delay spread are associated with the power delay profile $\varphi_c(\tau)$ of the radio channel. In fact, assuming that $\varphi_c(\tau)$ represents a probability distribution function, mean excess delay corresponds to the mean value of the delay, in which a very narrow pulse is subjected. Furthermore, the rms delay spread gives the spread of the delays around the mean delay of the pulse.

Frequency selectivity is characterized by a parameter called coherence bandwidth of the channel $[\Delta f]_C$. This is defined as the bandwidth within which the channel frequency response is approximately flat. This implies that for $\Delta f \geq [\Delta f]_C$, $\varphi_C(\Delta f) \approx 0$ and consequently the channel responses for frequency difference exceeding the coherence bandwidth are uncorrelated.

As frequency correlation function $\varphi_C(\Delta f)$ is the Fourier transform of power delay profile $\varphi_c(\tau)$, the following relation holds between rms delay spread and coherence bandwidth:

$$(\Delta f)_{\text{coh}} \cong \frac{k_\tau}{\sigma_\tau} \tag{1.98}$$

where k_τ depends on the shape of $\varphi_c(\tau)$ and the value at which we use for correlation. Most of the times we use $k_\tau = 1$.

Figure 1.11 shows the shape of a power delay profile and its Fourier transform from where the coherence bandwidth can be calculated.

Dispersion in Doppler Frequency and Time Selectivity

In analogy with time delay, $S_C(\nu)$ is used to determine the spread in Doppler frequency. More specifically, when there is Doppler effect due to movement of the mobile unit, a single carrier f_C transmitted through the channel, produces a Doppler spectrum occupying a frequency band $[f_C - \nu_D, f_C + \nu_D]$, where ν_D indicates the maximum Doppler frequency. The U-shaped Doppler is an example of this distribution. The mean Doppler frequency and rms Doppler spread are given respectively by:

$$m_\nu \equiv \int_{-\infty}^{+\infty} \nu S_C(\nu) d\nu \tag{1.99}$$

$$\sigma_\nu \equiv \sqrt{\int_{-\infty}^{+\infty} (\nu - m_\nu)^2 S_C(\nu) d\nu} \tag{1.100}$$

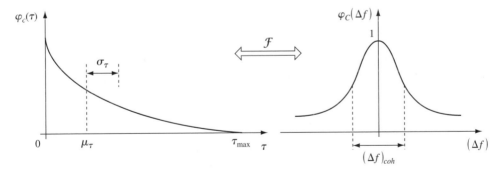

Figure 1.11 RMS delay spread and coherence bandwidth depicted on power delay profile and its autocorrelation function

In practice, the rms delay spread ΔF_D gives the bandwidth over which $S_C(v)$ is not close to zero.

To quantify how fast the radio channel changes with time, the notion of coherence time is introduced, which represents the time over which the time correlation function $\varphi_C(\Delta t)$ is not close to zero:

$$T_C : \varphi_C(\Delta t) \approx 0 \quad \text{for } \Delta t \geq T_C \tag{1.101}$$

More specifically, coherence time T_C indicates the time interval during which the channel impulse response does not change significantly. Coherence time T_C and rms Doppler spread are connected with an inverse relation of the form:

$$T_C \approx \frac{k}{\sigma_v} \tag{1.102}$$

Figure 1.12 shows $\varphi_C(\Delta t)$ and $S_C(v)$ and graphically depicts parameters ΔF_D and T_C.

1.4.3.4 Characterization of Small-scale Fading

Based on the above parameters of selectivity in frequency and time, small-scale fading can be classified in four categories. Two of them concern frequency selectivity and the other two are associated with selectivity in time.

Frequency Selectivity

By comparing the bandwidth W_{BP} of the bandpass signal to the coherence bandwidth of the channel $(\Delta f)_{coh}$, we classify fading in flat fading and frequency selective fading, given by the following criteria:

$$W_{BP} \ll (\Delta f)_{coh} \Rightarrow \text{Flat Fading} \tag{1.103}$$

$$W_{BP} \gg (\Delta f)_{coh} \Rightarrow \text{Frequency Selective Fading} \tag{1.104}$$

In the case of flat fading the relation between the input and output (transmitted and received) baseband signals is very simple and is given by:

$$r_L(t) = C(0;t) \int_{-\infty}^{+\infty} S_L(f)e^{j2\pi ft} df = C(0;t)s_L(t) \tag{1.105}$$

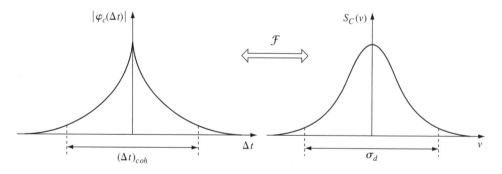

Figure 1.12 Coherence time and Doppler spread estimated from channel autocorrelation function and Doppler power spectrum

where

$$C(0;t) = \sum_n a_n(t)e^{-j2\pi f_C \tau_n(t)} \qquad (1.106)$$

In addition, in the time domain, the two conditions are expressed in terms of the rms delay spread and the signalling interval T ($T = 1/W_{BP}$):

$$T \gg \sigma_\tau \Rightarrow \text{Flat Fading} \qquad (1.107a)$$

$$T \ll \sigma_\tau \Rightarrow \text{Frequency Selective Fading} \qquad (1.107b)$$

It is important to note that the above relations indicate that in a frequency-selective channel, due to the relation $T_S \ll \sigma_\tau$, the channel introduces ISI.

Finally, we must note that the characterization of fading as frequency selective or not depends only on the bandwidth of the transmitted signal, compared with the channel frequency response.

Time Selectivity

In this case, the symbol interval T is compared with coherence time T_C. If the channel impulse response does not change significantly within the symbol interval, we have slow fading. In the opposite case, fading is characterized as fast fading. These two are expressed as:

$$T \ll (\Delta t)_{\text{coh}} \Rightarrow \text{Slow Fading} \qquad (1.108a)$$

$$T \gg (\Delta t)_{\text{coh}} \Rightarrow \text{Fast Fading} \qquad (1.108b)$$

Furthermore, in terms of Doppler spread we have:

$$W_{BP} \gg \sigma_\nu \Rightarrow \text{Slow Fading} \qquad (1.109a)$$

$$W_{BP} \ll \sigma_\nu \Rightarrow \text{Fast Fading} \qquad (1.109b)$$

As with frequency selectivity, characterizing the channel as slow fading or fast fading mostly depends on the transmitted signal bandwidth. However, in this case, since the channel statistics can change due to change in Doppler frequency, which depends on the change in Doppler, the channel can be transformed from slow fading to fast fading and vice versa.

Figure 1.13 depicts graphically the radio channel categorization in terms of frequency selectivity and time selectivity as discussed above.

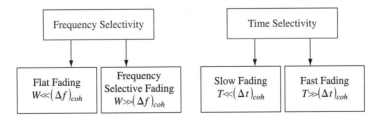

Figure 1.13 Categorization of channel properties with respect to selectivity in frequency and time

1.4.3.5 The Tapped Delay-line Channel Model

For a baseband transmitted signal of bandwidth W the received signal is given by:

$$r_L(t) = \sum_{n=1}^{L_{taps}} c_n(t)s_l\left(t - \frac{n-1}{W}\right)$$ (1.110)

It can be shown that [Proakis02] the coefficients are given by:

$$c_n(t) = \frac{1}{W}c\left(\frac{n-1}{W};t\right)$$ (1.111a)

The number of taps is given by:

$$L_{taps} = \lceil \tau_{\max} W \rceil \text{ where } \lceil x \rceil = N_x, N_x \le x < N_x + 1$$ (1.111b)

Finally, it should be noted that the impulse response $c(\tau;t)$ as a function of the tap values is:

$$c(\tau;t) = \sum_{n=1}^{L_{taps}} c_n(t)\delta\left(t - \frac{n-1}{W}\right)$$ (1.112)

Coefficients $\{c(\tau;t)\}$ are uncorrelated and follow a complex Gaussian distribution. When there is no line-of-sight (N-LOS), the magnitude of $\{c(\tau;t)\}$ follows a Rayleigh distribution with uniform distribution in phase.

1.5 Nonlinearity and Noise in Radio Frequency Circuits and Systems

An RF system is mainly plagued by two shortcomings produced by respective nonidealities. These are nonlinearity and noise. These two factors greatly affect its performance and define the region of useful operation of the RF front-end.

1.5.1 Nonlinearity

Let $x(t)$ represent the input of a system which by means of an operator $L\{\cdot\}$ produces output $y(t)$:

$$y(t) = L\{x(t)\}$$ (1.113)

This system is linear if the superposition principle is satisfied at its output. The linear combination of two input signals $x_1(t)$ and $x_2(t)$ produces the linear combination of the corresponding outputs:

$$x(t) = C_1 x_1(t) + C_2 x_2(t)$$
$$y(t) = C_1 L\{x_1(t)\} + C_2 L\{x_2(t)\}$$ (1.114)

Furthermore, a system is memoryless if at a time instant t its output depends only on the current value of the input signal $x(t)$ and not on values of it in the past:

$$y(t) = Cx(t)$$ (1.115)

In this subsection we present the effects of nonlinear memoryless systems that can be described by the general transfer function:

$$y(t) = \sum_{n=0}^{N} \alpha_n x^n(t) \tag{1.116}$$

The above equation specifies a system with nonlinearities of order N.

1.5.1.1 Output Saturation, Harmonics and Desensitization

We assume the input of a memoryless nonlinear system of order $N = 3$ is a simple sinusoid $x(t) = A \cos \omega t$. The output is given by:

$$y(t) = \sum_{n=1}^{3} a_n (A \cos \omega t)^n = \frac{1}{2} a_2 A^2 + \left(a_1 A + \frac{3}{4} a_3 A^3 \right) \cos(\omega t)$$

$$+ \frac{1}{2} a_2 A^2 \cos(2\omega t) + \frac{1}{4} A^3 \cos(3\omega t) \tag{1.117}$$

If we had an ideal linear system the output would be:

$$y(t) = a_1 x(t) \tag{1.118}$$

Coefficient a_1 constitutes the small-signal gain of the ideal system.

Hence, the above equation differs from the ideal in that it contains terms with frequencies 2ω and 3ω. These terms constitute harmonics of the useful input. There exist harmonics of all orders up to the order of the nonlinearity (in this case second and third).

In addition, from the above equation we see that the coefficient of the useful term $\cos(\omega t)$ contains one more term $(3 a_3 A^3 / 4)$ compared with the ideal. Assuming that the nature of nonlinearity does not change with the input signal, a_3 will have a constant value. Hence, the overall, nonideal gain is now a nonlinear function of the input amplitude A:

$$G(A) = A \left[a_1 + \frac{3}{4} a_3 A^2 \right] \tag{1.119}$$

Because, in most cases, $a_3 < 0$, when A starts increasing beyond a point, $G(A)$ stops increasing linearly and starts saturating. Figure 1.14 shows this variation. A usual measure to determine nonlinearity implicitly is the '1-dB compression point', which is defined as the input amplitude at which the output differs by 1 dB with respect to the ideally expected output amplitude. Using Equation (1.117) we get:

$$20 \log \left| \frac{a_1 + \frac{3}{4} a_3 A_{1\,\text{dB}}^2}{a_1} \right| = -1\,\text{dB} \quad \text{or} \quad A_{1\,\text{dB}} = \sqrt{0.145 \left| \frac{a_1}{a_3} \right|} \tag{1.120}$$

Desensitization concerns the impact of a high power interfering signal. To quantify the effect we assume that the input $x(t)$ of the system consists of a useful signal with amplitude A_1 at frequency f_1 and a strong interfering signal with amplitude A_2 at frequency f_2.

$$x(t) = A_1 \cos(\omega_1 t) + A_2 \cos(\omega_2 t) \tag{1.121}$$

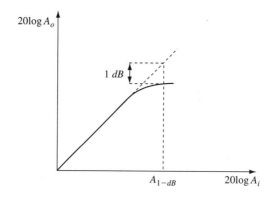

Figure 1.14 Determination of 1 dB compression point in nonlinear transfer function

Taking into account that $A_2 \gg A_1$, a good approximation of the output $y(t)$ is given as [Razavi98]:

$$y(t) = A_1 \left[a_1 + \frac{3}{2} a_3 A_2^2 \right] \cos \omega_1 t + \cdots \tag{1.122}$$

It is easy to realize that if $a_3 < 0$ the amplitude factor $[a_1 + (3a_3 A_2^2)/2]$, which represents the gain at the output of the system, will keep reducing until it gets close to zero. In this case, the receiver is desensitized by the strong interferer and the weaker useful signal is 'blocked'.

1.5.1.2 Distortion due to Intermodulation Products

Intermodulation (IM) products are created by the mixing operation applied on two signals appearing at the input of the receiver at different frequencies. The mixing operation is due to nonlinearities and produces signals falling within the demodulation bandwidth of the receiver. Assuming again that the input is given by two signals of comparable amplitudes as in Equation (1.121), third-order nonlinearity will give an output:

$$y(t) = \sum_{n=1}^{3} a_n (A_1 \cos \omega_1 t + A_2 \cos \omega_2 t)^n \tag{1.123}$$

Expanding Equation (1.123) and using trigonometric identities will give intermodulation products and harmonics the number and strength of which is depicted in Table 1.1.

The most important intermodulation products are, depending on the receiver architecture, the third order and second order. Most wireless communication systems have considerable numbers of user channels (10–100) close to each other. Neighbouring channels are located at distances equal to the channel bandwidth of the system. Hence, in such a system, two neighbouring channels, ch1 and ch2, with strong amplitudes, can create serious problems to the third successive channel ch3 because the third-order products will fall into its useful band. This is depicted in Figure 1.15. If the receiver is tuned to ch3, the neighbouring channels ch1 and ch2 will be treated as interferers and they will deteriorate the SINR (signal-to-interference and-noise ratio).

Table 1.1 Intermodulation products and harmonics and their respective amplitudes

Frequency	Component amplitude
ω_1	$a_1 A_1 + \dfrac{3}{4} a_3 A_1^3 + \dfrac{3}{2} a_3 A_1 A_2^2$
ω_2	$a_1 A_2 + \dfrac{3}{4} a_3 A_2^3 + \dfrac{3}{2} a_3 A_2 A_1^2$
$\omega_1 \pm \omega_2$	$a_2 A_1 A_2$
$2\omega_1 \pm \omega_2$	$\dfrac{3 a_3 A_1^2 A_2}{4}$
$2\omega_2 \pm \omega_1$	$\dfrac{3 a_3 A_1 A_2^2}{4}$

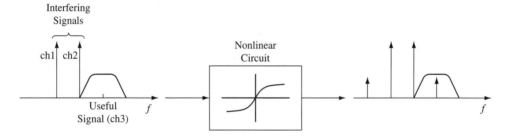

Interfering
Signals

ch1 ch2

Useful
Signal (ch3)

f

Nonlinear
Circuit

f

Figure 1.15 Useful signal and two interfering signals passing through a nonlinear circuit

To characterize third-order IM products, the parameter 'third-order intercept point' is used, also known as IP3. To evaluate IP3 we assume that the input $U_{IM}(t)$ of the nonlinear device creating IM distortion is the summation of two equal amplitude sinusoids tuned at ω_1 and ω_2:

$$U_{IM}(t) = A \cos \omega_1 t + A \cos \omega_2 t \qquad (1.124)$$

Using Table 1.1, the corresponding output, taking into account only the terms at frequencies ω_1, ω_2, $2\omega_1 - \omega_2$ and $2\omega_2 - \omega_1$, is given as:

$$U_{IM\text{-}O}(t) = \left[a_1 + \frac{9}{4} a_3 A^2 \right] U_{IM}(t) + \frac{3}{4} a_3 A^3 [\cos (2\omega_1 - \omega_2)t + \cos (2\omega_2 - \omega_1)t] \qquad (1.125)$$

The input IP3 is defined as the input amplitude for which the components at ω_1, ω_2 and $2\omega_1 - \omega_2$, $2\omega_2 - \omega_1$ of the output signal, have equal amplitudes:

$$\left[a_1 + \frac{9}{4} a_3 A_{IP3}^2 \right] A_{IP3} = \frac{3}{4} a_3 A_{IP3}^3 \qquad (1.126)$$

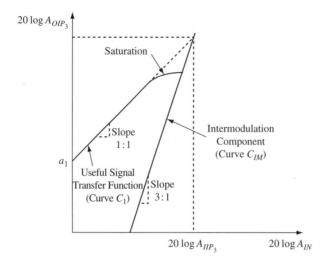

Figure 1.16 Intersection of ideal transfer function and third-order IM component

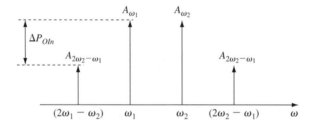

Figure 1.17 Output spectrum of nonlinear circuit when input consists of two tones at frequencies ω_1 and ω_2

Assuming $a_1 \gg \frac{9}{4}a_3 A_{\text{IP3}}^2$, the input and output IP3 are given as:

$$A_{\text{IP3}} = \sqrt{\frac{4}{3}\left|\frac{a_1}{a_3}\right|}, \quad A_{\text{OIP3}} = a_1 A_{\text{IP3}} \tag{1.127}$$

However, as we mentioned before, when the concept of saturation was presented, the above assumption for α_1 is not valid for high input amplitude and the $U_O(t)$ exhibits saturation.

Figure 1.16 shows geometrically the relations between the useful transfer function curve C_1 and the IM curve C_{IM} plotted in logarithmic scale. The above equations indicate that the slope of the latter is three times the slope of C_1 (3 dB/dB vs. 1 dB/dB). Furthermore, due to saturation the intercept point cannot be measured experimentally. However, the measurement can be done indirectly by applying the two tones at the input and noting the difference $\Delta P_{\text{O-IM}}$ at the output of the components at frequencies ω_1, ω_2 and $2\omega_1 - \omega_2$, $2\omega_2 - \omega_1$.

Figure 1.17 shows this relation graphically. Elementary calculations using the above formulas (1.125) and (1.126) show that this relation is [Razavi98]:

$$20 \log A_{\text{IP3}} = \frac{1}{2} 20 \log \left(\frac{A_{\omega_1}}{A_{2\omega_1 - \omega_2}}\right) + 20 \log A_{\text{in}} \tag{1.128}$$

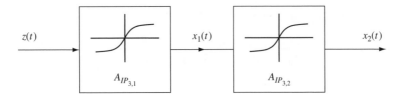

Figure 1.18 Cascading of two nonlinear stages with given IP_3

1.5.1.3 The Third-order Intercept Point of Cascaded Stages

Assume we have two successive stages with order of nonlinearity equal to 3 at the receiver, as illustrated in Figure 1.18. The input $z(t)$ of the first stage produces an output $x_1(t)$ which is used as the input of the next stage giving $x_2(t)$ at its output. The signals $x_1(t)$ and $x_2(t)$ are given as:

$$x_1(t) = \sum_{n=1}^{3} a_n z^n(t) \tag{1.129}$$

$$x_2(t) = \sum_{n=1}^{3} b_n x_1^n(t) \tag{1.130}$$

It is easy to express $x_2(t)$ as a function of $z(t)$ by eliminating $x_1(t)$:

$$x_2(t) = \sum_{n=1}^{6} K_n z^n(t) \tag{1.131}$$

Calculations show that K_1 and K_3 are given by [Razavi98]:

$$K_1 = a_1 b_1, \quad K_3 = a_3 b_1 + 2a_1 a_2 b_2 + a_1^3 b_3 \tag{1.132}$$

A_{IP3} is given by dividing K_1 by K_3 to get:

$$A_{IP3} = \sqrt{\frac{4}{3} \left| \frac{a_1 b_1}{a_3 b_1 + 2a_1 a_2 b_2 + a_1^3 b_3} \right|} \tag{1.133}$$

and consequently:

$$\frac{1}{A_{IP3}^2} = \frac{1}{A_{IP3,1}^2} + \frac{3a_2 b_2}{2b_1} + \frac{a_1^2}{A_{IP3,2}^2} \tag{1.134}$$

The terms $A_{IP3,i}^2$, $i = 1, 2$ give the input IP3 of the two successive stages 1 and 2 respectively. Finally, considering low values for second-order coefficients a_2, b_2 we get:

$$\frac{1}{A_{IP3}^2} \approx \frac{1}{A_{IP3,1}^2} + \frac{a_1^2}{A_{IP3,2}^2} \tag{1.135}$$

When there are more than two stages we have [Razavi98]:

$$\frac{1}{A_{\text{IP3}}^2} \approx \frac{1}{A_{\text{IP3,1}}^2} + \frac{a_1^2}{A_{\text{IP3,2}}^2} + \frac{a_1^2 b_1^2}{A_{\text{IP3,3}}^2} + \cdots \tag{1.136}$$

1.5.2 Noise

As mentioned in the beginning of this section, apart from nonlinearities noise is another factor that has considerable impact on the circuit and system performance. The aim of this subsection is to identify the noise sources in electronic elements (active and passive) and, based on these, to examine the parameters and techniques that will give the noise behaviour of complete circuits and systems.

1.5.2.1 The Noise Model of Bipolar Transistors

The fluctuation of the collector current due to the crossing of holes and electrons through the PN base–collector junction of a bipolar transistor, creates shot noise on the collector current and its spectral density is given by:

$$\overline{i_C^2} = 2qI_C\Delta f \tag{1.137}$$

A similar noise component (shot noise) is created at the base of the transistor originating from recombination at the base and carrier-crossing through the base–emitter junction. Apart from that, it was shown experimentally that two other sources of noise (flicker and burst) contribute to the overall noise at the base of the transistor. Flicker noise has a $1/f$ dependence on frequency whereas burst noise has $1/[1 + (f/f_C)^2]$ frequency dependence. Hence, the base noise spectral density for a bandwidth Δf is given as:

$$\overline{i_B^2} = 2qI_B\Delta f + K_{\text{FN}}\frac{I_B^{\text{f}}}{f}\Delta f + K_{\text{BN}}\frac{I_B^{\text{b}}}{1 + (f/f_C)^2}\Delta f \tag{1.138}$$

where K_{FN}, K_{BN} are multiplication constants and f, b are exponents associated with the base current I_B for flicker and burst noise, respectively.

Finally, there is thermal noise associated to physical resistance at the three terminals, base, collector and emitter, of the transistor:

$$\overline{i_{rb}^2} = \frac{4kT}{r_b}\Delta f, \quad \overline{i_{rc}^2} = \frac{4kT}{r_c}\Delta f, \quad \overline{i_{re}^2} = \frac{4kT}{r_e}\Delta f \tag{1.139}$$

where, k is the Boltzman constant and T is the absolute temperature (usually 290 K). We must note that $\overline{i_{rb}^2}$ is the most important because the other current densities have very low values.

Taking into account the above presentation of noise sources, Figure 1.19 depicts the noise model of bipolar transistors. Note that r_π, r_μ and r_o represent resistances in the noise model without contributing any thermal noise.

1.5.2.2 The Noise Model of MOS Transistors

The major contribution of noise in FETs comes from the channel that exhibits thermal noise due to its resistive behaviour [Gray01]. This is due to modulation of the channel by V_{GS}.

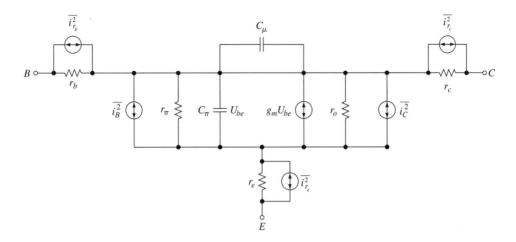

Figure 1.19 Noise model of bipolar transistor

In addition, experiments have shown that a flicker noise term can be included in the drain current spectral density $\overline{i_d^2}$:

$$\frac{\overline{i_d^2}}{\Delta f} = 4kT\gamma g_{do} + K_C \frac{I_d^m}{f}, \quad m = 0.5 - 2 \tag{1.140}$$

where g_{do} is the drain conductance of the device under zero-bias conditions and I_d is the drain current. γ is a factor taking values depending on the bias and the type of the device:

$$\left.\begin{array}{ll} \dfrac{2}{3} \le \gamma \le 1 & \text{for long channel devices} \\[2mm] 2 \le \gamma \le 3 & \text{for short channel devices (L} \le 0.7\,\mu\text{m)} \end{array}\right\} \tag{1.141}$$

Furthermore, a significant noise contributor in MOS devices is the induced gate noise current which is due to fluctuations of the channel charge originating from the drain noise current $4kT\gamma g_{do}$ [Shaeffer97]. This is given by:

$$\frac{\overline{i_g^2}}{\Delta f} = 4kT\delta g_g, \quad g_g = \frac{\omega^2 C_{gs}^2}{5 g_{do}} \tag{1.142}$$

and δ is a coefficient associated with gate noise.

Because induced gate noise is related to the drain noise as seen from the above expressions, the two noise factors are partially correlated:

$$\frac{\overline{i_g^2}}{\Delta f} = 4kT\delta g_g (1 - |c|^2) + 4kT\delta g_g |c|^2 \tag{1.143}$$

$$|c| = \left| \frac{\overline{i_g i_d^*}}{\sqrt{\overline{i_g^2} \cdot \overline{i_d^2}}} \right| \tag{1.144}$$

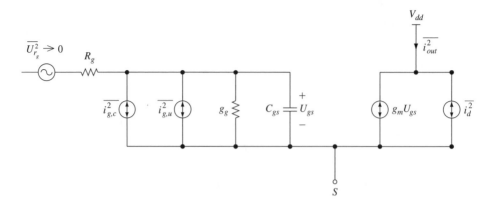

Figure 1.20   MOS transistor noise model

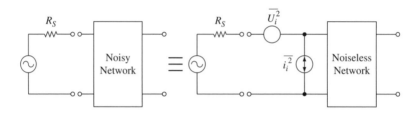

Figure 1.21   Transformation of noisy into a noiseless network with noise sources at the input

Another noise factor, similar to that of bipolar transistors, is shot noise current at the gate:

$$\frac{\overline{i_{gs}^2}}{\Delta f} = 2qI_G \tag{1.145}$$

This is due to gate leakage current and it has been experimentally found that its contribution to the gate noise current is significant at sub-GHz frequencies [Scholten03]. However, the gate noise current is dominated by the induced gate noise as in Equation (1.142) at frequencies above approximately 1 GHz.

Finally, there is a noise voltage due to the distributed gate resistance R_g. Taking into account all of the above, a reasonably accurate equivalent noise model of a MOS device for $f \geq 1$ GHz is as in Figure 1.20. Note that the shot noise component is not included for the reasons presented above and $\overline{U_{R_g}^2}$ is negligible.

1.5.2.3 Noise Performance of Two-port Networks and Noise Figure

According to the two-port noisy network theorem, any two-port noisy network is equivalent to a noiseless network with two noise sources, a voltage and a current noise source, connected at its input, as shown in Figure 1.21. When correlation of the two noise sources $\overline{U_{in}^2}$ and $\overline{i_{in}^2}$ is taken into account, the model is valid for any kind of source impedance.

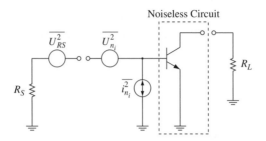

Figure 1.22 Equivalent two-port noise model of bipolar transistor

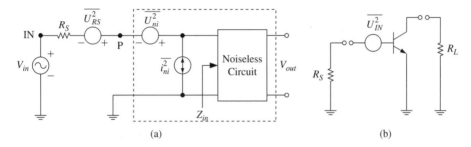

(a)   (b)

Figure 1.23 (a) Two-port network including all noise sources for NF calculation, (b) representation of overall noise (including noise from Rs) with $\overline{U_{IN}^2}$

The above theorem can be applied to a bipolar transistor to produce the equivalent two-port noise model, as shown in Figure 1.22, with input noise voltage and current given by the following expressions [Meyer94]:

$$\overline{U_{ni}^2} = 4kT[r_b + 1/(2g_m)]\Delta f, \quad \overline{i_{ni}^2} = 2q\left[I_B + \frac{I_C}{|\beta(j\omega)|^2}\right]\Delta f \qquad (1.146)$$

where flicker noise has been neglected in the calculation of noise current.

The SNR is defined as the ratio between signal power and noise power at any point along the receiver chain. The noise figure (*NF*) or noise factor (*F*) of a circuit (or system) along the receiver is defined as the amount by which SNR deteriorates when passing through it. Quantitatively *F* is expressed as:

$$F = \frac{\text{SNR}_{in}}{\text{SNR}_{out}} \qquad (1.147)$$

The noise figure is given by $NF = 10\log(F)$.

Figure 1.23(a) shows a two-port network including all noise sources, which is used to find a suitable expression for the noise factor. The source resistance R_S produces noise equal to:

$$\overline{V_S^2} = 4kTR_S \qquad (1.148)$$

We assume that the voltage gain is A_1 from point IN to point P and A_2 from point P to the output. The input SNR is measured at point P and is equal to the ratio of signal power produced by an input signal of amplitude V_{in} to the noise due to the source resistance R_S. Consequently, the expressions for SNR_{in} and SNR_{out} are as follows [Razavi98]:

$$SNR_{in} = \frac{A_1^2 V_{in}^2}{A_1^2 V_{Rs}^2} \tag{1.149}$$

$$SNR_{out} = \frac{V_{in}^2}{\left[\overline{V_{Rs}^2} + \overline{(U_{ni} + I_{ni}R_S)^2} \right] A_1^2 A_2^2} \tag{1.150}$$

Further calculations produce the following expression for F, useful for measurement and simulation:

$$F = \frac{\left[\overline{V_{Rs}^2} + \overline{(U_{ni} + I_{ni}R_S)^2} \right]}{(A_1 A_2)^2} \frac{1}{\overline{V_{Rs}^2}}$$

$$= \left(\frac{\text{Overall output noise power}}{\text{Voltage gain from IN to OUT}} \right) \Big/ (\text{Noise due to } R_S) = \frac{\overline{V_{n,o}^2}}{(A_1 A_2)^2} \frac{1}{4kTR_S} \tag{1.151}$$

As an example of how to use the above considerations to determine the noise figure we use a simple common emitter amplifier followed by subsequent stages, as shown in Figure 1.23(b). Elimination of $(A_1 A_2)^2$ in the first expression of Equation (1.151) shows that the noise factor can also be calculated by dividing the overall equivalent noise at the input $\overline{V_{n,In}^2} = \overline{V_{Rs}^2} + \overline{(U_{ni} + I_{ni}R_S)^2}$ by the noise due to R_S. Assuming U_{ni}, I_{ni} are statistically independent, $\overline{V_{n,In}^2}$ becomes:

$$\overline{V_{n,In}^2} = \overline{V_{Rs}^2} + \overline{U_{ni}^2} + \overline{I_{ni}^2} R_S^2 \tag{1.152}$$

Using the expressions for $\overline{U_{ni}^2}, \overline{I_{ni}^2}$ from Equation (1.146), we can finally obtain:

$$F = \frac{\overline{V_{n,In}^2}}{\overline{V_{Rs}^2}} = \frac{4kTR_S + 4kT[r_b + 1/(2g_m)] + 2q[I_B + I_C/|\beta(j\omega)|^2]R_S}{4kTR_S}$$

$$= 1 + \frac{r_b}{R_S} + \frac{1}{2g_m R_S} + \frac{g_m R_S}{2\beta(0)} + \frac{g_m R_S}{2|\beta(j\omega)|^2} \tag{1.153}$$

$\beta(0)$ is the value of β at DC. It must be noted that the load resistance appearing at the collector due to subsequent stages was not taken into account in this calculation.

1.5.2.4 Noise Figure of N Cascaded Stages

Let N stages at the input of the receiver after the antenna are connected in cascade, as shown in Figure 1.24. Let A_i denote the power gain of the ith stage assuming conjugate matching at both input and output of the corresponding stage. By replacing in the cascaded structure each stage by a noiseless two-port network with noise voltage and current connected at its input, as

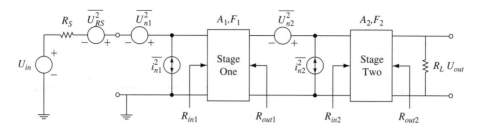

Figure 1.24 Noise model of two receiver stages connected in cascade

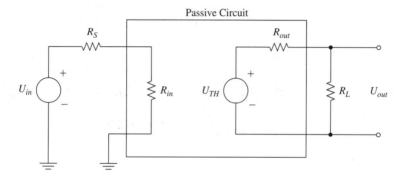

Figure 1.25 Noise model of passive circuit

shown in Figure 1.21, it can be shown after mathematical manipulation that the overall noise factor is given by [Razavi98]:

$$F_T = 1 + (F_1 - 1) + \frac{F_2 - 1}{A_1} + \frac{F_3 - 1}{A_1 A_2} + \cdots + \frac{F_n - 1}{A_1 A_2 \ldots A_{N-1}} \qquad (1.154)$$

This is the Friis noise formula and computes the overall noise factor for N cascaded stages along the receiver chain. Noise factors at each stage are determined by taking into account the impedance appearing at the input of the corresponding stage.

1.5.2.5 Noise Figure of Passive Circuits

A passive circuit is connected to a voltage input V_{in}, with R_S and R_L being the source and load resistances. R_{in} and R_{out} represent the input and output resistances respectively, as shown in Figure 1.25. First, we determine the output noise voltage and the overall voltage gain and subsequently we use Equation (1.151) to calculate the noise figure. This is done by employing the Thevenin equivalent of the output circuit. It is then easy to show that noise figure is equal to the losses of the passive circuit [Razavi98]:

$$F = \text{Losses} = (4kTR_{out}) \left(\frac{V_{in}}{V_{Th}} \right)^2 \left(\frac{1}{4kTR_S} \right) \qquad (1.155)$$

1.6 Sensitivity and Dynamic Range in Radio Receivers

1.6.1 Sensitivity and Dynamic Range

When the receiver is matched in impedance at its input, the noise power delivered to the receiver per unit bandwidth is equal to kT [Razavi98]. Consequently, for a receiver with information bandwidth B, the minimum detectable signal (MDS) represents the lowest possible power of the signal that just exceeds the noise threshold and is given by:

$$N_{\text{TH}} = N_{\text{MDS}} = kTB \qquad (1.156a)$$

For a specific application, the sensitivity of the receiver is defined as the minimum received power for which we can have satisfactory detection. This means that the SNR (or equivalently E_b/N_0), should have the minimum value to guarantee a required bit error rate (BER). For example, a GSM receiver needs approximately 9 dB in SNR in order to achieve satisfactory error rate. In addition, the noise figure of the receiver must be taken into account because, by its nature, it is the principle factor of deterioration of the SNR at the input of the receiver. Consequently, the sensitivity is given in decibels as follows:

$$N_S = 10 \log (kT) + 10 \log B + 10 \log (E_b/N_o) + NF \qquad (1.156b)$$

Another factor that can cause deterioration is the implementation losses L_{imp} associated with implementation of the RF receiver and the modem. For example, synchronization subsystems can contribute an SNR loss in the order of 0.5–2 dB due to remaining jitter of the carrier and time synchronizers. Implementation losses can worsen the receiver sensitivity by a few decibels.

On the other hand, there is a strongest allowable signal that the receiver is able to handle. Above that level, distortion becomes dominant and receiver performance deteriorates rapidly. The difference between the strongest allowable signal and the noise floor is defined as the dynamic range (DR) of the receiver. The highest allowable signal at the input is set as the input power at which the third-order IM products become equal to the system noise floor [Razavi98], [Parssinen01]. Figure 1.26 shows the fundamental and IP3 curves along with the noise floor. Point A is the point representing the strongest allowable signal and consequently DR is equal to the length AB. From the similar triangles it can be easily shown that DR is [Razavi98], [Parssinen01]:

$$DR = \frac{2}{3}(IIP_3 - N_{\text{TH}} - NF) \qquad (1.157)$$

For example, a GSM-900 system has a bandwidth of 200 kHz and requires an SNR of 9 dB. If in addition the receiver has noise figure and IP3 equal to NF = 8 dB and IIP3 = −10 dBm, the DR is 68.7 dB.

1.6.2 Link Budget and its Effect on the Receiver Design

Link budget refers to the analysis and calculation of all factors in the transceiver system in order to ensure that the SNR at the input of the digital demodulator is adequate for achieving satisfactory receiver performance according to the requirements of the application. In doing that, transmitted power and antenna gain must be taken into account in the transmitter. The

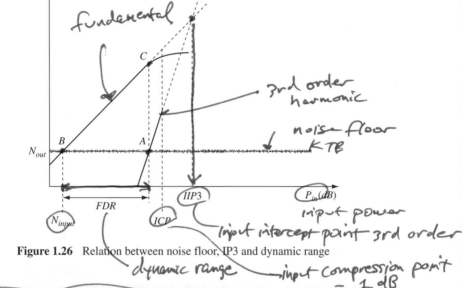

Figure 1.26   Relation between noise floor, IP3 and dynamic range

radio channel losses must be also determined. At the receiver side, sensitivity, noise, distortion and implementation losses have to be carefully determined.

The required sensitivity is a figure determined by taking into account the transmitted power, the gain of the transmitter and receiver antennas, as well as the average expected propagation losses between the transmitter and the receiver for the particular system application:

$$N_{req} = f(EIRP, G_r, L_{prop}) \qquad (1.158)$$

The overall system design and network planning sets the transmitted power levels, the antenna gains and the expected average propagation losses. An additional parameter is the fade margin associated with signal attenuation due to multipath fading. Fade margin is usually a factor of 2–4 dB and can be accommodated within the average propagation loses L_{prop} factor. Taking all these into account at the right-hand side of Equation (1.158), the required sensitivity N_{req} is determined and dictated by the system specification requirements.

Consequently, Equation (1.156) is used to determine the maximum allowable noise figure NF by setting: $N_{req} > N_S$. As an example we mention GSM900, requiring a reference sensitivity of -104 dBm for a 9 dB E_b/N_0. Taking into account that the information bandwidth is 200 kHz and assuming zero implementation losses, Equation (1.156b) gives a maximum allowable NF of 8 dB for the overall receiver.

Table 1.2 illustrates, in the form of an example for multiband UWB [Aiello03] system, the procedure of evaluating the basic quantities in the link budget. On the other hand, regarding linearity, the system requirements usually refer to a maximum level of a neighbouring interfering signal under which the useful signal can still be demodulated with satisfactory SNR. This maximum interferer power lets us calculate the input IP3 of the overall receiver. Hence here we do the inverse compared with the calculation of DR above. We use the required specification to calculate the receiver IIP3. For example, in GSM we should have satisfactory demodulation while two interfering signals of power -43 dBm are present, located 0.8 and 1.6 MHz

Table 1.2 Link Budget calculation for multiband UWB

Bit rate (R_b)	112 Mb/s
Transmitted power	−8.3 dBm
Tx, Rx antenna gains G_T, G_R	0
Path loss at 1 m	44.5 dB
Path loss at 10 m	20 dB
Rx power (at 10 m) $= P_r = P_t + G_T + G_R - L_1 - L_2$	−72.8 dBm
Rx noise figure at antenna terminal	7 dB
Noise power per bit [$N_{Th} = -174 + 10\log{(R_b)} + NF$]	−86.5 dBm
Minimum E_b/N_0	3.6 dB
Implementation losses (IL)	3 dB
Code rate	0.5
Raw bit rate	224 Mb/s
Link margin at 10 m	7.1 dB

away from the useful carrier. Input IP3 can be calculated by using $IIP_3 = P_{INT} - IM_3/2$. In our example, $P_{INT} = -43$ dBm and $IM_3 = -104 - 4 - (-43) = -65$ dB. Consequently we get $IIP_3 = -10.5$ dB.

From the above considerations it is easy to realize that the link budget involves a balancing procedure resulting in specific requirements for noise and linearity for the overall receiver. The overall noise and linearity requirements can then be translated into noise and linearity specifications for each circuit in the receiver chain (LNA, filters, mixers, etc.), taking into account the Friis formula for noise figure calculation of the cascaded receiver stages and the corresponding formula for overall IP3 calculation presented in Section 1.5.

1.7 Phase-locked Loops

1.7.1 Introduction

The phase-locked loop is one of the most frequently used subsystems in communications for detection, synchronization and frequency generation. We briefly present below the basic principles and operation of linear and hybrid PLLs. Linear PLLs are the ones using a linear analogue multiplier as a phase detector (PD), whereas 'hybrid' refers to the PLLs that use both digital and analogue components. Sometimes they are also called 'mixed-signal' PLLs [Best03].

Linear and hybrid PLLs represent the great majority of PLLs used in most applications today. All-Digital PLLs (ADPLL) are mostly used in purely digital subsystems where jitter requirements are not very stringent.

1.7.2 Basic Operation of Linear Phase-locked Loops

Figure 1.27 illustrates the basic structure of a linear PLL. Let us assume that its input and output are expressed as follows:

$$y_i(t) = A \cdot \sin{[\omega_i t + \varphi_i(t)]}$$
$$y_o(t) = B \cdot \cos{[\omega_o t + \varphi_o(t)]} \tag{1.159}$$

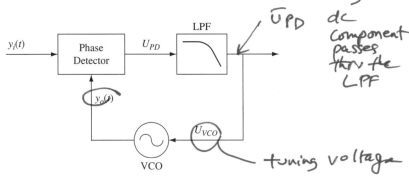

U_{PD} dc component passes thru the LPF

tuning voltage

Figure 1.27 Phase-locked loop block diagram

In more detail, it is a feedback system in which the incoming signal $y_i(t)$ and the output signal of the VCO $y_o(t)$ are used as the inputs for a phase detector, which compares the phase quantities $[\omega_i t + \varphi_i(t)]$ and $[\omega_o t + \varphi_o(t)]$ of the two signals. The PD generates an output signal U_{PD} which is proportional to the phase difference of y_i and y_o. The lowpass filter following the PD passes only the DC component $\overline{U}_{PD}$. Subsequently, the voltage-controlled oscillator (VCO) is commanded by $\overline{U}_{PD}$. We use U_{VCO} to represent the tuning voltage at the input of the VCO, which changes its output frequency and phase such that its frequency becomes equal to the frequency of the input signal $y_i(t)$, whereas its phase locks to the phase of $y_i(t)$. The above equations give the state of the system while the feedback path is still open. By closing it, $y_o(t)$ changes its phase and frequency until it locks to the input. At that time the output signal is given by:

$$y_o(t) = B \cdot \cos[\omega_i t + \psi_o] \tag{1.160}$$

Before the system reaches its steady-state condition (final locking), the output of the phase detector can be represented by the low frequency component of the product of y_i and y_o:

$$\overline{U}_{PD} = K_{PD} \cos[(\omega_i - \omega_o(t)) + \varphi_i - \varphi_o(t)] \tag{1.161}$$

where K_{PD} is the sensitivity constant of the phase detector and φ_o is a function of time representing the dynamic change of the output phase and frequency of the VCO during the locking procedure.   *positive increase in freq*

Figure 1.28 shows the procedure of locking when a step Δf occurs at the frequency of the input signal y_i at $t = t_0$. The frequency difference is translated into a phase difference $\theta_e = \Delta f \cdot t$. Hence, in terms of phase, in the beginning the phase error keeps increasing. However, application of this DC voltage at the input of the VCO forces it to increase its output frequency which has as a consequence the continuous reduction of Δf and, therefore, of the corresponding phase error. Hence, as depicted in the figure, after an initial increase of the phase error and $\overline{U}_{PD}$, at t_1 and t_2, the resulting increase of the VCO frequency will decrease the phase error (see time instants t_3, t_4) until the system becomes locked.   *ω*

The equation of operation of the VCO is given in terms of angular frequency as:

$$\omega_{inst} = \omega_o + K_{VCO} U_{VCO} = \frac{d}{dt}[\omega_o(t) t + \varphi_o(t)], \qquad \frac{d\varphi_o(t)}{dt} = K_{VCO} U_{VCO} \tag{1.162}$$

where K_{VCO} is the sensitivity constant of the VCO in Hz/V or rad/sec/V.

tuning voltage
"U" is used for voltage !

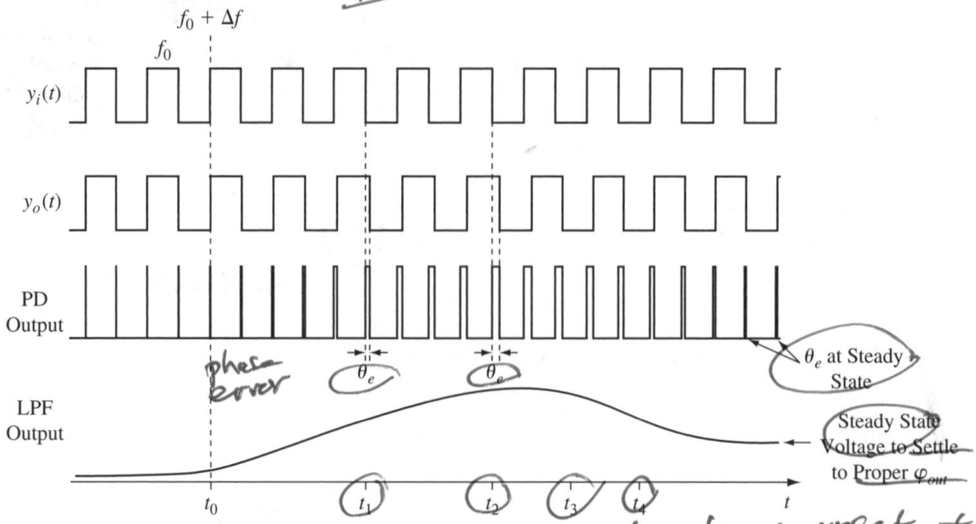

[handwritten top: "PD output → LPF → LPF output"]

[handwritten annotations on figure: "phase error", "θₑ", "t₃ t₄ represent time when the phase error has decreased"]

Figure 1.28 PLL's basic behaviour in the time domain when a frequency step of the input signal occurs

If $f(t)$ and $F(j\omega)$ are the impulse response and transfer function of the LPF respectively, the output of the LPF and input of the VCO is given by:

[handwritten: "the tuning voltage"]   *[handwritten: "convolution see wikipedia"]*

$$U_{VCO}(t) = U_{PD}(t) * f(t) \qquad (1.163)$$

We assume, for purposes of simplicity, that the frequency of the two signals y_i and y_o is the same. By combining the above equations, we obtain:

$$\frac{d\varphi_0(t)}{dt} = K_{PD}K_{VCO} \cdot \left\{ \sin\left[\varphi_i(t) - \varphi_0(t)\right] * f(t) \right\} \qquad (1.164)$$

1.7.3 The Loop Filter

The loop filter, apart from eliminating possible high-frequency components at the output of the phase detector, also affects the stability of the feedback system. In most cases, three types of filters are used.

(1) The simple RC filter (which we call S-RC) with transfer function

$$F(j\omega) = \frac{1}{1 + j\omega\tau_1}, \qquad \tau_1 = RC \qquad (1.165)$$

Figure 1.29(a) shows the circuit and the magnitude of the transfer function.

(2) When the capacitor in parallel is replaced by a combination of a capacitor and a resistor we have what we call the lowpass with phase-lead filter (LP-PL). Its transfer function is given by:

$$F(j\omega) = \frac{1 + j\omega\tau_2}{1 + j\omega\tau_1}, \qquad \tau_2 = R_2C, \quad \tau_1 = (R_1 + R_2)C \qquad (1.166)$$

[handwritten: "τ₁ includes R₁ and R₂"]

[handwritten: "if τ₁ = R₂C then F jw = (1 + jωτ₂)/(1 + jω(τ₂ + τ₂)) as shown in Fig 1.29(b)"]

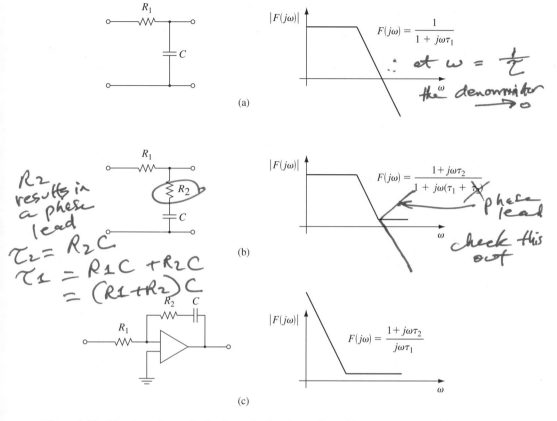

The handwritten notes read:

(on figure a) $\therefore$ at $\omega = \frac{1}{\tau}$ the denominator $\to 0$

$F(j\omega) = \frac{1}{1 + j\omega\tau_1}$

(on figure b) $F(j\omega) = \frac{1+j\omega\tau_2}{1 + j\omega(\tau_1 + \tau_2)}$

phase lead — check this out

R_2 results in a phase lead

$\tau_2 = R_2 C$

$\tau_1 = R_1 C + R_2 C = (R_1 + R_2)C$

(on figure c) $F(j\omega) = \frac{1+j\omega\tau_2}{j\omega\tau_1}$

Figure 1.29 Circuits and magnitude of transfer functions of PLL filters: (a) simple RC filter, (b) lowpass filter with phase lead, (c) active filter

(handwritten) due to $R_2 C$ time constant $\tau_2 = R_2 C$

Figure 1.29(b) depicts the circuit and transfer function for this type of filter.

(3) The third choice is an active filter, the circuit and transfer function of which are illustrated in Figure 1.29(c). By analysing the circuit we can easily obtain the corresponding equation for the transfer function:

$$F(j\omega) = -G\frac{1+j\omega\tau_2}{1+j\omega\tau_1}, \quad \tau_2 = R_2 C, \ \tau_1 = (R_1 + GR_1 + R_2)C \qquad (1.167)$$

If the gain of the filter G is high, the above transfer function can be approximated by:

$$F(j\omega) \approx -G\frac{1+j\omega\tau_2}{j\omega\tau_1} \qquad (1.168)$$

The denominator corresponds to the integrator function and consequently we call this filter integrator with phase lead filter (I-PL).

1.7.4 Equations and Dynamic Behaviour of the Linearized PLL

We use Equation (1.164) to determine the linear equation for PLL and its transfer function for all three types of filters presented above. For this purpose, we consider that phase difference $(\varphi_i(t) - \varphi_0(t))$ is small and therefore $\sin[\varphi_i(t) - \varphi_0(t)] \approx \varphi_i(t) - \varphi_0(t)$. Hence, the resulting equation becomes:

$$\frac{d\varphi_0(t)}{dt} = K_{PD}K_{VCO} \cdot \{[\varphi_i(t) - \varphi_0(t)] * f(t)\} \tag{1.169}$$

We assume that the PLL remains locked. By taking Fourier transforms, we obtain the PLL transfer function and error transfer function in terms of the overall open loop gain K after elementary manipulations [Blanchard78]:

$$H(j\omega) = \frac{\Phi_0(j\omega)}{\Phi_i(j\omega)} = \frac{KF(j\omega)}{j\omega + KF(j\omega)}, \quad H_e(j\omega) = \frac{j\omega}{j\omega + KF(j\omega)} \tag{1.170}$$

By inserting in the above equations the expressions for $F(j\omega)$ presented in the previous section, we can determine expressions for the PLL transfer function in terms of K and the time constants of the filters.

In the case where there is no filter in the loop $[F(j\omega) = 1]$, the PLL transfer function is given by:

$$H(j\omega) = \frac{K}{j\omega + K} \tag{1.171}$$

Similarly, for the PLL with filter $F(j\omega) = (1 + j\omega\tau_2)/j\omega\tau_1$, the transfer function becomes:

$$H(j\omega) = \frac{K + j\omega K\tau_2}{K - \omega^2\tau_1 + j\omega K\tau_2} \tag{1.172}$$

Replacing the $j\omega$ operator by the Laplace operator s, we have:

$$H(s) = \frac{(K\tau_2)s + K}{\tau_1 s^2 + (K\tau_2)s + K} \tag{1.173}$$

In similar fashion, for the other two lowpass filters (S-RC and LP-PL) we obtain:

$$H(s) = \frac{K}{\tau_1 s^2 + s + K} \tag{1.174}$$

$$H(s) = \frac{(K\tau_2)s + K}{\tau_1 s^2 + (K\tau_2 + 1)s + K} \tag{1.175}$$

Taking into account the feedback systems theory we observe that, by elementary manipulations, the denominator in all cases can be expressed in the form $s^2 + 2\zeta\omega_n s + \omega_n^2$. After straightforward calculations we obtain the expressions presented in Table 1.3 for the three different filters.

By using Fourier transform properties in the time domain, we can easily derive the differential equations for first-order PLLs (no filter) and second-order PLLs (using one of the previously presented LPFs). For these cases the corresponding equations are [Blanchard78]:

$$\frac{d\varphi_0(t)}{dt} + K\varphi_0(t) = K\varphi_i(t) \tag{1.176}$$

Table 1.3 Transfer function and loop parameters for second order PLL with different filters

Filter type	$H(s)$	$\omega_n^2, 2\zeta\omega_n$
$1/(1+j\omega\tau_1)$	$\dfrac{\omega_n^2}{s^2+2\zeta\omega_n s+\omega_n^2}$	$\omega_n^2=\dfrac{K}{\tau_1},\quad 2\zeta\omega_n=\dfrac{1}{\tau_1}$
$\dfrac{(1+j\omega\tau_2)}{j\omega\tau_1}$	$\dfrac{2\zeta\omega_n s+\omega_n^2}{s^2+2\zeta\omega_n s+\omega_n^2}$	$\omega_n^2=\dfrac{K}{\tau_1},\quad 2\zeta\omega_n=\dfrac{K\tau_2}{\tau_1}$
$\dfrac{(1+j\omega\tau_2)}{(1+j\omega\tau_1)}$	$\dfrac{(2\zeta\omega_n-\omega_n^2/K)s+\omega_n^2}{s^2+2\zeta\omega_n s+\omega_n^2}$	$\omega_n^2=\dfrac{K}{\tau_1},\quad 2\zeta\omega_n=\dfrac{1+K\tau_2}{\tau_1}$

$$\tau_1\frac{d^2\varphi_o(t)}{dt^2}+K\tau_2\frac{d\varphi_o(t)}{dt}+K\varphi_o(t)=K\tau_2\frac{d\varphi_i(t)}{dt}+K\varphi_i(t) \tag{1.177}$$

Observing that the first equation is a first-order differential equation (DE) and the second is a second-order DE, we define the loop without filter as a first-order PLL, whereas the loop with one of the LPFs presented above is called a second-order PLL.

It is of great interest to determine the dynamic behaviour of the PLL under various kinds of excitation such as an abrupt change in phase or frequency of the input signal. The behaviour of the system to such changes finds application to phase and frequency demodulation and carrier recovery techniques, as we shall see in Section 1.7.8. By using standard Laplace transform techniques, the resulting time domain function can be obtained for each kind of excitation.

As an example, let us assume that we have a unit step change at the phase of the input signal $y_i(t)$ which in time and Laplace domains is given by:

$$\varphi_i(t)=\Delta\varphi u(t),\quad \Phi_i(s)=\frac{\Delta\varphi}{s} \tag{1.178}$$

where $u(t)$ represents the unit step function.

Using the expression for the error function $H_e(s)$ from Equation (1.170) we can obtain expressions for $\Phi_e(s)$ and subsequently for $\varphi_e(t)$. For example, for the integrator with phase-lead filter we get:

$$\Phi_e(s)=\frac{\tau_1 s\Delta\varphi}{\tau_1 s^2+K\tau_2 s+K}=\frac{s\Delta\varphi}{s^2+2\zeta\omega_n s+\omega_n^2} \tag{1.179}$$

By splitting the denominator into a product of the form $(s+\alpha)(s+\beta)$ and taking the inverse Laplace transform, we obtain the following expression for $\varphi_e(t)$:

$$\varphi_e(t)=\begin{cases} \Delta\varphi\exp[-\zeta\omega_n t]\left\{\cosh\left[\omega_n\sqrt{\zeta^2-1}\,t\right]-\dfrac{\zeta}{\sqrt{\zeta^2-1}}\sinh\left[\omega_n\sqrt{\zeta^2-1}\,t\right]\right\}, & \zeta>1 \\[2mm] \Delta\varphi\exp[-\omega_n t]\,(1-\omega_n t), & \zeta=1 \\[2mm] \Delta\varphi\exp[-\zeta\omega_n t]\left\{\cos\left[\omega_n\sqrt{1-\zeta^2}\,t\right]-\dfrac{\zeta}{\sqrt{1-\zeta^2}}\sin\left[\omega_n\sqrt{1-\zeta^2}\,t\right]\right\}, & \zeta<1 \end{cases}$$

$$\tag{1.180}$$

Figure 1.30 shows the normalized response of the phase error $\varphi_e/\Delta\varphi$ as a function of normalized (with respect to natural frequency ω_n) time. These equations and corresponding

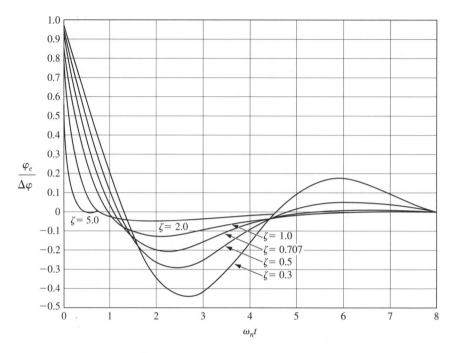

Figure 1.30 Resulting phase error as a response to a phase step $\Delta\varphi$ (from 'Phaselock Techniques', 2nd edn, F. Gardner, copyright © 1979 by John Wiley & Sons)

diagrams are very useful to determine the time needed for the system to settle to the new phase or frequency.

Sometimes it is of interest to know only the steady-state value of the phase or frequency in order to realize whether the system will finally lock or not. In that case, it is only necessary to use the final value theorem of Laplace transform:

$$\lim_{t\to\infty}[\varphi_e(t)] = \lim_{s\to 0}[s\Phi_e(s)] \tag{1.181}$$

For example, if a step change $\Delta\omega \cdot u(t)$ of the input frequency is applied at the input of the PLL, the steady-state error for a second-order loop is given by:

$$\lim_{t\to\infty}[\varphi_e(t)] = \lim_{s\to 0}\left[s\frac{\Delta\omega/s^2}{1 + \frac{K}{s(1+s\tau_1)}}\right] = \frac{\Delta\omega}{K} \tag{1.182}$$

At this point it is helpful to define lock range, hold range and lock time.

Lock range is defined as the frequency range within which the PLL will lock, while initially unlocked, within one cycle. Hold range represents the frequency range within which the PLL will remain locked while it is already in lock.

Straightforward analysis shows that the hold range $\Delta\omega_{HR}$ is the maximum value that $\Delta\omega$ takes in the following equation [Gardner05]:

$$\lim_{t\to\infty}[\sin\varphi_e(t)] = \frac{\Delta\omega}{K_{PD}K_{VCO}F(0)} \tag{1.183}$$

Since the maximum value of $\sin \varphi_e(t)$ is equal to 1 (for $\varphi_e = \pi/2$), the hold range is given by

$$\Delta\omega_{\mathrm{HR}} = K_{\mathrm{PD}}K_{\mathrm{VCO}}F(0) \tag{1.184}$$

The typical lowpass with phase lead filter with low frequency gain $F(0) = K_F$ will have a hold range of $K_{\mathrm{PD}}K_{\mathrm{VCO}}K_F$ whereas the perfect integrator filter I-PL theoretically exhibits infinite hold range. However, its hold range is limited by the tuning frequency range of the VCO.

On the other hand, in order to determine the lock range, we initially assume that the PLL is unlocked and the frequency applied at its input is $\omega_{iS} = \omega_0 + \Delta\omega$. Assuming that the phase detector is a simple multiplier and that the high frequency term is filtered by the LPF, the signal at the output of the LPF (input of the VCO) is:

$$U_F = U_{\mathrm{VCO}} \approx K_{\mathrm{PD}}|F(j\Delta\omega)| \sin(\Delta\omega t + \varphi_i(t)) \tag{1.185}$$

This represents a slow sinusoid modulating the VCO. Thus, the VCO output frequency from Equation (1.162) is given by:

$$\omega_i = \omega_0 + K_{\mathrm{VCO}}K_{\mathrm{PD}}|F(j\Delta\omega)| \sin(\Delta\omega t) \tag{1.186}$$

The above equation indicates that the VCO output frequency increases and decreases in a sinusoidal fashion with upper and lower limits:

$$\omega_i = \omega_0 \pm K_{\mathrm{VCO}}K_{\mathrm{PD}}|F(j\Delta\omega)| \tag{1.187}$$

In order for locking to take place within one period, the upper limit $\omega_{i\,\max}$ of the VCO frequency ω_i, should exceed the input frequency ω_{iS} and consequently:

$$K_{\mathrm{VCO}}K_{\mathrm{PD}}|F(j\Delta\omega)| \geq \Delta\omega \tag{1.188a}$$

The upper limit of $\Delta\omega$ in Equation (1.188a) gives the lock range $\Delta\omega_L$:

$$K_{\mathrm{VCO}}K_{\mathrm{PD}}|F(j\Delta\omega_L)| = \Delta\omega_L \tag{1.188b}$$

To solve this nonlinear equation, approximate expressions for $F(j\Delta\omega_L)$ must be used. These give approximately the same value for all types of filters [Best03]:

$$\Delta\omega_L \approx 2\zeta\omega_n \tag{1.189}$$

To find the lock-in time the transient response of the second-order PLL must be determined. It is shown that the transient response is confined to the steady-state value within a small percentage of it, in approximately one signal period. Hence, the lock in time can be expressed as:

$$T_L = \left(\omega_n/2\pi\right)^{-1} \tag{1.190}$$

1.7.5 Stability of Phase-locked Loops

The requirement for stability is that the phase of the open-loop transfer function $G(j\omega)$ at the gain crossover frequency ω_{CO} (at which $|G(j\omega)| = 1$) is higher than $-180°$:

$$\angle G(j\omega_{\mathrm{CO}}) \geq -180° \tag{1.191}$$

Taking into account that the open-loop gain is expressed as:

$$G(s) = \frac{K_{\mathrm{PD}}K_{\mathrm{VCO}}F(s)}{s} \tag{1.192}$$

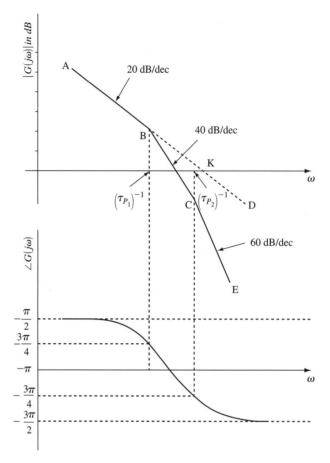

Figure 1.31 Magnitude and phase Bode plots of G(s) for loop with no filter (ideal and lowpass behaviour of VCO and PD)

we distinguish the following cases:

(1) *No filter.* In that case, the Bode diagrams of the magnitude and phase for $G(s)$ are shown in Figure 1.31, indicated by the ABD straight line and a constant $-\pi/2$ phase. It is obvious that in this case the system is stable. However, the filter is necessary after the PD for proper operation. Even if such a filter is absent, the phase detector and VCO exhibit lowpass behaviour at the respective outputs. This behaviour can be modelled as an LPF at the output of the respective components with poles at $1/\tau_{P1}$ and $1/\tau_{P2}$, where τ_{P1} and τ_{P2} represent the time constants of the PD and VCO respectively. Hence, the final open-loop transfer function will become:

$$G(j\omega) = \frac{K_{PD}K_{VCO}K_F}{j\omega(1+j\omega\tau_{P1})(1+j\omega\tau_{P2})} \qquad (1.193)$$

Figure 1.31 shows hypothetical but possible Bode plots for such a system. It is easy to realize that depending on the values of τ_{P1}, τ_{P2} and K, the system can become unstable.

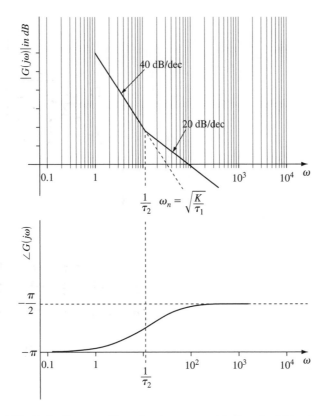

Figure 1.32   Bode diagram with perfect integrator with phase lead filter $(\omega_n > 1/\tau_2)$

(2) *Perfect integrator with phase lead filter.* Figure 1.32 shows the Bode diagrams for the case that $\omega_n > 1/\tau_2$, for which stability can be very satisfactory. The condition for stable operation is: $\omega_n \tau_2 > 2\zeta > 1$. Hence, high values of ζ guarantee good stability.

(3) *Filter with LPF and phase-lead correction.* This filter has an open-loop transfer function given by:

$$G(j\omega) = \frac{K(1 + j\omega\tau_2)}{j\omega(1 + j\omega\tau_1)} \tag{1.194}$$

Because of its two time constants, this filter permits to independently choose values for τ_1 (and subsequently ω_n) and τ_2 and ζ, which is a function of τ_2. Hence, no matter how low the value for ω_n is chosen, increasing τ_2 provides better stability [Blanchard78].

1.7.6 Phase Detectors

The phase detector is one of the principle elements of the PLL as it serves as the component closing the feedback loop through its second input coming from the VCO. The most popular phase detectors are the analogue multiplier and the digital PDs based on flip-flops and logic circuitry. The analogue multiplier performs direct multiplication of the two input signals y_i and y_o resulting in a useful term of the form $\sin\left[((\omega_i - \omega_o(t)) + \varphi_i - \varphi_o(t)\right]$ as given previously

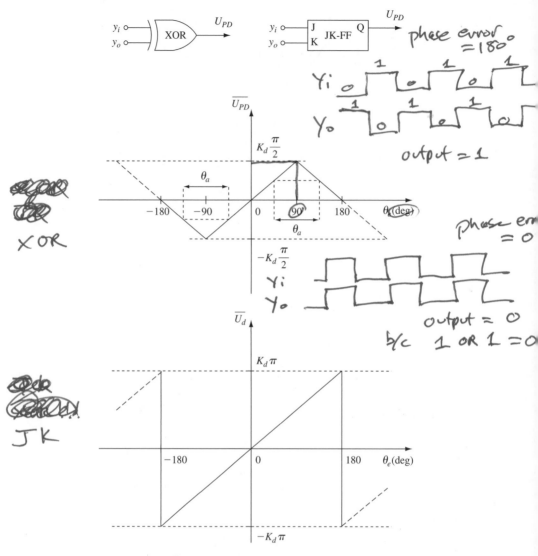

Figure 1.33 Exclusive-OR and JK flip-flop phase detectors with their characteristics

in Equation (1.161). Disregarding the frequency difference, the output signal is a nonlinear function of the phase difference $\sin[\varphi_i - \varphi_o(t)]$, resulting in a major disadvantage. However, when the input signal y_i contains considerable noise, this PD exhibits superior performance compared with the digital detectors. On the other hand, owing to the advances of digital technology, digital PDs have dominated most applications excluding those involving very high-speed analogue signals where the analogue detector seems irreplaceable [Gardner05]. The most popular digital PD are the exclusive-OR PD, the Edge-triggered JK-flip flop (ET-JK) detector and the phase-frequency detector (PFD).

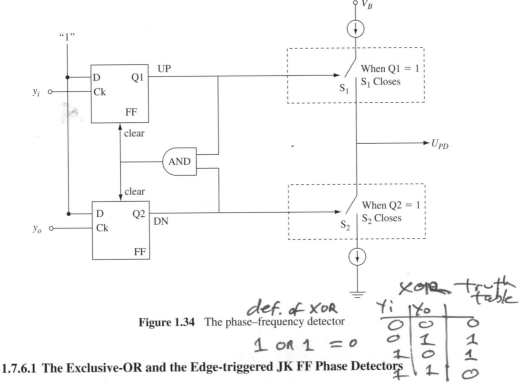

Figure 1.34   The phase–frequency detector

handwritten: def. of XOR
handwritten: 1 OR 1 = 0

handwritten: XOR truth table

Y_i	Y_o	
0	0	0
0	1	1
1	0	1
1	1	0

1.7.6.1 The Exclusive-OR and the Edge-triggered JK FF Phase Detectors

Figure 1.33 shows both detectors with inputs and output y_i, y_o and U_{PD} respectively. The exclusive-OR phase detector gives a high-level (logic '1') output when the two input signals have different levels and a logic '0' when they both have high or low levels. Based on that, the exclusive-OR characteristic is also given in Figure 1.33.

However, when one of the signals has a duty cycle different from 50%, the characteristic curve exhibits flattening. This happens because the detector is insensitive to phase changes during time periods so that the high level of one of the inputs is not crossed by the falling or the rising edge of the other.

The edge-triggered-JK flip-flop produces a high level at the output when a rising edge occurs at input y_i whereas it gives a low-level when a rising edge takes place at the second input y_o. Figure 1.33 also shows the JK PD characteristic exhibiting a linear range that is twice as high as that of the exclusive-OR detector. The edge-triggered-JK flip-flop does not have the same problem with that of exclusive-OR detector since the logic levels '1' and '0' at the output result from occurrences of rising edges and not from durations of logic states '1' and '0' at the two inputs.

1.7.6.2 The Phase-frequency Detector

Figure 1.34 depicts the PFD consisting of two D flip-flops with their clock inputs connected to the PD input signals y_i and y_o. Their Q-outputs represent the UP and DN PFD digital outputs the combination of which gives the current state of PFD. Three combinations of logic levels for UP and DN are possible, whereas the fourth (UP = DN = '1') is prohibited through the AND gate. Table 1.4 illustrates all possible states and transitions, which actually can be represented

Table 1.4 Phase-frequency detector states and transitions

Current state	Output signal U_{PD}	Next state for rising edge of y_i	Next state for rising edge of y_o
State $= '-1'$: UP $= '0'$, DN $= '1'$	$U_{PD} = 0$ Volts	'0'	'−1'
State $= '0'$: UP $= '0'$, DN $= '0'$	High impedance	'+1'	'−1'
State $= '+1'$: UP $= '1'$, DN $= '0'$	$U_{PD} = V_B$ Volts	'+1'	'0'

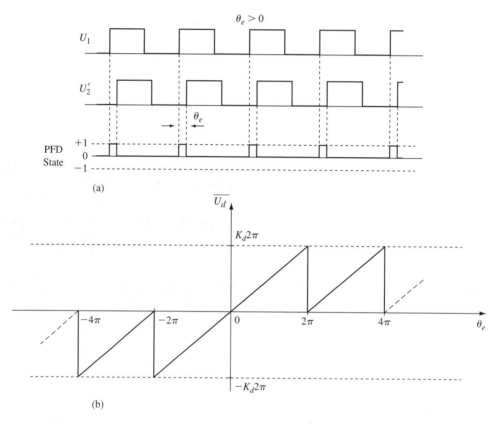

(a)

(b)

Figure 1.35 (a) Output states of the PFD with the first input leading the second, (b) characteristics of the PFD

graphically by a state diagram [Best03]. It is easy to realize that the three possible states correspond to three states for the output U_{PD} of the PD as shown at the table, taking into account that switches S_1 and S_2 are controlled by UP and DN respectively as shown in Figure 1.34.

Figure 1.35(a) shows the output states of the PFD when y_i leads in phase compared with y_o, whereas Figure 1.35(b) illustrates the characteristic of the PD for the average $\overline{U}_{PD}$. The value of $\overline{U}_{PD}$ depends on the average occupancy in time of a particular state. However, when the frequency of the two signals y_i and y_o is different, the average occupancy of a particular state will be higher than the other states. Consequently, $\overline{U}_{PD}$ changes as a function of frequency making the PFD sensitive in frequency variations, as well.

1.7.7 PLL Performance in the Presence of Noise

Let the input y_i at the phase detector be corrupted by additive noise:

$$y_i(t) = V_S \sin(\omega_i t + \varphi_i) + n(t) \tag{1.195a}$$

The output of the VCO and second input of the PD is given by:

$$y_o(t) = V_o \cos(\omega_i t + \varphi_o) \tag{1.195b}$$

where θ_o can be initially treated as time invariant for reasons of convenience. It can be shown [Gardner79] that the output signal of the PD can be expressed as:

$$U_{PD} = K[\sin(\varphi_i - \varphi_o)] + n'(t) \tag{1.196}$$

where K is a constant representing the product $V_S V_O K_{PD}$. The variance $\sigma_{n'}^2$ of the equivalent noise $n'(t)$ as a function of input SNR, SNR_i, is given by:

$$\sigma_{n'}^2 = \frac{\sigma_n^2}{V_S^2} = \frac{P_n}{2P_S} = \frac{1}{2SNR_i} \tag{1.197}$$

where P_S and P_n are the signal and noise power at the input of the PLL, respectively. The noise at the output of the PD $n'(t)$ could also be created by a phase disturbance $\sin \varphi_{ni}(t) = n'(t)$. When the signal-to-noise ratio at the input is high, the disturbance can be linearized due to small $\varphi_{ni}(t)$ [Gardner79]. In this case, it can be considered that the noise is created by an input phase variance $\overline{\varphi_{ni}^2}$ which represents the jitter of the input signal due to phase noise.

Calculation of the spectral density of the noise $n'(t)$ gives [Gardner79]:

$$\Phi_{n'}(f) = \frac{2N_0}{V_S^2} \tag{1.198}$$

Consequently, since the PLL transfer function is $H(j\omega)$ the spectral density of the VCO output phase noise $\Phi_{no}(f)$ and the corresponding variance are given by:

$$\Phi_{no}(f) = \Phi_{n'}(f)|H(j\omega)|^2 \tag{1.199}$$

$$\overline{\varphi_{no}^2} = \int_0^\infty \Phi_{n'}(f)|H(j2\pi f)|^2 df \tag{1.200}$$

When the noise density of $n(t)$ at the input is white in the band of interest (corresponding to the bandwidth of a bandpass filter in front of the PLL), then $\Phi_n(f) = N_0$ and the VCO output phase noise is:

$$\overline{\varphi_{no}^2} = \frac{2N_0}{V_S^2} \int_0^\infty |H(j2\pi f)|^2 df \tag{1.201}$$

As noted in a previous section, the above integral represents the equivalent noise bandwidth:

$$B_L = \int_0^\infty |H(j2\pi f)|^2 df \tag{1.202}$$

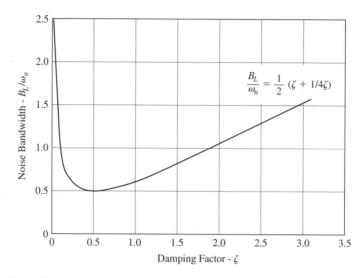

Figure 1.36 Normalized noise bandwidth of a second order PLL (from 'Phaselock Techniques', 2nd edn, F. Gardner, copyright © 1979 by John Wiley & Sons, Inc.)

$\overline{\varphi_{no}^2}$ represents the phase jitter at the output of the VCO and, like the noise at the input, it can be associated with the signal-to-noise ratio of the loop SNR$_L$:

$$\overline{\varphi_{no}^2} = \frac{1}{2SNR_L} \tag{1.203}$$

Consequently, the PLL improves the SNR of the signal at its input as follows:

$$SNR_L = SNR_i \cdot \frac{B_i}{2B_L} \tag{1.204}$$

where B_i is the bandwidth of a bandpass filter at the input of the PLL.

Figure 1.36 illustrates the noise bandwidth of a second-order PLL, normalized to the natural angular frequency ω_n. One can realize that there is an optimum value for ζ close to 0.5 resulting in the minimum noise bandwidth B_L.

1.7.8 Applications of Phase-locked Loops

Phase-locked loops have a wide variety of applications in radio communications ranging from frequency synthesizers to modulators and demodulators of analogue and digital signals. Furthermore, they constitute important building blocks in carrier recovery and synchronization in coherent receivers.

The principle of frequency synthesizers is based on deriving an output frequency *fvco*, which in the most general case could be the linear combination of a number of reference frequencies. In the most usual case, *fvco* is an integer multiple of an input frequency *f$_{in}$* used as reference frequency. This is widely known as the integer-N frequency synthesizer. The synthesizer should be able to generate a wide range of frequencies necessary for down-conversion in the receiver or up-conversion in the transmitter. The number of frequencies and frequency

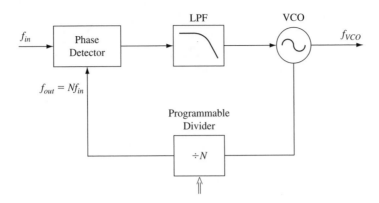

Figure 1.37 Typical integer-N frequency synthesizer

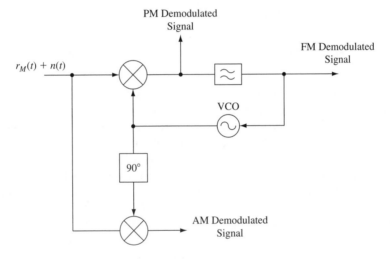

Figure 1.38 FM demodulator using PLL

resolution (minimum frequency step) of the synthesizer are dictated by the application. Phase noise generated by the synthesizer is an important issue as it has a considerable impact on the receiver performance (this will be examined later in Chapter 4). It can be shown that in a simple integer-N configuration the phase noise of the output signal f_o of the VCO follows the phase noise of the reference frequency (which is usually low) within the loop bandwidth. This rule aids in compromising frequency resolution for phase noise. Figure 1.37 illustrates a typical integer-N synthesizer. One can notice that it is a classical PLL with a programmable divider ($\cdot/N$) inserted at the feedback loop at the output of the VCO.

When the input at the phase detector is an FM/FSK modulated signal (including noise) $r_M + n(t)$, the output of the LPF (and input of the VCO) produces the noisy modulated signal (Figure 1.38). This is because the VCO of the locked system follows the frequency variation of the input signal r_M. In order for the VCO to follow the frequency variation, its input is a time-varying voltage corresponding to the information signal. Improved demodulator performance is achieved by designing the loop to have an increased output SNR.

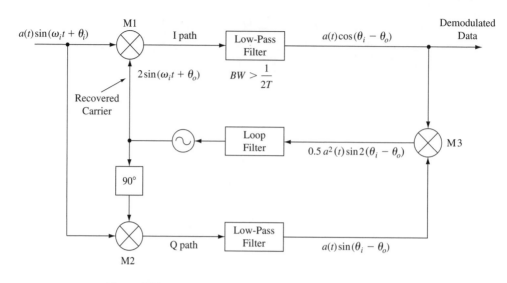

Figure 1.39   Costas loop for carrier recovery and demodulation

Because PLLs can implement maximum likelihood (ML) phase or frequency estimators of feedback nature [Proakis02], they find wide use in carrier and data synchronizers. For example, the Costas loop illustrated in Figure 1.39 is widely known and applicable in BPSK carrier recovery and demodulation. Indeed, by carefully looking at it, if a BPSK modulated signal is used at the input, we see that the output $0.5a^2(t)\sin 2(\theta_i - \theta_0)$ of the multiplier M3 due to the factor of 2 within the phase argument has eliminated phase modulation and consequently the output of the VCO produces the recovered carrier in phase and frequency. In addition, by designing the I-path LPF as a matched filter with bandwidth equal to the information bandwidth, the output of the I-arm constitutes the BPSK data demodulator [Gardner05]. Other forms of the Costas loop can be applied for carrier recovery in M-PSK modulation. The Costas loop can be designed as a digital system for carrier and clock recovery [Mengali97] in modern communication systems. Commercial digital implementations [HSP50210] find wide use in digital modem design and development.

References

[Aiello03]: R. Aiello, 'Challenges for ultra-wideband (UWB) CMOS integration', IEEE Radio Frequency Integrated Circuits Symp., RFIC'2003, pp. 497–500.

[Best03]: R. Best, 'Phase-Locked Loops: Design, Simulation and Applications', 5th edn, McGraw-Hill, New York, 2003.

[Blanchard78]: A. Blanchard, 'Phase-locked Loops', John Wiley & Sons Inc., New York, 1978.

[Chugh05]: M Chugh, D. Bhatia, P. Balsara, 'Design and implementation of configuable W-CDMA rake receiver architectures on FPGA', IEEE Int. Parallel and Distributed Processing Symposium (IPDPS'05).

[Copani05]: T. Copani, S. Smerzi, G. Girlando, G. Palmisano, 'A 12-GHz silicon bipolar dual-conversion receiver for digital satellite applications', IEEE J. Solid St. Circuits, vol. 40, June 2005, pp. 1278–1287.

[Euro-COST 231-1991]: 'Urban Transmission Loss models for Mobile Radio in the 900 and 1800 MHz bands', September 1991.

[Fleury00]: B. Fleury, 'First- and second-order characterization of dispersion and space selectivity in the radio channel', IEEE Trans. Inform. Theory, vol. 46, September 2000, pp. 2027–2044.

[Gardner79]: F. Gardner, 'Phaselock Techniques', 2nd edn, John Wiley & Sons Inc., New York, 1979.

[Gardner05]: F. Gardner, 'Phaselock Techniques', 3rd edn, John Wiley & Sons Inc., New York, 2005.

[Geer06]: D. Geer, 'UWB standardization effort ends in controversy', Computer, July 2006, pp. 13–16.

[Goldsmith05]: A. Goldsmith, 'Wireless Communications', Cambridge University Press, Cambridge, 2005.

[Gray01]: P. Gray, P. Hurst, S. Lewis, R. Meyer, 'Analysis and Design of Analog Integrated Circuits', John Wiley & Sons Inc., New York, 2001.

[Hata80]: M. Hata, 'Empirical formula for propagation loss in land mobile radio services', IEEE Trans. Vehicular Technol., vol. VT-29, no. 3, August 1980, pp. 317–325.

[Hawwar06]: Y. Hawwar, E. Farag, S. Vanakayala, R. Pauls, X. Yang, S. Subramanian, P. Sadhanala, L. Yang, B. Wang, Z. Li, H. Chen, Z. Lu, D. Clark, T. Fosket, P. Mallela, M. Shelton, D. Laurens, T. Salaun, L. Gougeon, N. Aubourg, H. Morvan, N Le Henaff, G. Prat, F. Charles, C. Creach, Y. Calvez, P. Butel, '3G UMTS wireless system physical layer: baseband proccessing hardware implementation perspective', IEEE Commun. Mag., September 2006, pp. 52–58.

[HSP50210]: 'Digital Costas Loop', Data sheet, 2 July 2008, Intersil.

[Ibnkahla04]: M. Ibnkahla, Q. M. Rahman, A. I. Sulyman, H. A. Al-Asady, J. Yuan, A. Safwat, 'High-speed satellite mobile communications: technologies and challenges', IEEE Proc., vol. 92, February 2004, pp. 312–339.

[Jo04]: G-D Jo, K-S Kim, J-U Kim, 'Real-time processing of a software defined W-CDMA modem', IEEE Int. Conf. on Vehicular Technology (VTC04), pp. 1959–1962.

[Mengali97]: U. Mengali, A. D'Andrea, 'Synchronization Techniques for Digital Receivers', Plenum Press, London, 1997.

[Meyer94]: R. Meyer, W. Mack, 'A 1-GHz BiCMOS RF Front-End IC', IEEE J. Solid-St. Circuits, vol. 29, March 1994, pp. 350–355.

[Muhammad05]: K. Muhammad, 'Digital RF processing: towards low-cost reconfigurable radios', IEEE Commun. Mag., August 2005, pp. 105–113.

[Okumura68]: T. Okumura, E. Ohmori, K. Fukuda, 'Field strength and its variability in VHF and UHF land mobile service,' Rev. Elect. Commun. Lab., vol. 16, nos 9–10, September–October 1968, pp. 825–873.

[Parssinen01]: A. Parssinen, 'Direct Conversion Receivers in Wide-Band Systems', Kluwer Academic, Dordrecht, 2001.

[Pra98]: R. Prasad, 'Universal Wireless Personal Communications', Artech House, Norwood, MA, 1998.

[Proakis02]: J. Proakis, M. Salehi, 'Communication Systems Engineering', 2nd edn, Prentice Hall, Englewood Cliffs, NJ, 2002.

[Rappaport02] T. Rappaport, 'Wireless Communications: Principles and Practice', 2nd edn, Prentice Hall, Englewood Cliffs, NJ, 2001.

[Razavi97]: B. Razavi, 'Design considerations for direct conversion receivers', IEEE Trans. Circuits Systems II, vol. 44, June 1997, pp. 428–435.

[Razavi98]: B. Razavi, 'RF Microelectronics', Prentice Hall, Englewood Cliffs, NJ, 1998.

[Scholten03]: A. Scholten, L. Tiemeijer, R. van Langevelde, R. Havens, A. Zegers-van Duijnhoven, V. Venezia, 'Noise modeling for RF CMOS circuit simulation', IEEE Trans. Electron Devices, vol. 50, March 2003, pp. 618–632.

[Schreier02]: R. Schreier, J. Lloyd, L. Singer, D. Paterson, M. Timko, M. Hensley, G. Patterson, K. Behel, J. Zhou, 'A 10-300-MHz F-digitizing IC with 90–105-dB dynamic range and 15–333 kHz bandwidth', IEEE J. Solid St. Circuits, vol. 37, December 2002, pp. 1636–1644.

[Shaeffer97]: D. Shaeffer, T. Lee, 'A 1.5-V, 1.5-GHz CMOS low noise amplifier', IEEE J. Solid St. Circuits, vol. 32, May 1997, pp. 745–759.

[Tolson99]: N. Tolson, 'A novel receiver for GSM TDMA radio', IEEE Int. Proc. Vehicular Technlogy, 1999, pp. 1207–1211.

[Van Nee99]: R. van Nee, G. Awater, M. Morikura, H. Takanashi, M. Webster, K. Halford, 'New high-rate wireless LAN standards', IEEE Commun. Mag., December 1999, pp. 82–88.

[Wheeler07]: A. Wheeler, 'Commercial applications of wireless sensor networks using ZigBee', IEEE Commun. Mag., April 2007, pp. 71–77.

[Wu00]: Y.-C. Wu, T.-S. Ng, 'New implementation of a GMSK demodulator in linear-software radio receiver', IEEE Personal, Indoor Radio Commun. Conf. (PIMRC'00), pp. 1049–1053.

[Zervas01]: N. Zervas, M. Perakis, D. Soudris, E. Metaxakis, A. Tzimas, G. Kalivas, K. Goutis, 'low-power design of direct conversion baseband DECT receiver', IEEE Trans. Circuits Systems II, vol. 48, December 2001.

[Zhang05]: X. Zhang, G. Riley, 'Energy-aware on-demand scatternet formation and routing for Bluetooth-based wireless sensor networks', IEEE Commun. Mag., July 2005, pp. 126–133.

2

Digital Communication Principles

The objective of this chapter is to give to the reader the principles of digital modem design for radio transmission and reception. As it is one of the introductory chapters of this book, we tried to keep it as short as possible. On the other hand, all topics necessary to develop the concept of digital radio modem design needed to be included. Furthermore, it was necessary to cover the subjects in sufficient depth and detail to make the material of this chapter useful in modem analysis and design. To keep a balance, details that would distract the reader (like mathematical derivations) from the objectives of the book were omitted.

To present the material in sufficient depth and concise way, [Proakis02] and [Proakis04] were used. Furthermore, we have followed [Simon05] in the sections concerning detection formulation and receiver performance of QPSK, QAM, coherent detection with nonideal synchronization and noncoherent and differentially coherent detection. Moreover, [Simon05] was used to briefly present performance in fading channels. Finally [Goldsmith05] was used in providing final formulas for error performance in coded systems.

2.1 Digital Transmission in AWGN Channels

2.1.1 Demodulation by Correlation

As presented in Chapter 1, all possible transmitted signals belonging to a set $\{s_m(t), \ m = 1, \ldots, M\}$ can be expressed as a linear combination of N basis functions $\psi_n(t)$. We assume that the demodulation consists of a parallel bank of correlators, as shown in Figure 2.1(a). Then, at the output of the kth branch we have:

$$r_k = \int_0^T r(t)\psi_k(t)\mathrm{d}t = \int_0^T [s_m(t) + n(t)]\psi_k(t)\mathrm{d}t$$

$$r_k = s_{mk} + n_k, \qquad k = 1, 2, \ldots, N \qquad (2.1)$$

Digital Radio System Design   Grigorios Kalivas
© 2009 John Wiley & Sons, Ltd

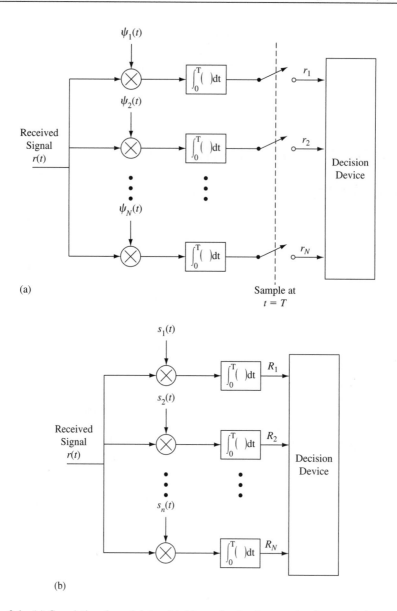

Figure 2.1 (a) Correlation demodulator. (b) Alternative implementation for correlation receiver

where s_{mk} are components of a vector $\mathbf{s}_m$, and n_k are random variables associated with the input noise at the receiver $n(t)$. They are given as follows:

$$s_{mk} = \int_0^T s_m(t)\psi_k(t)\,dt, \quad k = 1, 2, \ldots, N$$

$$n_k = \int_0^T n(t)\psi_k(t)\,dt, \quad k = 1, 2, \ldots, N \qquad (2.2)$$

Noise components n_k are zero mean random variables with variance equal to $\sigma_n^2 = N_0/2$, where N_0 represents the spectral density of the noise.

Furthermore, the received signal can be written as a linear combination of the basis functions $\psi_n(t)$ [Proakis02] within the integration interval $0 \le t \le T$:

$$r(t) = \sum_{k=1}^{N} s_{mk} \psi_k(t) + \sum_{k=1}^{N} n_k \psi_k(t) + n'(t) = \sum_{k=1}^{N} r_k \psi_k(t) + n'(t) \tag{2.3a}$$

$$n'(t) = n(t) - \sum_{k=1}^{N} n_k \psi_k(t) \tag{2.3b}$$

It can be shown [Proakis02] that $n'(t)$ does not influence the decision of the receiver on the transmitted signal and it is only necessary to use the outputs r_k from the correlators to demodulate the received signal. Furthermore, these signal components are random variables with mean and variance as shown below:

$$E[r_k] = E[s_{mk} + n_k] = s_{mk}, \quad \sigma_{r_k}^2 = \sigma_n^2 = \frac{N_0}{2} \tag{2.4}$$

It can be proved [Peebles87] that the decision variables can be formulated as:

$$R_k = \int_0^T r(t) s_k(t) dt \tag{2.5}$$

In this case, an alternative implementation of the correlation receiver is as illustrated in Figure 2.1(b), where the signalling waveforms $s_m(t)$ are used as the multiplying functions.

Coming back to the first receiver implementation, it can be shown that the conditional probability density functions (p.d.f.) of signal vector $\mathbf{r} = (r_1, r_2, \ldots r_N)^{\mathrm{T}}$ are only a function of r_k and s_{mk} [Proakis02]:

$$f(\mathbf{r}|\mathbf{s_m}) = \frac{1}{(\pi N_0)^{N/2}} \exp\left[\frac{-\sum_{k=1}^{N} (r_k - s_{mk})^2}{N_0}\right] \tag{2.6}$$

Furthermore, the components of the output vector $\mathbf{r}$ of the correlator are sufficient statistics to decide which signal, out of the M possible waveforms, was transmitted, because $n'(t)$ is uncorrelated to the elements of $\mathbf{r}$.

2.1.2 Demodulation by Matched Filtering

Instead of correlation, the received signal can go through a filter $h(t)$, which is designed such that it maximizes the signal to noise ratio at the output. Therefore, if the composite signal (y_S) plus noise (y_n) waveform at the output is:

$$y(t_0) = y_S(t_0) + y_n(t_0) \tag{2.7}$$

we seek the impulse response $h(t)$ such that we get a maximum as expressed below:

$$\max\left(\frac{S}{N}\right)_{\text{out}} = \max\left(\frac{y_S^2(t_0)}{E[y_n^2(t_0)]}\right) \tag{2.8}$$

In addition, the optimum time instant t_0 at which the output signal $y(t_0)$ assumes the maximum value, must be determined.

If the transmitted signalling waveform is $s(t)$, it can be shown that the matched filter impulse response is:

$$h(t) = s(t_0 - t) \tag{2.9}$$

In this way, the receiver can take the form of Figure 2.2(a) for which the signal at the output of the receiver is the correlation function of the received signal $r(t)$ with the signalling waveform $s(t)$:

$$y(t) = \int_0^t r(\tau) \cdot s(t_0 - t + \tau)d\tau \tag{2.10}$$

To demonstrate that the form of the matched filter is as in Equation (2.9), the ratio given at the right hand of Equation (2.8) must be calculated. Indeed, this will give the following expression:

$$\left(\frac{S}{N}\right)_{out} = \frac{\left[\int_0^\infty h(\tau)s(t_0 - \tau)d\tau\right]^2}{\frac{N_0}{2}\int_0^\infty h^2(t_0 - t)dt} \tag{2.11}$$

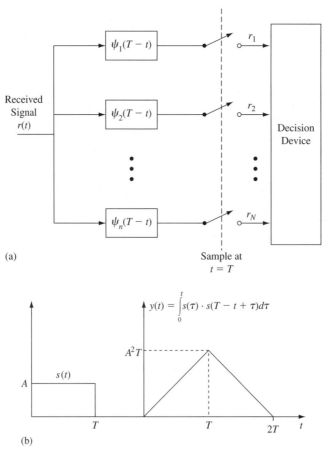

(a)

(b)

Figure 2.2 (a) Matched filter receiver. (b) Unit pulse transmitted signal waveform and resulting matched filter output

Having in mind that, for finite energy signals $s(t)$ and $h(t)$, the Cauchy–Swartz inequality holds:

$$\left[\int_0^\infty h(t)s(t_0 - t)dt \right]^2 \leq \int h^2(t)dt \int s^2(t_0 - t)dt \tag{2.12}$$

it is easy to see that the two terms of the above become equal when $h(t) = Ks(t_0 - t)$. In addition, when this condition is satisfied the resulting maximum SNR is:

$$\left(\frac{S}{N} \right)_{out} = \frac{2}{N_0} \int_0^{t_0} s^2(t)dt \tag{2.13}$$

Since this is directly related to the energy of the signal $s(t)$, it can be deduced that the optimum time instant t_0 at which the output of the matched filter is sampled is the one at which the signalling waveform ends. This is easy to realize when a specific shape is used for $s(t)$. Figure 2.2(b) shows the case for a rectangular unit pulse signal ending at T. In this case we notice that the autocorrelation of $s(t)$ is maximized at T, at which the output of the matched filters must be sampled. In general the integration intervals are usually set equal to the duration of the signalling waveform.

2.1.3 The Optimum Detector in the Maximum Likelihood Sense

To proceed further, at this stage, it is necessary to find a detector, which will use the observed vector $\mathbf{r} = (r_1, r_2, \ldots, r_N)^T$ to decide on the transmitted waveform by maximizing the probability of correct decision. To do this, the maximum a posteriori probability (MAP) criterion must be used, according to which, the a posteriori probability given below must be maximized:

$$P(\mathbf{s}_m|\mathbf{r}) = P(\mathbf{s}_m \text{was transmitted}|\mathbf{r}) \tag{2.14a}$$

From the Bayes rule we get:

$$P(\mathbf{s}_m|\mathbf{r}) = \frac{p(\mathbf{r}|\mathbf{s}_m) \cdot P(\mathbf{s}_m)}{p(\mathbf{r})} = \frac{p(\mathbf{r}|\mathbf{s}_m) \cdot P(\mathbf{s}_m)}{\sum_{m=1}^M p(\mathbf{r}|\mathbf{s}_m)P(\mathbf{s}_m)} \tag{2.14b}$$

Assuming that the M possible transmitted signals are equally probable $[P(\mathbf{s}_m) = 1/M]$ and taking into account that the denominator value is independent of the transmitted signal, the maximization criterion is:

$$\max_{\mathbf{s}_m} \{P(\mathbf{s}_m|\mathbf{r})\} = \max_{\mathbf{s}_m} \{p(\mathbf{r}|\mathbf{s}_m)\} \tag{2.15}$$

Hence the conditional p.d.f. $p(\mathbf{r}|\mathbf{s}_m)$ and its natural logarithm $\ln \{p(\mathbf{r}|\mathbf{s}_m)\}$ can both be chosen as the likelihood function. The criterion used to maximize the likelihood function is defined as the maximum likelihood criterion. We must note here that the MAP and ML criteria are exactly the same when all possible M transmitted signals have the same a priori probability $P(\mathbf{s}_m)$. As we will see in Chapter 5, a similar situation exists for MAP and ML criteria regarding other receiver subsystems such as those used for synchronization.

Examining the case of AWGN and taking into account Equation (2.6), the log-likelihood function becomes

$$\ln [f(\mathbf{r}|\mathbf{s}_m)] = -\frac{N \ln(\pi N_0)}{2} - \frac{\sum_{k=1}^N (r_k - s_{mk})^2}{N_0} \tag{2.16}$$

As the first term in the previous equation is independent of the transmitted signal, the ML criterion becomes:

$$ML|_{\text{AWGN}} = \min\left[D(\mathbf{r}, \mathbf{s}_m)\right] = \min\left[-2\mathbf{r} \cdot \mathbf{s}_m + |\mathbf{s}_m|^2\right] \tag{2.17}$$

where $D(\mathbf{r}, \mathbf{s}_m)$ is defined as the distance metrics and expresses the intuitively obvious consideration that the transmitted signal is the one which corresponds to the minimum distance from the received vector:

$$D(\mathbf{r}, \mathbf{s}_m) = \sum_{n=1}^{N} r_n^2 - 2\sum_{n=1}^{N} r_n s_{mn} + \sum_{n=1}^{N} s_{mn}^2 = |\mathbf{r}|^2 - 2\mathbf{r} \cdot \mathbf{s}_m + |\mathbf{s}_m|^2, \quad m = 1, 2, \ldots, M \tag{2.18}$$

The last two terms of the above equation, represent the correlation metric:

$$C(\mathbf{r}, \mathbf{s}_m) = 2\mathbf{r} \cdot \mathbf{s}_m - |\mathbf{s}_m|^2 \tag{2.19}$$

2.1.3.1 Example – Special Case: Probability of Error for M Orthogonal Signals

Assuming $\mathbf{s}_1$ was transmitted, the received signal vector is:

$$\mathbf{r} = (r_1, r_2, \ldots, r_M) = \left(\sqrt{E_S} + n_1, n_2, \ldots, n_M\right) \tag{2.20}$$

where n_i, $i = 1, \ldots, M$ represents noise terms characterized as mutually independent Gaussian random variables with zero mean and variance $\sigma_{ni}^2 = N_0/2$. E_S represents the signal energy.

According to the above, the following correlation metric must be maximized:

$$C(\mathbf{r}, \mathbf{s}_m) = \mathbf{r} \cdot \mathbf{s}_m = \sum_{k=1}^{M} r_k s_{mk}, \quad m = 1, 2, \ldots, M \tag{2.21}$$

The elements of the above metric are:

$$\begin{aligned}
C(\mathbf{r}, \mathbf{s}_1) &= \sqrt{E_S}\left(\sqrt{E_S} + n_1\right) \\
C(\mathbf{r}, \mathbf{s}_i) &= \sqrt{E_S}\, n_i, \quad i = 2, 3, \ldots, M
\end{aligned} \tag{2.22}$$

Dividing all correlator outputs by $\sqrt{E_S}$ and taking into account the nature of n_i, $i = 1, \ldots, M$, the p.d.f.s of random variables r_i, $i = 1, \ldots, M$ are:

$$f_{r1}(x_1) = \frac{1}{\sqrt{\pi N_0}} \exp\left[\frac{-\left(x_1 - \sqrt{E_S}\right)^2}{N_0}\right] \tag{2.23}$$

$$f_{rm}(x_m) = \frac{1}{\sqrt{\pi N_0}} \exp\left[\frac{-(x_m)^2}{N_0}\right], \quad m = 2, 3, \ldots, M \tag{2.24}$$

The probability of symbol error P_{SER} is:

$$P_{SER} = \frac{1}{M} \sum_{j=1}^{M} P(e|s_j)$$

$$= \frac{1}{M} \sum_{j=1}^{M} \left[1 - \int dr_j \int_{-\infty}^{r_j} \cdots \int_{-\infty}^{r_j} p(\mathbf{r}|s_j) dr_1 dr_2, \ldots, dr_{j-1} dr_{j+1}, \ldots, dr_M \right] \quad (2.25)$$

where $P(e|s_j)$ is the conditional probability of error.

The joint p.d.f. of the vector $\mathbf{r}$ is expressed as [Cooper86]:

$$p(\mathbf{r}|s_j) = \frac{1}{(2\pi)^{M/2}|\mathbf{Q}|^{1/2}} \exp\left[-0.5\mathbf{r}_z^T \mathbf{Q}^{-1}\mathbf{r}_z\right] \quad (2.26)$$

where $\mathbf{r}_z$ represents the received signal vector, from each element of which the correlation to the transmitted signal $s_j(t)$ $[\int_0^T s_j(t)s_i(t)dt]$ has been subtracted. $\mathbf{Q}$ is a matrix with elements proportional to the correlation of the noise components at the output of the matched filters, $q_{ij} = \sigma_{ni}^2 E[n_i n_j]$. Since the noise components n_i, $i = 1, \ldots, M$ are mutually independent zero-mean Gaussian random variables, we have that $\mathbf{Q} = [N_0 E/2] \cdot \mathbf{I}$, where $\mathbf{I}$ is the identity matrix.

Consequently, the conditional p.d.f. [in the square brackets of Equation (2.25)] can now be expressed as the product of the p.d.f. of each element r_i:

$$p(r|s_j) = \prod_{i=1}^{M} \frac{1}{\sqrt{2\pi}\sigma} \exp\left[-(r_i - m_{ij})^2/2\sigma^2\right] \quad (2.27)$$

where m_{ij} represents the mean of the corresponding p.d.f.s.

Some mathematical manipulations can now give the conditional probability of error:

$$P(e|s_j) = 1 - \int_{-\infty}^{\infty} \left[1 - Q\left(r + \sqrt{\frac{2E}{N_0}}\right)\right]^{M-1} \frac{1}{\sqrt{2\pi}} \exp\left(-r^2/2\right) dr \quad (2.28)$$

Since $P(e|s_j)$ is independent of j, the above equation gives:

$$P_{SER} = \frac{1}{M} \sum_{j=1}^{M} P(e|s_j) = P(e|s_j) \quad (2.29)$$

2.1.3.2 Decision Regions

As noted in the introductory chapter, the basis functions $\psi_n(t)$ are used to map the vectors s_m onto the N-dimensional signal space. In turn, the signal space can be divided into M different regions. For equally probable transmitted signals, the boundaries of a particular region I_{ms} are determined such that all vectors $\mathbf{r}$ belonging to that region are closer to the point s_{ms} than any other point. These regions I_m ($m = 1, \ldots, M$) are defined as the decision regions for the particular signal space. Figure 2.3 shows the four decision regions for a two-dimensional signal space with $M = 4$, for which signal vectors s_1, s_2, s_3, s_4 are located at the edges of a trapezoid. To determine the boundaries for equally probable transmitted signals, we draw perpendicular

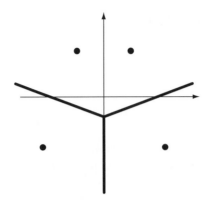

Figure 2.3 Decision regions of signal space with $M = 4$

lines bisecting the lines that connect the signal points. Based on that, a general formula for the error probabilities based on decision regions can be given as follows:

$$P[e|s_j] = 1 - P[r_j \geq r_k|s_j] = 1 - \int dr_j \int \cdots \int p(\mathbf{r}|s_j)d\mathbf{r}' \quad \text{for all } k \neq j \qquad (2.30)$$

where $d\mathbf{r}' = dr_1 dr_2, \ldots, dr_{j-1} dr_{j+1}, \ldots, dr_M$.

2.1.4 Techniques for Calculation of Average Probabilities of Error

In the following sections we will seek simplified expressions for the probability of symbol error, P_{SER}, and the probability of bit error, P_{BER}, for a variety of modulation techniques in radio communications. As we shall see, common terms in all these expressions will be the Gaussian Q-function and the Marcum Q-function (see Appendix A). In an effort to present final expressions for the calculation of P_{BER} and P_{SER} without having to get involved in complex and time-consuming mathematical manipulations, we will follow [Simon05] in the presentation of such probabilities. For this reason we will only give final expressions (except from the cases where there is a reason to give more details). The interested reader is directed to [Simon05], where all derivations, along with the necessary mathematical tools are presented in great detail.

Most P_{BER} and P_{SER} for coherent detection involve the following integral:

$$I_G = \int_0^\infty Q(a \cdot \sqrt{\gamma})p_\gamma(\gamma)d\gamma \qquad (2.31)$$

The integrand of I_G is the product of a Q-function and of the p.d.f. $p_\gamma(\gamma)$. Variable γ represents the instantaneous SNR of a slowly fading signal. The argument of the Q-function is the product of $\sqrt{\gamma}$ and a constant a, which is associated with the modulation/detection technique.

In most cases a compact form of I_G can be used which includes the moment generating function (MGF) of γ [Simon05]:

$$I_G = \frac{1}{\pi} \int_0^{\pi/2} M_\gamma \left(-\frac{a^2}{2\sin^2\theta} \right) d\theta \qquad (2.32)$$

It is much easier to compute I_G using the above expression because M_γ represents the Laplace transform of the SNR p.d.f. $p_\gamma(\gamma)$:

$$M_\gamma(s) = \int_0^\infty \exp(s\gamma)p_\gamma(\gamma)d\gamma \tag{2.33}$$

Regarding the probability of error for differential and noncoherent communication systems, the following integral is very useful involving the Marcum Q-function of order l and double argument:

$$I_M = \int_0^\infty Q_l(a\sqrt{\gamma}, b\sqrt{\gamma})p_\gamma(\gamma)d\gamma \tag{2.34}$$

In a similar way as above, I_M can be simplified by expressing it as a function of M_γ:

$$I_M = \frac{1}{2\pi}\int_{-\pi}^{\pi}\left[\frac{c(\theta;\varsigma,l)}{1+2\varsigma\sin\theta+\varsigma^2}\right] \times M_\gamma\left(-\frac{b^2}{2}f(\theta;\varsigma)\right)d\theta$$

$$f(\theta;\varsigma) \equiv 1+2\varsigma\sin\theta+\varsigma^2, \quad \varsigma = \frac{a}{b}$$

$$c(\theta;\varsigma,l) \equiv \varsigma^{-(l-1)}\left\{\cos\left[(l-1)\left(\theta+\frac{\pi}{2}\right)\right]-\varsigma\cos\left[l\left(\theta+\frac{\pi}{2}\right)\right]\right\}, \quad 0^+ \le \varsigma < 1 \tag{2.35}$$

Similarly,

$$I_M = \frac{1}{2\pi}\int_{-\pi}^{\pi}\left[\frac{c(\theta;\varsigma,-(l-1))}{1+2\varsigma\sin\theta+\varsigma^2}\right] \times M_\gamma\left(-\frac{a^2}{2}f(\theta;\varsigma)\right)d\theta$$

$$\varsigma = \frac{b}{a}, \quad 0^+ \le \varsigma < 1 \tag{2.36}$$

2.1.5 M-ary Pulse Amplitude Modulation (PAM)

2.1.5.1 Transmitted Waveforms and their Properties

This scheme involves M possible transmitted signals located on a straight line (one-dimensional signal space), which take the following values:

$$s_m = \sqrt{E_g} \cdot a_m = \sqrt{E_g} \cdot (2m-1-M), \quad m = 1, 2, \ldots, M \tag{2.37}$$

Defining the decision thresholds as the points between successive signal vectors s_i, s_{i+1}, we have the corresponding decision regions II, as indicated in Figure 2.4(a). The distance between neighbouring constellation points is $d = 2\sqrt{E_g}$. Employing the decision metrics as presented previously, for all signal vectors except those at the end (s_{1-M}, s_{M-1}), we have:

$$P_{SER}(E|s_m) = P[\hat{s} \neq s_m|s = s_m] = P(|r - s_m| > \sqrt{E_g})$$

$$= P\left[\left\{r < (a_m-1)\frac{d}{2}\right\} \cup \left\{r > (a_m+1)\frac{d}{2}\right\}|s = s_m\right] = 2Q\left(\frac{d}{2\sigma}\right) \tag{2.38}$$

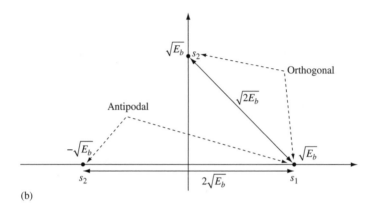

Figure 2.4 (a) M-PAM signaling vectors and decision regions. (b) Binary PAM (antipodal and orthogonal)

Similarly, for the edge signal vectors we have:

$$P_{\text{SER}}[E|s_{M-1}] = P_{\text{SER}}[E|s_{1-M}] = P(\hat{s} \neq s_{M-1}|s = s_{M-1})$$

$$= P[r > (1 + (1 - M))\sqrt{E_g}] = Q\left(\frac{d}{2\sigma}\right) \qquad (2.39)$$

Consequently the overall probability of error is:

$$P_{\text{SER}}(E) = \frac{1}{M}\left[(M-2)P_{\text{SER}}[E|\underset{m \neq \pm(M-1)}{s_m}] + P_{\text{SER}}[E|s_{M-1}] + P_{\text{SER}}[E|s_{1-M}]\right] \qquad (2.40)$$

In terms of average power P_{av} we have:

$$P_{\text{SER}}(E) = \frac{2(M-1)}{M}Q\left(\sqrt{\frac{6P_{\text{av}}T_S}{(M^2-1)N_0}}\right) \qquad (2.41)$$

2.1.5.2 Special Case: Binary PAM

In this case, we have two signal waveforms s_1 and s_2. The signal points at the constellation map can be positioned in two ways, as shown in Figure 2.4(b): (a) s_1 and s_2 at $+\sqrt{E_b}$ and $-\sqrt{E_b}$, respectively on the real axis – this is called antipodal signalling; (b) s_1 and s_2 on the x and y axes, respectively, at distances $\sqrt{E_b}$ from the origin. This represents orthogonal signalling.

(1) *Antipodal signalling* – the received signal from the optimum (correlator-matched filter) demodulator is:

$$r_i = s_i + n = \begin{cases} \sqrt{E_b} + n \text{ when } s_1(t) \text{ is transmitted} \\ -\sqrt{E_b} + n \text{ when } s_2(t) \text{ is transmitted} \end{cases} \tag{2.42}$$

In this case it is easy to calculate P_{BER} by setting the threshold of the two decision regions at zero. Let $s_1(t)$ be the transmitted signal. In this case P_{BER} is just the probability of r being less than zero:

$$P_{BER}(E) = P_{BER}[E|s_1] = P(r < 0) = \int_{-\infty}^{0} p(r|s_1)dr$$

$$= \int_{-\infty}^{0} \frac{1}{\sqrt{\pi N_0}} \exp\left[-(r - \sqrt{E_b})^2/N_0\right]dr = Q\left[\sqrt{\frac{2E_b}{N_0}}\right] \tag{2.43}$$

(2) *Orthogonal signalling* – in this case we use vector notation and the transmitted vector is:

$$\mathbf{s_1'} = (\sqrt{E_b}, 0), \quad \mathbf{s_2'} = (0, \sqrt{E_b})$$

Assuming that $\mathbf{s_1'}$ was transmitted, the received vector is:

$$\mathbf{r} = (\sqrt{E_b} + n_1, n_2)$$

Using correlation metrics from Equation (2.19) we get [Proakis02]:

$$P_{BER}(E|s_1) = P[C(\mathbf{r}, \mathbf{s_2}) > C(\mathbf{r_1}, \mathbf{s_1})] = P[n_2 - n_1 > \sqrt{E_b}] = Q\sqrt{\frac{E_b}{N_0}} \tag{2.44}$$

The resulting poorer performance (by 3 dB) of orthogonal signalling with respect to the antipodal is due to the fact that the distances between s_1 and s_2 are $2E_b$ and $4E_b$ respectively.

2.1.6 Bandpass Signalling

Transmission of digitally modulated signals through channels with bandpass frequency response (radio channel, wireline channel) can be achieved using a carrier frequency f_C located within the passband of the channel. The carrier frequency is generated in the local oscillator (LO) section and mixing operation is performed to up-convert the baseband information signal at the transmitter. In turn, downconversion at the receiver transforms the received signal into its baseband form for further processing and detection. The radio channel creates delay and distortion, which have to be compensated for at the receiver. Furthermore, the receiver must remove frequency and phase shifts for proper coherent detection. Consequently, there are two major aspects that must be considered for passband reception: synchronization and channel distortion. The objective of this section is two-fold. The first is to demonstrate, under perfect synchronization, the equivalence of baseband and passband reception. The second aspect is to give a general framework for representing all major modulations using complex signals.

This is very effective when detection techniques are examined in detail for most modulation methods in the subsequent sections of this chapter.

Before we give a more general representation for signals as complex quantities along the transceiver chain, let us use real-valued signals to present the practical aspects of passband transmission and reception. The baseband signal waveforms s_m formulated as above in Equation (2.37) can be replaced by the following up-converted waveform for transmission:

$$u_m(t) = s_m(t)\cos(2\pi f_C t) = A_m g_T(t) \cos(2\pi f_C t) \qquad (2.45)$$

where g_T is a pulse formulation function for signal conditioning.

This sinusoidal modulation of the digital information, transforms the PSD of the baseband signal (of bandwidth W) to passband at $\pm f_C$, as illustrated in Figure 2.5.

The composite baseband signal consists of an infinite sequence and is given as:

$$U(t) = \sum_{-\infty}^{\infty} a_n g_T(t - nT) \qquad (2.46)$$

whereas the transmitted passband signal is expressed as follows:

$$u(t) = U(t)\cos(2\pi f_C t) \qquad (2.47)$$

The energy of the passband waveforms E_m can be shown to be half of that of the baseband waveforms [Proakis02]:

$$E_m = \int_{-\infty}^{\infty} u_m^2(t)\mathrm{d}t \approx \frac{A_m^2}{2} \int_{-\infty}^{\infty} g_T^2(t)\mathrm{d}t = \frac{A_m^2}{2} \cdot E_g \qquad (2.48)$$

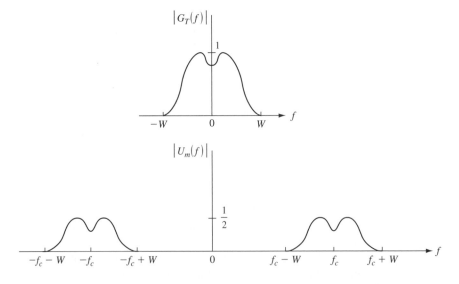

Figure 2.5   Spectral density of baseband and passband modulated signals

2.1.6.1 Complex Representation, Demodulation and Carrier Synchronization of Passband Signals

For optimum demodulation, the received signal $r(t)$ is correlated with the basis function $\psi(t)$:

$$r(t) = u_m(t) + n(t) = A_m g_T(t) \cos(2\pi f_C t) + n(t) \tag{2.49}$$

where $n(t)$ is the bandpass noise process (see Chapter 1), whereas $\psi(t)$ is considered a real-valued signal also and is given as [Proakis02]:

$$\psi(t) = \sqrt{\frac{2}{E_g}} g_T(t) \cos(2\pi f_C t) \tag{2.50}$$

Correlation will give:

$$R_C(t) = \int_{-\infty}^{\infty} r(t)\psi(t)\mathrm{d}t = A_m \sqrt{\frac{2}{E_g}} \int_{-\infty}^{\infty} g_T^2(t) \cos^2(2\pi f_C t)\mathrm{d}t$$

$$+ \int_{-\infty}^{\infty} n(t)\psi(t)\mathrm{d}t = A_m \sqrt{E_g/2} + n_C \tag{2.51}$$

In this subsection we assume an ideal bandpass channel in which the channel does not affect in any other way the received signal apart from adding additive white Gaussian noise. In the next subsection we will examine the effect of a nonideal bandpass channel.

In general the passband communication system can have the form of Figure 2.6. Although considering real-valued signals is useful to grasp the practical aspects of the transceiver, for the rest of this chapter we adopt a unified formulation for signals and waveforms which helps to examine in a more systematic way the various modulation schemes and their performance. To do this, signals are represented as complex quantities (similar to [Simon05]), along the transmitter and receiver chain, as indicated in Figure 2.7. The three basic modulation categories are quadrature amplitude modulation (QAM), phase shift keying (PSK) and frequency shift keying (FSK). The receiver structure in Figure 2.7 corresponds to PSK and QAM. These two schemes can be visualized by mapping the signalling waveforms on the complex (two-dimensional) plane. This is called a constellation mapping. Figure 2.8 illustrates the two cases. When signal points vary in both amplitude and phase we have QAM modulation. When only the phase is changed we have PSK. Capital $S(t)$ denotes the baseband transmitted waveform at time instant t, whereas lower case $s(t)$ denotes the passband transmitted waveform. $R(t)$ and $r(t)$ represent the baseband and passband received waveforms at time instant t, respectively. Finally, by assigning different frequency increments corresponding to the transmitted symbols, we have FSK modulation. The baseband FSK receiver is somewhat different in that the multiplication by the possible transmitted waveforms $s_m(t)$ is done prior to integration.

Taking into account the above discussion the mathematical formulation for all necessary signals is as follows:

$$S(t) = \begin{cases} Aa(t), & \text{QAM modulation} \\ A\exp[j\theta(t)], & \text{PSK modulation} \\ A\exp[j2\pi f(t)t], & \text{FSK modulation} \end{cases} \tag{2.52}$$

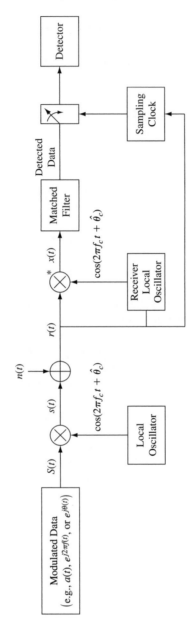

Figure 2.6 General form of transmitter and coherent receiver

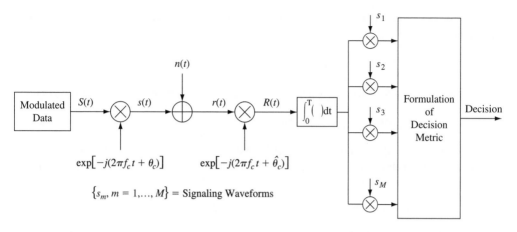

Figure 2.7 General form of complex transceiver for QAM and M-PSK modulation

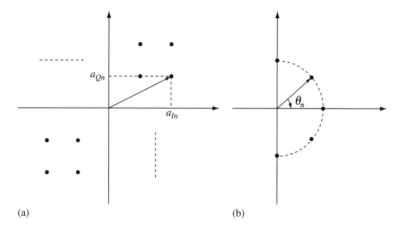

(a)   (b)

Figure 2.8 Constellation mapping for QAM and M-PSK

Considering the same quantities at the nth symbol interval, the expressions become:

$$S(t_n) = \begin{cases} A(a_{In} + ja_{Qn}), & \text{QAM modulation} \\ A\exp[j\theta_n], & \text{PSK modulation} \\ A\exp[j2\pi f_n(t - nT_S)], & \text{FSK modulation} \end{cases} \tag{2.53}$$

The transmitted passband signal is obtained after up-conversion by multiplication with the local oscillator, which has the form of an exponential function to anticipate for complex operations:

$$s(t) = S(t)\exp[j(2\pi f_C t + \theta_C)] \tag{2.54}$$

The corresponding received passband signal $\varphi_c(\tau_1; \Delta t)$ is:

$$r(t) = \alpha s(t) + n(t) \tag{2.55}$$

where the Greek letter α stands for the channel attenuation.

At the receiver, the received signal $r(t)$ is used to obtain carrier synchronization in the block 'receiver LO'.

$$c_{LO}(t) = \exp[-j(2\pi f_C t + \hat{\theta}_C)] \tag{2.56}$$

The baseband signal $R(t)$ after downconversion at the receiver by multiplying with $c_{LO}(t)$ is:

$$R(t) = r(t)\exp[-j(2\pi f_C t + \hat{\theta}_C)] = \alpha S(t)\exp j(\theta_C - \hat{\theta}_C) + n(t)\exp[-j(2\pi f_C t + \hat{\theta}_C] \tag{2.57}$$

The term θ_C is a random phase introduced by the transmitter LO and $\hat{\theta}_C$ its estimate. Ideal coherent reception entails perfect carrier synchronization in frequency and phase $(\hat{\theta}_C = \theta_C)$ at the receiver and transmitter LO:

$$R(t) = \alpha S(t) + n(t)\exp[-j(2\pi f_C t + \hat{\theta}_C] \tag{2.58}$$

It is easy to see that, for passband transmission and perfect carrier recovery, optimum reception is similar to that performed at baseband (after downconversion), as presented in the above sections.

Furthermore, optimum demodulation and decision can be performed in the passband instead of baseband. In this case, the system has the form of Figure 2.9. Looking at a practical system (real-valued signals instead of complex), the received signal is multiplied by $g_T(t)\cos(2\pi f_C t - \hat{\theta}_C)$ when bandpass correlation is employed, or by $g_T(T - t)\cos\{2\pi f_C(T - t) - \hat{\theta}_C\}$ where bandpass matched filtering is used at the receiver. This can be achieved using PLLs when feedback techniques are used for that. The output of the PLL-based LO system where carrier recovery has taken place is $c_{LO}(t) = \cos(2\pi f_C t - \hat{\theta}_C)$. For perfect carrier synchronization $(\hat{\theta}_C = \theta_C)$ detection is done using Equation (2.51), as before. Furthermore, the optimum detection, like in the baseband system model, can be based on finding the minimum distance using correlation metrics:

$$C(\mathbf{r}, \mathbf{s}_m) = 2\mathbf{r}\mathbf{s}_m - \mathbf{s}_m^2 \tag{2.59}$$

In the sections below, we use $y(t)$ or y_n to represent the decision variables at the output of the matched filter in order not to create any confusion with $r(t)$, which is used to represent the passband received signal.

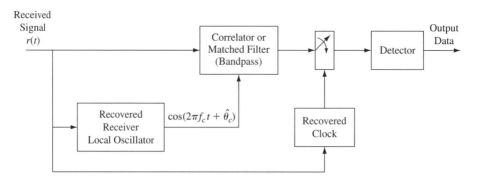

Figure 2.9 Passband demodulation

The quantities $s(t)$, $S(t)$, $r(t)$ and $R(t)$ represent transmitted and received symbols at a specific time instant. The transmitted waveform consisting of a sequence of symbols is given as follows:

$$u(t) = \sum_{-\infty}^{\infty} A_n g_T(t - nT_S) \tag{2.60}$$

where A_n represents one of the three expressions in Equation (2.53), depending on the modulation. For example, for QAM, $A_n = A(a_{In} + ja_{Qn})$. The time function g_T represents a unit pulse function designed so as to limit the bandwidth of the transmitted signal. Most of the modulation schemes along with their variants can be represented using the above expressions. Possible variants can be detection with nonideal carrier recovery, partially coherent detection, FSK with continuous phase and differentially encoded modulations.

2.1.6.2 Effect of Bandpass Channel on System Design

In radio transmission the propagation channel is of bandpass nature. In this subsection we include a nonideal bandpass channel and we briefly examine its effect on the system. The main issue here is to demonstrate that a passband channel can be converted into its equivalent baseband channel. As a result, the techniques outlined in Section 2.1.3 can be directly applied. Figure 2.10 illustrate the passband and equivalent baseband systems. The relations connecting the frequency responses of the bandpass and baseband channels $C_p(f)$ and $C_{bb}(f)$ are as follows [Proakis02]:

$$C_{bb}(f - f_C) = \begin{cases} 2C_p(f), & f > f_C \\ 0, & f < f_C \end{cases}, \quad C_{bb}^*(-f - f_C) = \begin{cases} 0, & f > -f_C \\ 2C_p^*(-f), & f < -f_C \end{cases} \tag{2.61}$$

$$C_p(f) = \frac{1}{2}\left[C_{bb}(f - f_C) + C_{bb}^*(-f - f_C)\right] \tag{2.62}$$

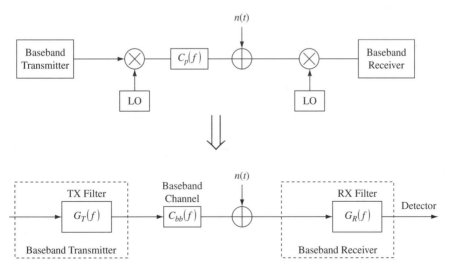

Figure 2.10 Passband and equivalent baseband transceiver and the role of filtering

Regarding impulse responses we have:

$$c_p(t) = \mathrm{Re}\{c(t)\exp[j(2\pi f_C t)]\} \tag{2.63}$$

To anticipate for the effects of the channel, transmitter and receiver filters must be designed. Regarding the design of these filters we have two cases:

(1) The frequency response $C_{bb}(f)$ of the channel is known – in this case the transmitter and receiver filter responses $G_T(f)$, $G_R(f)$ are designed such that they satisfy the relation:

$$G_T(f)C_{bb}(f)G_R(f) = X_{rc}(f)\exp[-j\pi f t_0], \quad |f| \le W \tag{2.64}$$

where $X_{rc}(f)$ represents a filtering function designed to eliminate intersymbol interference and has usually the form of a raised-cosine function. This is discussed in much detail in subsequent sections when techniques are considered for elimination of channel distortion using equalizers. The delay t_0 is used to anticipate the delay of the transmitted signal in reaching the receiver.

(2) The frequency response $C_{bb}(f)$ is unknown – in this case transmitter and receiver filters are designed to satisfy:

$$|G_T(f)||G_R(f)| = X_{RC}(f) \tag{2.65}$$

The above approach would give ideal reception in the case of an ideal channel [$C_{bb}(f)=1$ in the passband]. However, owing to the nonideal channel, there will be intersymbol interference. To eliminate it, an equalizer can be designed with frequency response $G_{Eq}(f)$ which will compensate for the channel transfer function $C_{bb}(f)$. Detailed presentation of equalization techniques will be given in a subsequent section.

2.1.7 M-ary Phase Modulation

2.1.7.1 Transmitted Waveforms and their Properties

As mentioned before, in M-phase shift keying (M-PSK) modulation the transmitted symbols are located on the circumference of a unit circle at equal angular distances. Each data symbol s_k is located at an angle $\theta_k = 2\pi k/M$. Figure 2.11 gives the mapping (constellations) of the

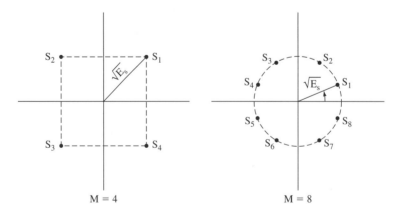

Figure 2.11   Constellations for 4-PSK and 8-PSK modulations

PSK-modulated signals for $M = 4$ and $M = 8$. For the case of M-PSK we will briefly present two approaches for detection in order to show that both techniques employing decision regions and optimum receiver principle based on distance metrics are equivalent.

Phase modulation of the band-pass signal is done by digitally modulating the phase of the signal waveforms $u(t)$ using phase values θ_k as follows:

$$u_m(t) = g_T(t) \exp\left(2\pi f_C t + \frac{\theta_m}{M} + \theta_C\right), \quad \theta_m = \frac{2\pi m}{M} + \frac{\pi}{M}, \quad m = 0, 1, 2 \ldots, (M-1) \quad (2.66)$$

When $g_T(t)$ is a rectangular pulse it has the form:

$$g_T(t) = \sqrt{2E_S/T}, \quad 0 \leq t \leq T$$

By using the real part of the transmitted signal $u_m(t)$ we can easily show how a PSK-modulated signal can be represented as two-dimensional vector:

$$\mathrm{Re}[u_m(t)] = A_{mC}g_T(t)\cos(2\pi f_C t) + A_{mS}g_T(t)\sin(2\pi f_C t) \quad (2.67)$$

where:

$$A_{mC} = \cos\left(\frac{2\pi m}{M} + \frac{\pi}{M}\right), \quad A_{mS} = \sin\left(\frac{2\pi m}{M} + \frac{\pi}{M}\right), \quad m = 0, 1, \ldots, M\text{-}1 \quad (2.68)$$

Therefore the signal waveforms consist of two-dimensional signal vector $\mathbf{s}_m$ as follows:

$$\mathbf{s}_m = \left(\sqrt{E_S}\cos\left(\frac{2\pi m + \pi}{M}\right), \sqrt{E_S}\sin\left(\frac{2\pi m + \pi}{M}\right)\right) \quad (2.69)$$

The transmitted bandpass signal consists of a sequence of information symbols and can be expressed in a more general form as follows:

$$u(t) = \mathrm{Re}[\exp(j2\pi f_C t) \cdot U(t)] = \mathrm{Re}\left[\exp(j2\pi f_C t) \cdot \sum_{n=-\infty}^{\infty} \exp(j\theta_n)g_T(t - nT)\right] \quad (2.70)$$

By standard manipulations, this expression can give the power spectrum of the transmitted signal by calculating the autocorrelation function and taking in turn the Fourier transform.

2.1.7.2 Special Case 1: Power Spectrum of M-PSK

The first step is to obtain an expression for the average autocorrelation $\overline{R}_{U(\tau)}$ of $U(t)$. Subsequently, by using the Fourier transform, we arrive at the power spectrum, which is as follows [Proakis02]:

$$S_u(f) = \frac{1}{4}[S_U(f - f_C) + S_U(-f - f_C)] \quad (2.71)$$

where $S_U(f)$ is the PSD of the corresponding low-pass signal $U(t)$:

$$S_U(f) = \frac{S_{an}(f)}{T}|G_T(f)|^2 \quad (2.72)$$

$S_{an}(f)$ is the PSD of the transmitted information sequence $a_n \equiv \{a_{In} + ja_{Qn}\}$ with autocorrelation $R_{an}(\tau)$:

$$R_{an}(\tau) = E\left[a_k^* a_{k+n}\right] = E[\exp(-j\theta_k)\exp(j\theta_{k+n})] \quad (2.73)$$

$$S_{an}(f) = \sum_{n=-\infty}^{\infty} R_{an}(n)\exp\left[-j2\pi f n T\right] \tag{2.74}$$

For a rectangular PSK pulse $G_T(t)$ we have:

$$|G_T(f)|^2 = 2E_S T \left[\frac{\sin(\pi f T)}{\pi f T}\right]^2 \tag{2.75}$$

Since the autocorrelation of equally probable transmitted data is equal to 1, the PSD of $S_u(f)$ in this case is:

$$S_u(f) = \frac{E_S}{2}\left[\frac{\sin\pi(f - f_C)T}{\pi(f - f_C)T}\right]^2 + \frac{E_S}{2}\left[\frac{\sin\pi(f + f_C)T}{\pi(f + f_C)T}\right]^2 \tag{2.76}$$

This is the power spectrum of M-PSK and it is illustrated in Figure 2.12. When we look at the PAM spectrum in Section 2.4 (for band-limited channels), we will note that it is obtained using the same steps as for the power spectrum of M-PSK. This can be easily realized by taking into account that the baseband M-PSK signal $U(t)$ from the above equation can be written as follows:

$$U(t) = \sum_{n=-\infty}^{\infty} g_T(t - nT)\cdot(a_{In} + ja_{Qn}) \tag{2.77}$$

The corresponding equation for PAM is:

$$U(t) = \sum_{n=-\infty}^{\infty} a_n g_T(t - nT) \tag{2.78}$$

The only difference between the two expressions is that in M-PSK the information sequence $a_{In} + ja_{Qn} = a_n$ is complex whereas for PAM it is real.

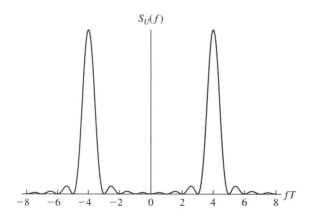

Figure 2.12 Power spectrum of M-PSK modulation with $f_C T = 4$

2.1.7.3 Coherent Detection of M-PSK

To implement the optimum receiver, we consider the correlation metric as given in Equation (2.21) which is repeated below:

$$C(\mathbf{r}, \mathbf{s}_m) = \mathbf{r} \cdot \mathbf{s}_m \tag{2.21}$$

The elements of $\mathbf{s}_m$ in complex form are given as:

$$s_m = \exp[j\beta_m], \quad \beta_m = \frac{2m\pi}{M} + \frac{\pi}{M}, \quad m = 0, 1, \ldots, M - 1 \tag{2.79a}$$

and the time-domain transmitted signal is:

$$s(t) = A_C \exp[j(2\pi f_C t + \theta_n)] \tag{2.79b}$$

The data chosen is that giving the highest value for the correlation metric in Equation (2.21). Hence the following equations represent the decision variables:

$$y_{nk} = A_C T_S \exp[j(\theta_n - \beta_k)] + \exp(-j\beta_k)n_{xx}, \quad k = 1, 2 \ldots, M$$

$$n_{xx} = \int_{nT_S}^{(n+1)T_S} n(t)dt \tag{2.79c}$$

The decision rule is:

$$\text{Choose } k: \ \operatorname{Re}[y_{nk}] = \max_i\{\operatorname{Re}[y_{ni}]\}, \quad i = 1, \ldots, M \tag{2.80}$$

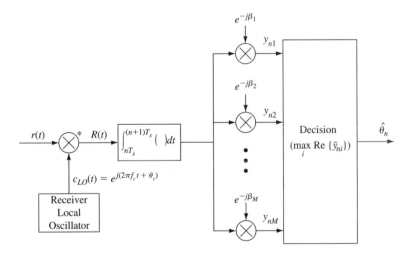

Figure 2.13 Optimum coherent detection for M-PSK modulation [adapted from Simon05]

The above expressions are mapped in the optimum receiver, illustrated in Figure 2.13. To calculate the error probability we can extract expressions based on the formulation of decision regions as follows [Benvenuto02]:

$$P_{SER}(E) = 1 - P(\text{Correct}|s_n) = 1 - P(\mathbf{r} \in I_n|s_n) = 1 - \iint_{I_n} p_{\mathbf{r}}(r_I, r_Q|s_n)dr_I dr_Q$$

$$= 1 - \iint_{I_n} \frac{1}{\pi N_0} \exp\left\{-\frac{1}{N_0}\left[\left(r_I - \sqrt{E_S}\cos\theta_n\right)^2 + \left(r_Q - \sqrt{E_S}\sin\theta_n\right)^2\right]\right\}dr_I dr_Q$$

$$(2.81)$$

where I_n represents angular sectors formed by drawing lines bisecting the distances connecting neighbouring vectors (s_m, s_{m+1}). Figure 2.14 illustrates the region I_n corresponding to an angle $2\pi/M$ and a received vector $\mathbf{r}$ located in the same decision region. Vector $\mathbf{w}$ represents the noise at the receiver.

Transformation of the above expression into polar coordinates will help to express both the integrand and the limits of the integral in a more compact form. After some mathematical manipulations the result is:

$$P_{SER}(E) = 1 - \int_{-\pi/M}^{\pi/M} p_{\Theta}(\theta)d\theta \tag{2.82}$$

$$p_{\Theta}(\theta) = \frac{\exp(-E_S/N_0)}{2\pi}\left\{1 + 2\sqrt{\frac{\pi E_S}{N_0}}\exp\left[(E_S/N_0)\cos^2\theta\right]\cos\theta\left[1 - Q\left(\sqrt{\frac{2E_S}{N_0}}\cos\theta\right)\right]\right\}$$

$$(2.83)$$

The random variable θ takes values in the interval $[-\pi, \pi]$.

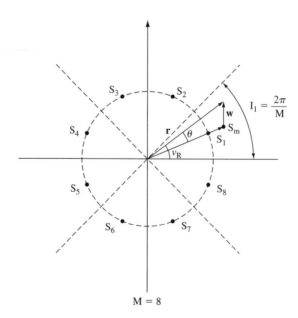

M = 8

Figure 2.14 Partition into decision regions I_n for 8-PSK modulation

The same expression is derived if we use an equivalent metric Θ_r as in [Proakis02]:

$$\Theta_r = \tan^{-1}\left(\frac{r_Q}{r_I}\right) \tag{2.84}$$

Because of the circular symmetry of the signal space, the transmitted signals s_m have the same energy and the correlation metric leads to the calculation of Θ_r. In this case, after some manipulations, it can be shown that the p.d.f. of Θ_r will be similar to Equation (2.83) above. However, the derivation of exact expressions for the probability of bit error P_{BER} or symbol error P_{SER} is quite involved.

It can be shown that P_S can be expressed as [Simon05]:

$$P_{SER}(E) = Q\left(\sqrt{\frac{2E_S}{N_0}}\right) + \frac{2}{\sqrt{\pi}}\int_0^\infty \exp\left[-\left(u - \sqrt{\frac{E_S}{N_0}}\right)^2\right]Q\left(\sqrt{2}u\tan\frac{\pi}{M}\right)du \tag{2.85}$$

Following [Simon05], [Pawula82] and [Craig91], a simple expression can be obtained for the symbol error probability:

$$P_{SER}(E) = \frac{1}{\pi}\int_{-\pi/2}^{\pi/2-\pi/M} \exp\left[-\frac{E_S}{N_0}\frac{g_{PSK}}{\sin^2\theta}\right]d\theta = \frac{1}{\pi}\int_0^{(M-1)\pi/M} \exp\left[-\frac{E_S}{N_0}\frac{g_{PSK}}{\sin^2\theta}\right]d\theta \tag{2.86}$$

where $g_{PSK} = \sin^2(\pi/M)$.

An alternative expression is [Simon05]:

$$P_{SER}(E) = 2Q\left(\sqrt{2\gamma_S}\sin\frac{\pi}{M}\right) - Q\left[\sqrt{2\gamma_S}\sin\frac{\pi}{M}, \sqrt{2\gamma_S}\sin\frac{\pi}{M}; -\cos\frac{2\pi}{M}\right] \tag{2.87}$$

where $\gamma_S = E_S/N_0$ represents the symbol SNR. For $\gamma_S \gg 1$ only the first term is significant.

The factor $1/\sin^2\theta$ is eliminated from the integrand of Equation (2.86) by replacing it with its minimum value. In this way Equation (2.86) can be replaced by an upper bound [Simon05]:

$$P_{SER}(E) \leq \frac{M-1}{M}\exp\left(-\frac{E_S}{N_0}g_{PSK}\right) = \frac{M-1}{M}\exp\left[-\frac{E_S}{N_0}\sin^2\left(\frac{\pi}{M}\right)\right] \tag{2.88}$$

A frequently used union bound for the P_{SER} is [Simon05], [Benvenuto02]:

$$P_{SER}(E) \leq 2Q\left(\sqrt{\frac{2E_S}{N_0}}\sin(\pi/M)\right) \tag{2.89}$$

Plotting Equations (2.86) and (2.88) in Figure 2.15 gives very close results, demonstrating the accurateness of the Union bound. The exact calculation of the probability of bit error P_{BER} for M-PSK can be quite tedious [Simon05]. Some recent results [Lu99] give approximate expressions of reasonable accuracy for any value of M and are valid for all values of SNR:

$$P_{BER}(E) \cong \frac{2}{\max(\log_2 M, 2)} \cdot \sum_{i=1}^{\max(M/4,1)} Q\left(\sqrt{\frac{2E_b\log_2 M}{N_0}} \cdot \sin\frac{(2i-1)\pi}{M}\right) \tag{2.90}$$

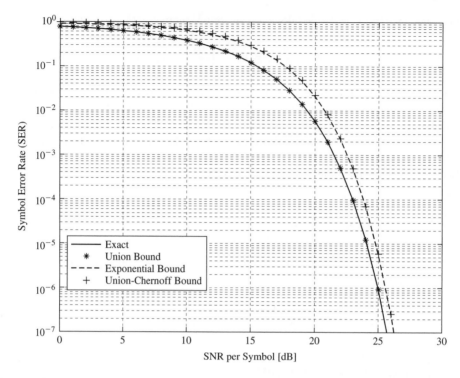

Figure 2.15 Probability of symbol error for M-PSK including Union, Exponential and Union-Chernoff bounds (reprinted from M. Simon, 'Digital Communication over Fading Channels', 2nd edn, copyright © 2005 by John Wiley & Sons, Inc.)

In a similar fashion as with P_{SER}, for high E_b/N_0 we have:

$$P_{\text{BER}}(E) \cong \frac{2}{\log_2 M} Q\left(\sqrt{\frac{2E_b \log_2 M}{N_0}} \sin \frac{\pi}{M}\right) \tag{2.91}$$

2.1.7.4 Special Case 2: QPSK

A special case of M-PSK is quadrature PSK (QPSK), which along with some of its modified versions (such as OQPSK, π/4QPSK) has found many applications in wireless communications. Since QPSK constellation consists of four points located on the periphery of a unit circle at angles 45, 135, 225 and 315°, it can also be considered as a hybrid amplitude-phase modulation. The general format of this family of modulations is defined as quadrature amplitude modulation and will be presented in the next section. Consequently, the QPSK modulated bandpass signal can be expressed as:

$$u_m(t) = [a_{I,m}(t) + ja_{Q,m}(t)] \exp[j(2\pi f_C t + \theta_C)] \tag{2.92}$$

where $a_{I,m}(t)$, $a_{Q,m}(t)$ are the information-bearing symbols taking the values ± 1:

$$a_{I,m}(t) = \sum_{-\infty}^{\infty} a_{I,m} g_T(t - mT_S), \quad a_{Q,m}(t) = \sum_{-\infty}^{\infty} a_{Q,m} g_T(t - mT_S) \tag{2.93}$$

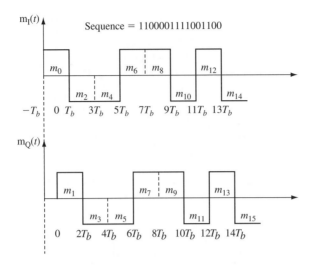

Figure 2.16   In-phase and quadrature bit streams in OQPK modulation

As discussed in previous sections, $g_T(t)$ is a pulse shaping function of unit duration ($t = T_S$). In most cases (including QPSK) this pulse has a rectangular shape. Using Equation (2.85) for $M = 4$, it can be easily shown that:

$$P_{SER}(E) = 2Q(\sqrt{\gamma_S}) - Q^2(\sqrt{\gamma_S}) \tag{2.94}$$

where, as noted before, $\gamma_S = E_S/N_0$ represents the symbol SNR.

2.1.8 Offset QPSK

From the previous presentation of QPSK it can be realized that, at the mth transmission interval, both a_{Im} and a_{Qm} change abruptly causing fluctuations in the ideally constant envelop of QPSK [Rappaport02]. If we delay the quadratic data stream by $T_S/2$, this problem can be surpassed. This represents the offset-QPSK (OQPSK) modulation and can be expressed as:

$$a_{Im}(t) = \sum_{-\infty}^{\infty} a_{I,m} g_T(t - mT_S), \quad a_{Qm}(t) = \sum_{-\infty}^{\infty} a_{Q,m} g_T(t - mT_S - T_S/2) \tag{2.95}$$

Figure 2.16 shows the effect of the $T_S/2$ delay on the relative positions of the I and Q data streams. From the above equation and figure it turns out that at mth time interval, the transmitted passband signal in complex form is:

$$s(t) = \begin{cases} A(a_{I,m} + ja_{Q,m-1}) \exp[j(2\pi f_C t + \theta_C)], & mT_S \leq t \leq mT_S + T_S/2 \\ A(a_{I,m} + ja_{Q,m}) \exp[j(2\pi f_C t + \theta_C)], & mT_S + T_S/2 \leq t \leq mT_S + T_S \end{cases} \tag{2.96}$$

Figure 2.17 gives the optimum receiver for OQPSK. After matched filtering and hard-limiting, the decision variables $y_{I,m}$ and $y_{Q,m}$ for the I and Q branches are obtained as follows [Simon05]:

$$y_{I,m} = Aa_{I,m}T_S + n_{I,m} \qquad y_{Q,m} = Aa_{Q,m}T_S + n_{Q,m} \tag{2.97}$$

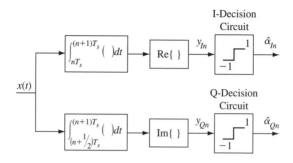

Figure 2.17 Optimum receiver for OQPSK modulation

$$n_{I,m} = \text{Re}\left[\int_{mT_S}^{(m+1)T_S} n(t)\mathrm{d}t\right], \qquad n_{Q,m} = \text{Im}\left[\int_{mT_S+T_S/2}^{(m+1)T_S+T_S/2} n(t)\mathrm{d}t\right] \qquad (2.98)$$

Although there is a time offset of $T_S/2$ between the I and Q data streams, from the expressions of the decision variables it is easy to deduce that ideal coherent detection will give the same P_{BER} with that of QPSK.

However, due to this offset of I and Q components, the possibility to have simultaneous transitions in phase is eliminated. Consequently, the sidelobes of the power spectrum of OQPSK are considerably lower compared to these of QPSK.

2.1.9 Quadrature Amplitude Modulation

2.1.9.1 Transmitted Waveforms and their Properties

By combining amplitude (PAM) and phase modulation (PSK) we can obtain the signal waveform for QAM at the mth time interval:

$$s_m(t) = A(a_{Im} + ja_{Qm})\exp\left[j(2\pi f_C t + \theta_C)\right] \qquad (2.99)$$

90° phase diff

The quadrature components a_{Im}, ja_{Qm} can take values from the in-phase and the quadrature components A_i and A_q respectively, in independent fashion:

$$A_i = 2i - 1 - \sqrt{M},\, i = 1, 2, \ldots, \sqrt{M} \text{ and } A_q = 2q - 1 - \sqrt{M},\, q = 1, 2, \ldots, \sqrt{M} \qquad (2.100)$$

Employing Cartesian coordinates the real-valued signal $s_{mR}(t)$ corresponding to $s_m(t)$ can be expressed as follows:

$$s_{mR}(t) = A(a_{Im}g_T(t)\cos 2\pi f_C t + a_{Qm}g_T(t)\sin 2\pi f_C t) \qquad (2.101)$$

Both representations lead to signal vectors of two dimensions as follows:

$$\mathbf{s}_m = (\sqrt{E_S}a_{Im},\ \sqrt{E_S}a_{Qm}) \qquad (2.102)$$

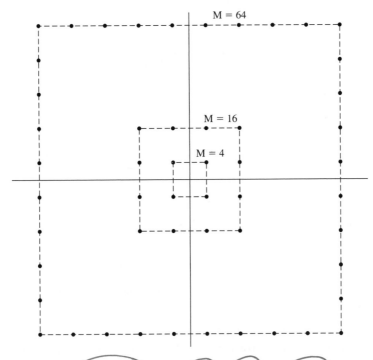

Figure 2.18 Signal constellations for QAM-4, QAM-16 and QAM-64 modulation

Figure 2.18 illustrates the signal constellation of 4-QAM, 16-QAM and 64-QAM modulation schemes. From the above, the transmitted band-pass signal is:

$$u(t) = \text{Re}[\exp{(j2\pi f_C t)} \cdot U(t)] = \text{Re}\left[\exp{(j2\pi f_C t)} \cdot \sum_{n=-\infty}^{\infty} A_n \exp{(j\theta_n)}g_T(t - nT)\right]$$
$$(2.103)$$

2.1.9.2 Coherent Detection of M-QAM

Representation according to Equation (2.101) above helps to transform the system into a two-dimensional M-PAM arrangement in order to calculate P_{BER} and P_{SER} more easily. At the receiver, after ideal downconversion and carrier synchronization, the baseband signal will have the form:

$$R(t) = (a_{Irn} + ja_{Qrn}) + n(t) \qquad (2.104)$$

Ideal downconversion and carrier synchronization assumes no errors in phase and frequency in the received baseband signal $R(t)$. The opposite case (coherent detection with nonideal carrier synchronization) will be examined briefly in the next section. Figure 2.19 shows the structure of the optimum QAM receiver [Simon05].

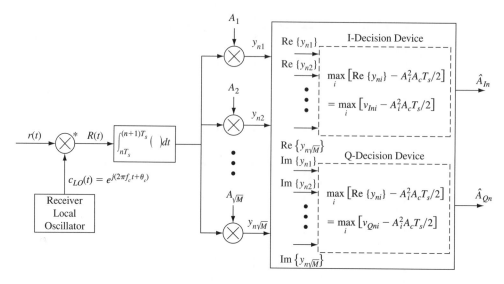

Figure 2.19 Optimum receiver for coherent QAM demodulation

Under the realistic assumption of independent I and Q components the received signal passes through a matched filter receiver to obtain the decision variables $y_{In}(k)$ and $y_{Qn}(k)$ [Simon05]:

$$y_{In}(k) = \text{Re}\{R(k)\} = A_k a_{Irn} T_S + A_k n_{In}, \quad k = 1, 2, \ldots, \sqrt{M}/2 \tag{2.105a}$$

$$y_{Qn}(k) = \text{Im}\{R(k)\} = A_k a_{Qrn} T_S + A_k n_{Qn} \tag{2.105b}$$

$$n_{In} = \text{Re}\left\{ \int_{nT_S}^{(n+1)T_S} n(t)\mathrm{d}t \right\}, \quad n_{Qn} = \text{Im}\left\{ \int_{nT_S}^{(n+1)T_S} n(t)\,\mathrm{d}t \right\} \tag{2.105c}$$

where k corresponds to the sampling instants t_k of the continuous received signal $R(t)$.

The optimum detector corresponds to separate maximization of the following quantities:

$$\max_i[D_I] = \max_i\left[y_{In}(k) - \frac{1}{2}A_i^2 T_S \right], \quad \max_i[D_Q] = \max_i\left[y_{Qn}(k) - \frac{1}{2}A_i^2 T_S \right] \tag{2.106}$$

The rectangular M-QAM constellations (16-QAM, 64-QAM, etc.) correspond to two independent $\sqrt{M}$-PAM arrangements if carrier synchronization is perfect (as will be explained later). Consequently, a correct decision in M-QAM is reached only in case that correct decisions are made for both in-phase and quadrature $\sqrt{M}$-PAM. Therefore, the probability of correct symbol decision $P_S(C)$ as a function of symbol error rate $P_{SER}(E)$ is:

$$P_S(C) = \left[1 - P_{SER}(E)\big|_{\sqrt{M}-\text{PAM}} \right]^2 = 2P_{SER}(E)\big|_{\sqrt{M}-\text{PAM}}\left\{ 1 - \frac{1}{2}P_{SER}(E)\big|_{\sqrt{M}-\text{PAM}} \right\} \tag{2.107}$$

The expression of P_{SER} in the square brackets corresponds to half the average power of the M-QAM constellation. Taking this into account and Equation (2.41) derived above for M-PAM we get [Simon05]:

$$P_{SER}(E) = 4\left(\frac{\sqrt{M}-1}{\sqrt{M}}\right)Q\left(\sqrt{\frac{3E_S}{N_0(M-1)}}\right) - 4\left(\frac{\sqrt{M}-1}{\sqrt{M}}\right)^2 Q^2\left(\sqrt{\frac{3E_S}{N_0(M-1)}}\right) \quad (2.108)$$

After mathematical manipulations using Gaussian Q- and Q^2-functions, the expression for P_{SER} becomes [Simon05]:

$$P_{SER}(E) = \frac{4}{\pi}\left(\frac{\sqrt{M}-1}{\sqrt{M}}\right)\int_0^{\pi/2}\exp\left[-\frac{3}{2(M-1)\sin^2\theta}\frac{E_S}{N_0}\right]d\theta$$
$$-\frac{4}{\pi}\left(\frac{\sqrt{M}-1}{\sqrt{M}}\right)^2\int_0^{\pi/2}\exp\left[-\frac{3}{2(M-1)\sin^2\theta}\frac{E_S}{N_0}\right]d\theta \quad (2.109)$$

Accurate expressions for bit-error probability P_{BER} were recently obtained for arbitrary M [Yoon02]. However we will only give an approximate formula which is valid for both high and low SNRs. It is necessary to have such simplified expressions for a complete range of SNR values because they are very useful to obtain P_{BER} for multipath fading channels later in this chapter. Consequently, [Lu99] derives the following approximate expression:

$$P_{BER}(E) \cong 4\left(\frac{\sqrt{M}-1}{\sqrt{M}}\right)\left(\frac{1}{\log_2 M}\right)\sum_{i=0}^{\frac{\sqrt{M}}{2}-1}Q\left((2i+1)\sqrt{\frac{E_b}{N_0}\frac{3\log_2 M}{M-1}}\right) \quad (2.110)$$

2.1.10 Coherent Detection for Nonideal Carrier Synchronization

As explained in Section 2.1.6, to achieve coherent reception the local oscillator $c_{LO}(t)$ at the receiver must be synchronized in frequency and phase with the transmitted carrier. Let us incorporate very small frequency errors between the two oscillators in the phase variable. In this way, if the transmitted carrier phase is θ_C, the phase of the received carrier (at LO) is an estimate $\hat{\theta}_C$ produced by recovery circuits such as PLLs. In this case the resulting phase error $\theta_E = \theta_C - \hat{\theta}_C$ will be a random variable with a p.d.f. $p(\theta_E)$.

First of all, a detailed presentation of the system model is in order. This will help to effectively present the resulting decision variables. Figure 2.20 illustrates the structure of a coherent receiver with nonideal carrier recovery (CR). The received signal $r(t)$ is multiplied by the local oscillator signal at the receiver, which produces an estimate of the transmitted carrier phase $\hat{\theta}_C$. In accordance with nomenclature in Section 2.1.6, we have for the transmitted and received passband signals $s(t)$ and $r(t)$ as well as the local oscillator signal $c_{LO}(t)$ and the baseband received signal $R(t)$ respectively:

$$s(t) = \begin{cases} A(a_{In} + ja_{Qn})\exp[j(2\pi f_C t + \theta_C)] + n(t), & \text{QAM} \\ A\exp[\theta_n]\exp[j(2\pi f_C t + \theta_C)] + n(t), & M\text{-PSK} \end{cases} \quad (2.111a)$$

$$r(t) = s(t) + N(t) \quad (2.111b)$$

$$c_{LO}(t) = \exp\left[j\left(2\pi f_C t + \hat{\theta}_C\right)\right] \quad (2.111c)$$

$$R(t) = AS(t)\exp[j(\theta_C - \hat{\theta}_C)] + n_b(t) \quad (2.111d)$$

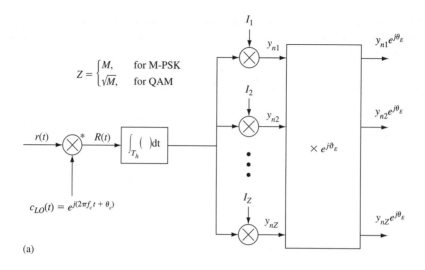

(a)

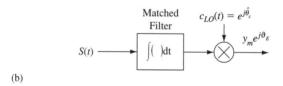

(b)

Figure 2.20 (a) Coherent receiver with non-ideal carrier recovery. One-branch baseband equivalent of coherent detection with non-ideal carrier recovery

The term $n_b(t)$ represents a baseband noise.

The downconverted signal $R(t)$ at the output of the multiplier goes through matched filter receiver which for M-PSK and QAM has the same structure as the corresponding coherent ideal receivers. For those two cases the multiplier operators in Figure 2.20(a) are given as:

$$I_i = \begin{cases} A_i, & i = 1, 2, ..\sqrt{M} \quad \text{QAM} \\ \exp\left[2\pi i/M\right], & i = 1, 2, ...M \quad M\text{-PSK} \end{cases} \tag{2.112}$$

Because we have a resulting phase error θ_E between the estimated and the actual phase of the carrier of the transmitted signal, there is an extra block multiplying all decision variables y_{ni} corresponding to ideal coherent detection by $\exp(j\theta_E)$:

$$y(\theta_E)_{ni} = y_{ni} \exp(j\theta_E) + N_n(t) \tag{2.113a}$$

$$N_n(t) = \int n(t) \cdot \exp(j\theta_E) dt \tag{2.113b}$$

The above considerations lead to an equivalent baseband representation of the receiver as illustrated in Figure 2.20(b).

Coming back to the distribution $p(\theta_E)$ of the random variable θ_E, it has been shown [Simon05], [Viterbi63] that it follows the Tikhonov model according to which:

$$p(\theta_E) = \frac{\exp(\text{SNR}_L \cos \theta_E)}{2\pi I_0(\text{SNR}_L)} \tag{2.114}$$

where SNR_L is the SNR of the feedback loop (PLL) used to perform carrier recovery, whereas I_0 represents the zero-order Bessel function of the first kind.

To assess coherent reception taking into account nonideal CR it is assumed that the detector has a coherent structure as if the LO carrier phase is perfectly known at the receiver. However, the nonideal nature of the CR circuitry is taken into account when the performance of the receiver is evaluated. To do so, we use the fact that $p(\theta_E)$ follows the Tikhonov distribution. Subsequently, the bit error probability P_B can be evaluated from the following integral:

$$P_{\text{BER}} = \int_{-K}^{K} P_{\text{BER}}(E; \theta_E) \cdot p(\theta_E) d\theta_E \tag{2.115}$$

By inserting appropriate values for integration limits and conditional probability $P_{\text{BER}}(E; \theta_E)$ in the above equation, P_{BER} can be determined for BPSK and QPSK.

2.1.10.1 BPSK and QPSK Performance

Using a Costas loop carrier recovery tracking the double phase error $2\theta_E$, we can get for BPSK [Simon05]:

$$P_{\text{BER}}(E; \theta_E) = Q\left[\sqrt{\frac{2E_b}{N_0}} \cdot \cos \theta_E\right]$$

$$K = \pi/2, \quad p(\theta_E) = \frac{\exp[\text{SNR}_{eq} \cos(2\theta_E)]}{\pi I_0(\text{SNR}_{eq})} \qquad 0 \leq |\theta_E| \leq \frac{\pi}{2} \tag{2.116}$$

where the equivalent loop SNR_{eq} and squaring loss S_L are given by:

$$\text{SNR}_{eq} = \frac{\text{SNR}_L S_L}{4}, \quad \text{SNR}_L = \frac{E_b/T_b}{N_0 B_L}, \quad S_L = \frac{1}{1 + 1/(2E_b/N_0)} \tag{2.117}$$

where B_L represents the equivalent noise bandwidth of the loop. Using the above formulas in Equation (2.115) it is easy to get:

$$P_{\text{BER}}(E) = \int_{-\pi/2}^{\pi/2} Q\left(\sqrt{\frac{2E_b}{N_0}} \cos \theta_E\right) \frac{\exp[\text{SNR}_{eq} \cos(2\theta_E)]}{\pi I_0(\text{SNR}_{eq})} d\theta_E \tag{2.118}$$

Simon reports a simplification in which the integrand is a function of the zeroth-order modified Bessel function:

$$P_{\text{BER}}(E) = \frac{1}{\pi} \int_0^{\pi/2} \exp\left[-\frac{E_b}{2N_0 \sin^2 \theta}\right] \cdot \frac{I_0\left[-\frac{E_b}{2N_0 \sin^2 \theta} + \text{SNR}_{\text{eq}}\right]}{I_0(\text{SNR}_{\text{eq}})} d\theta \qquad (2.119)$$

Employing a similar technique does not result in simplified P_{BEP} expressions for QPSK. To produce exact expressions for both BPSK and QPSK, another approach is used in which the local oscillator signal at the receiver $c_{\text{LO}}(t)$ is modelled such that the error in the phase estimate is replaced by additive noise [Simon05]:

$$c_{\text{LO}}(t) = \exp\left[j(2\pi f_C t + \theta_C)\right] + N_{\text{LO}}(t) \qquad (2.120)$$

In this case the bit error probability is:

$$P_{\text{BER}}(E) = \frac{1}{2}[1 - Q_1(\sqrt{b}, \sqrt{a}) + Q_1(\sqrt{a}, \sqrt{b})] \qquad (2.121)$$

where a and b are given for BPSK and QPSK as follows ([Simon05], [Fitz92]):

$$a_{\text{QPSK}} = \frac{E_b}{2N_0}\left(\sqrt{2G} - 1\right)^2, \quad b_{\text{QPSK}} = \frac{E_b}{2N_0}\left(\sqrt{2G} + 1\right)^2 \qquad (2.122)$$

G represents the SNR gain for this approach and is related to the previous Tikhonov phase distribution model by:

$$G = \frac{1}{B_L T_b} = \frac{\text{SNR}_L}{(E_b/N_0)} \qquad (2.123)$$

2.1.11 M-ary Frequency Shift Keying

2.1.11.1 Transmitted Waveforms and their Properties

In M-FSK the transmitted signal waveforms are given as:

$$s_k(t) = \sqrt{\frac{2E_S}{T_S}} \exp\left\{j[2\pi f_C t + 2\pi f_i(k)(t - kT_S) + \theta_C]\right\} \qquad (2.124)$$

where $f_i(k)$ represents the digital frequency modulation value in the interval $[kT_S, (k+1)T_S]$ which can take any of the following M values:

$$\nu_i = (2i - 1 - M) \cdot \Delta f/2, \quad i = 1, 2, \ldots, M \qquad (2.125)$$

In addition, the energy E_S per symbol and symbol interval are:

$$E_S = KE_b, \quad T_S = KT_b, \quad K = \log_2 M \qquad (2.126)$$

where K represents the number of bits per symbol and Δf is the frequency separation between successive frequency symbols. Δf is an important parameter because it determines the degree of similarity between the M symbols. This is quantified by calculating the correlation between two transmitted waveforms, the nth and the mth [Proakis02]:

$$R_{mn}(\Delta f) = \frac{1}{E_S} \int_0^{T_S} s_m(t)s_n(t)dt \qquad (2.127)$$

Using Equations (2.124) and (2.125) after some manipulations we get:

$$R_{mn}(\Delta f) = \frac{\sin 2\pi(m-n)\Delta f T_S}{2\pi(m-n)\Delta f T_S} \tag{2.128}$$

This function has the usual $\sin x/x$ form. Consequently, the orthogonality condition among transmitted waveforms is:

$$\Delta f = \frac{n}{2T_S}, \quad n = 1, 2, \ldots \tag{2.129}$$

Furthermore, Δf is associated with another parameter of M-FSK, the modulation index h which is given by $h = \Delta f T_S$.

2.1.11.2 Coherent Detection for M-FSK

The optimum M-FSK receiver for coherent detection can have the form of Figure 2.21. The ith receiver branch has a complex multiplier multiplying the incoming signal by the complex frequency carrier corresponding to the ith frequency increment $I_{ni} = \exp[j2\pi v_i(t-nT_S)]$. The resulting product passes through an integrator to obtain the decision variables y_{ni} given by:

$$y_{ni} = \int_{nT_S}^{(n+1)T_S} \exp[j2\pi(f_n - v_i)(t - nT_S)]dt + n_{ni} \tag{2.130a}$$

$$n_{ni} = \int_{nT_S}^{(n+1)T_S} \exp[-j2\pi v_i(t - nT_S)]n_b(t)dt \tag{2.130b}$$

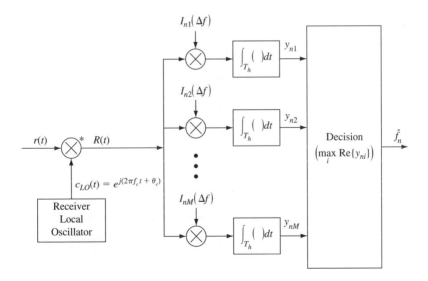

Figure 2.21 Optimum coherent receiver for M-FSK modulation

If the transmitted frequency increment is $f_n = v_k = (2k - 1 - M)\Delta f/2$ then the above integral gives:

$$y_{ni} = T_S \exp\left[j\pi(k-i)N/2\right]\frac{\sin\left[\pi(k-i)N/2\right]}{\pi(k-i)N/2} \tag{2.131a}$$

$$n_{ni} = \int_0^{T_S} \exp\left[\frac{-j\pi(2i-1-M)N}{2T_S}t\right]n_b(t+nT_S)dt \tag{2.131b}$$

The decision rule is to take $\max_i \text{Re}\left(y_{ni}\right)$.

Because it is extremely difficult to employ coherent detection for a modulation scheme such as M-FSK, we do not give calculations for error rate. For reasons of comparison we only give the probability of error for binary FSK:

$$P_{\text{BER}}(E) = Q\left(\sqrt{\frac{E_b}{N_0}}\right) = \frac{1}{\pi}\int_0^{\pi/2} \exp\left[-\frac{E_b}{2N_0}\frac{1}{\sin^2\theta}\right]d\theta \tag{2.132}$$

As will be demonstrated later, noncoherent detection is the most suitable approach for M-FSK demodulation because it considerably simplifies the receiver.

2.1.12 Continuous Phase FSK

As discussed above, the M-FSK modulated signal is generated by assigning one of the M possible values, $f(k) = (2i - 1 - M) \cdot \Delta f/2$ in the signalling interval $[kT_S, (k+1)T_S]$ to the frequency of the transmitted waveform. A straightforward way to implement it is by using M different oscillators with frequency difference $\Delta f/2$. In this way, the imposed frequency steps are abrupt and do not have any continuity. This discontinuity in the switching process results in high sidelobes in the transmitted spectrum. To avoid such an effect, a single oscillator can be used which must be modulated in frequency according to the above modulation pattern. In this case, phase continuity is preserved.

In general, there is a family of modulation techniques in which the phase of the modulated waveform changes in continuous manner. These are categorized as continuous phase modulation (CPM) techniques and are characterized by the following expression for the modulated waveform:

$$\text{Re}\{s(t)\} \equiv u(t) = \sqrt{\frac{2E_S}{T}}\cos\left[2\pi f_C t + \theta(t;\mathbf{a}) + \theta_C\right] \tag{2.133}$$

$$\theta(t;\mathbf{a}) = 2\pi \sum_{k=-\infty}^{n} \alpha_k h_k q(t - kT_S) \tag{2.134}$$

where $\{\alpha_k\}$ represents the sequence of information symbols, h_k are the modulation indices and $q(t)$ is an arbitrary waveform taking specific forms depending on the special cases of CPM.

Continuous-phase FSK is a special case of CPM in which $\theta(t;\mathbf{a})$ is given as:

$$\theta(t;\mathbf{a}) = 4\pi T_S f_d \int_{-\infty}^{t} U(\tau)d\tau \tag{2.135}$$

where f_d is the peak frequency deviation and $U(t)$ represents a baseband PAM modulated signal expressed as in Equation (2.46) (Section 2.1.6), which is repeated below:

$$U(t) = \sum_n \alpha_n g_T(t - nT_S) \tag{2.46}$$

A modulating waveform of this form gives two distinguishing characteristics: It provides an M-FSK modulated signal with distinct instantaneous frequencies equal to:

$$f_i(t) = f_C + 2T_S f_d \cdot U(t) \tag{2.136}$$

while at the same time this signal is continuous in phase.

Elementary manipulations give the following expression for $\theta(t; a)$ [Proakis02]:

$$\theta(t; a) = 2\pi f_d T \sum_{k=-\infty}^{n-1} \alpha_k + 2\pi(t - nT)f_d \alpha_n = \theta_n + 2\pi h \alpha_n q(t - nT) \tag{2.137}$$

where h, θ_n and $q(t)$ are given as follows:

$$h = 2f_d T, \quad \theta_n = \pi h \sum_{k=-\infty}^{n-1} \alpha_k, \quad q(t) = \begin{cases} 0, & t < 0 \\ t/2T, & 0 \le t \le T \\ 1/2, & t > T \end{cases} \tag{2.138}$$

Since the principle characteristic for CP-FSK is the change of phase in a continuous manner as a function of time, it gives more insight to consider the phase variable $\theta(t; a)$ as representing phase trajectories in time. Figure 2.22 shows the phase trajectories or the phase tree representing the values of $\theta(t; a)$ for various values of the information sequence $\{\alpha_k\}$ as a function of time. The straight lines connecting the points represent the actual trajectory for a unit pulse g_T of rectangular shape. Other types of pulses give different shapes for the trajectories [Aulin81]. Plotting the trajectories in modulo $[2\pi]$ format reduces the figure into a phase trellis like the one illustrated in Figure 2.23.

This form of representation is very useful when expressions for the probability of error are to be derived. This is due to the fact that MLS detection techniques can be used since CP-FSK is a modulation with memory.

2.1.12.1 Continuous Phase Modulation and Detection

We will give some more attention to the general class of phase modulation defined as CPM in the previous section, in order to suggest a receiver structure and determine the probability of error. In our presentation we consider only full response signals for which $g_T(t) = 0$ for $t > T$. Partial response signals for which the signalling pulse extends beyond T are mostly employed in applications where bandwidth limitation is the principle factor in the system design.

Usually the modulation index is fixed over all time periods and Equation (2.134) can be written as:

$$\theta(t; a) = 2\pi h \sum_{k=-\infty}^{n} \alpha_k q(t - kT) = \pi h \sum_{k=-\infty}^{n-1} \alpha_k + 2\pi h \alpha_n q(t - nT) = \theta_n + \Theta(t, \alpha_n),$$

$$\text{for } nT \le t \le (n+1)T \tag{2.139}$$

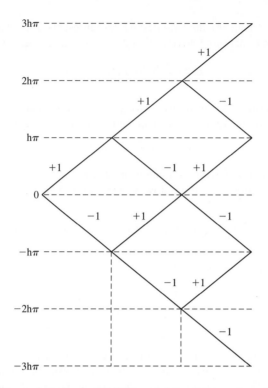

Figure 2.22   Phase trajectories for CP-FSK modulation [rectangular pulse shape $g_T(t)$]

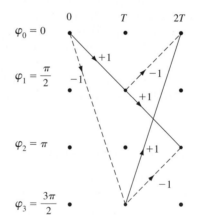

Figure 2.23   Phase-state trellis for binary CP-FSK

where

$$\theta_n \equiv \pi h \sum_{k=-\infty}^{n-1} \alpha_k, \quad \Theta(t, \alpha_n) \equiv 2\pi h \alpha_n q(t - nT) \qquad (2.140)$$

In CPM and in contrast to CP-FSK, the function $q(t)$ is given by $q(t) = \int_0^t g_T(\tau)d\tau$.

From the discussion on the optimum ML detector in Section 2.1.3 it was shown that the ML receiver must calculate one of the following metrics R_k from Equation (2.6) or $C(\mathbf{r}, \mathbf{s_m})$ from Equation (2.21):

$$R_k = \int r(t)s_k(t)\mathrm{d}t, \;\; C(\mathbf{r}, \mathbf{s_m}) = \mathbf{r} \cdot \mathbf{s_m} = \sum_{k=1}^{M} r_k s_{mk}, \;\;\; m = 1, 2, \ldots, M$$

Considering this, there are two aspects that must be resolved. The first is to define the structure of the ML receiver for a modulation scheme with memory. The second is to determine the probability of error for such a receiver.

The Viterbi Detector

In accordance with the previous equation, the ML receiver minimizes:

$$\int (r(t) - s(t,\mathbf{a}))\mathrm{d}t \tag{2.141}$$

in terms of an infinitely long sequence of data symbols $\{\mathbf{a}\}$. It must be noticed that the difference from (2.139) above is that now the transmitted signal terms $s_b(t, \mathbf{a})$ depend on the long sequence of symbols $\{\mathbf{a}\}$. Hence the following correlation must be maximized:

$$C(\mathbf{a}) = \int_{-\infty}^{+\infty} r(t)s(t,\mathbf{a})\mathrm{d}t \tag{2.142}$$

Because calculation of the above correlation must be performed with respect to a long sequence $\{\mathbf{a}\}$, it is convenient to break the correlation metric in two terms. The first is a cumulative term incorporating the previous decisions up to the $(n-1)$ time instant. The second is a metric referred to the current time instant and is usually called the incremental metric:

$$C_n(\mathbf{a}) = C_{n-1}(\mathbf{a}) + W_n(\mathbf{a}) \tag{2.143}$$

with C_n and W_n as follows:

$$C_n(\mathbf{a}) \equiv \int_{-\infty}^{(n+1)T} r(t)s(t,\mathbf{a})\mathrm{d}t \tag{2.144}$$

$$W_n(\mathbf{a}) = \int_{nT}^{(n+1)T} r(t)s(t,\mathbf{a})\mathrm{d}t \tag{2.145}$$

$W_n(\alpha)$ represents the incremental metric from the nth to the $(n+1)$th time period.

Figure 2.24 illustrates the structure of the receiver whereas the implementation procedure below outlines, in algorithmic fashion, its operation.

Implementation Procedure

(1) Calculate $W_n(\mathbf{a})$ for all possible modulated waveforms. The number of possible modulated waveforms depends on the number of previous bits on which it depends. For example, if it depends on the current and two previous bits, we have $2^3 = 8$ possible $W_n(\mathbf{a})$.

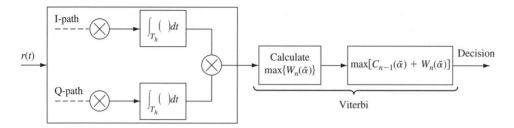

Figure 2.24 Receiver for continuous phase (CP) modulation

(2) There was a limited number of possible values for $C_{n-1}(a)$ (for example 4). For all these values calculate the summations $C_{n-1}(a) + W_n(a)$.
(3) Using step 2, decide upon the next four metrics $C_n(a)$ to be used in the next iteration.

Calculation of the Probability of Symbol Error

We will follow [Aulin81] and [Steele92] to derive the probability of error. Let us assume that the transmitted signal is $s_i(t)$ and that the ML receiver decides on $s_k(t)$ with probability $P_{SER}(k|i)$. This is the probability of erroneously deciding that $s_k(t)$ was transmitted because:

$$\|r(t) - s_k(t)\| \leq \|r(t) - s_i(t)\| \tag{2.146}$$

The conditional probability $P_E(k|i)$ that $\|r_1 - s_{i1}\| > d_{ik}/2$ is as follows:

$$P_{SER}(k|i) = \int_{d_{ik}/2}^{\infty} \frac{1}{\sqrt{\pi N_0}} \exp\left(-\frac{x^2}{N_0}\right) dx \tag{2.147}$$

It can be easily shown [Steele92] that:

$$P_{SER}(i) \leq \sum_{k \neq i} Q\left(\frac{\|s_i(t) - s_k(t)\|}{\sqrt{2N_0}}\right) \tag{2.148}$$

Equation (2.147) is true because the components $r_1 - s_{i1}$ are Gaussian-distributed zero-mean noise with variance $N_0/2$. The distance d_{ik} between signals $s_i(t)$ and $s_k(t)$ can be expressed as:

$$d_{ik}^2 = \|s_i(t) - s_k(t)\|^2 = \int (s_i(t) - s_k(t))^2 dt = \int_0^{NT} s_i^2(t)dt$$

$$+ \int_0^{NT} s_k^2(t)dt - 2\int_0^{NT} s_i(t)s_k(t)dt \tag{2.149}$$

Some trivial mathematics reveals that the first term of the above equation depends on the energy E of the transmitted signal and the inverse of carrier frequency $1/\omega_C$. The second term is also a function of the energy E. Only the cross-correlation term (the third one) is a function of the information and is the only one which will determine the distance d_{ik}:

$$\int_0^{NT} s_i(t)s_k(t)dt = \frac{2E}{T}\int_0^{NT} \cos\left[2\omega_C t + \varphi_i(t) + \varphi_k(t)\right]dt + \frac{2E}{T}\int_0^{NT} \cos\left[\varphi_i(t) - \varphi_k(t)\right]dt \tag{2.150}$$

Owing to the term of the first integral, only the second integral remains to determine d_{ik}^2:

$$d_{ik}^2 = \frac{2E}{T} \int_0^{NT} [1 - \cos \Delta\varphi(t)]dt, \quad \Delta\varphi \equiv \varphi_i(t) - \phi_k(t) \tag{2.151}$$

Normalizing the distance with respect to the energy per bit E_b we have:

$$d_N^2 \equiv \frac{d_{ik}^2}{2E_b} = \frac{\log_2 M}{T} \int_0^{NT} [1 - \cos \Delta\varphi(t)]dt \tag{2.152}$$

Consequently, Equation (2.148) becomes:

$$P_{SER}(i) \leq \sum_{k \neq i} Q\left(\sqrt{\frac{E_b}{N_0}} d[s_i(t), s_k(t)]\right) \tag{2.153}$$

Depending on the SNR, there is one or more paths associated to a minimum distance over the observation interval of length NT. These paths are the ones which determine the P_{SER} because they are the dominant terms in the summation of Equation (2.153) [Proakis02], [Steele92]. Consequently, if the number of minimum paths within NT is N_{dm}, the error probability is:

$$P_{SER} \cong N_{dm} Q\left(\sqrt{\frac{E_b}{N_0} d_{min}^2}\right) \tag{2.154}$$

[Aulin81] has computed an upper bound d_B^2 for d_{min}^2:

$$d_B^2(h) = \log_2 (M) \cdot \min_{1 \leq k \leq M-1} \left\{ 2 - \frac{1}{T} \int_0^{2T} \cos\left[2\pi h \cdot 2k(q(t) - q(t-T))\right]dt \right\} \tag{2.155}$$

Special case: CP-FSK. For M-ary CP-FSK the upper bound $d_B^2(h)$ becomes

$$d_B^2(h) = \log_2 (M) \cdot \min_{1 \leq k \leq M-1} 2\left(1 - \frac{\sin 2\pi kh}{2\pi kh}\right) \tag{2.156}$$

The minimum distance $d^2(h)$ for $N = 1,2,3,4$ for binary CP-FSK is illustrated in Figure 2.25 as a function of h. The dashed line depicts the bound $d_B^2(h)$. From this figure one can notice that for MSK ($h = 1/2$) $d_B^2(0.5) = 2$. MSK requires two-bit interval observations before a decision is made. From Figure 2.25 one can also deduce that the optimum h for CP-FSK is $h = 0.715$ for $N = 3$. This is translated to an SNR gain of 0.85 dB compared with MSK and PSK [Aulin81].

2.1.13 Minimum Shift Keying

Binary CP-FSK with modulation index $h = 1/2$ is defined as minimum shift keying (MSK) because the frequency separation between the corresponding symbols is $\Delta f = 1/2T_S$, which represents the minimum separation in frequency to achieve orthogonality when there are only two signal waveforms [Equation (2.129)].

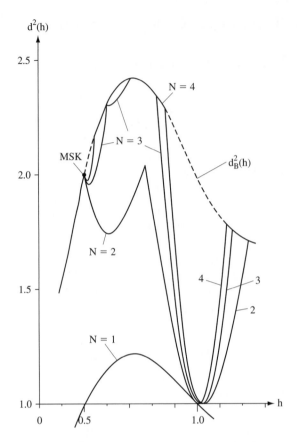

Figure 2.25 Minimum distance $d^2(h)$ as a function of h (reprinted from T. Aulin, C. Sundberg, 'Continuous phase modulation – Part I: full response signaling', IEEE Trans. Commun., vol. COM-29, March 1981, pp. 196–209, © 1983 IEEE)

For the above values of h and Δf it is easy to get the phase function and the corresponding waveforms for MSK:

$$\theta(t;\boldsymbol{a}) = \frac{\pi}{2} \cdot \sum_{k=-\infty}^{n-1} \alpha_k + \pi \alpha_n q(t - nT_S) = \theta_n + \frac{\pi}{2}\alpha_n \cdot \left(\frac{t - nT_S}{T_S}\right), \quad \text{for } nT_S \le t \le (n+1)T_S$$

$$(2.157)$$

$$u_m(t) = \sqrt{\frac{2E_b}{T_S}} \cos\left[2\pi\left(f_C + \frac{1}{4T_S}\alpha_m\right)t - \alpha_m\frac{m\pi}{2} + \theta_m\right] \qquad (2.158)$$

From the frequency argument in the above equation, it is easy to see that:

$$\Delta f = f_2 - f_1 = \left(f_C + \frac{1}{4T_S}\right) - \left(f_C - \frac{1}{4T_S}\right) = \frac{1}{2T_S} \qquad (2.159)$$

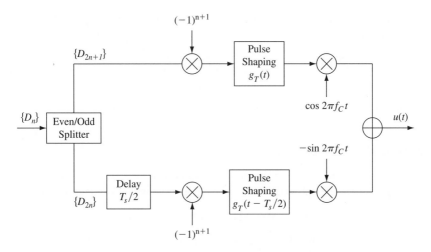

Figure 2.26 MSK transmitter using pulse-shaped OQPSK modulator

MSK modulation can be conceived like an offset QPSK with different pulse shaping function g_T. This is because Equation (2.158) can be written as:

$$u(t) = \sqrt{\frac{2E_b}{T_S}} \left\{ \left[\sum_{-\infty}^{\infty} a_{\mathrm{Im}} g_T(t - mT_S) \right] \cos 2\pi f_C t \right.$$
$$\left. + \left[\sum a_{Qm} g_T(t - mT_S - T_S/2) \right] \sin 2\pi f_C t \right\} \qquad (2.160)$$

where the unit pulse is shaped as follows:

$$g_T(t) = \begin{cases} \sin \dfrac{\pi t}{T_S} = \ldots, 0 \le t \le T_S \\ 0, \qquad \text{otherwise} \end{cases} \qquad (2.161)$$

If φ_C is the cumulative phase argument in $u(t)$, $\varphi_C(m) = \theta_m - a_m(m\pi/2)$, it is easy to see that the phase difference between two successive signalling intervals is:

$$\varphi_C(m) - \varphi_C(m - 1) = \frac{m\pi}{2} (a_{m-1} - a_m) \qquad (2.162)$$

From this and the fact that the data can be looked at as differentially encoded before transmission, $D_m = a_m D_{m-1}$, it can be verified that [Simon05]:

$$a_{\mathrm{Im}} = (-1)^{m+1} \cdot D_{2m+1}, \quad a_{Qm} = (-1)^{m+1} \cdot D_{2m} \qquad (2.163)$$

and that the data can be transmitted using a precoded-MSK format (Figure 2.26) such that they can be coherently detected using a pulse-shaped OQPSK optimum demodulator without employing differential decoding. All the above lead to the conclusion that precoded-MSK modulation, demodulation and BER performance are identical to those of pulse-shaped OQPSK.

From Section 2.1.12.1 above it can be easily deduced that, since $d_{\min}^2(MSK) = 2$, then the error probability is:

$$P_{\text{BER}} \approx Q\left(\sqrt{\frac{2E_b}{N_0}}\right) \tag{2.164}$$

This is identical to the probability of bit error for BPSK modulation.

2.1.14 Noncoherent Detection

When no information exists or can be obtained regarding the carrier phase at the receiver, noncoherent detection is employed. The received passband waveform can be expressed as:

$$r(t) = S(t)\exp[j(2\pi f_C t + \theta_C)] + n(t) \tag{2.165}$$

whereas the corresponding baseband received signal is:

$$R(t) = S(t)\exp[j\theta_C] + n(t)\exp[-j2\pi f_C t] \tag{2.166}$$

where $n(t)$ is passband white noise and θ_C, as explained before, is unknown and represents a uniformly distributed random variable within the interval $[-\pi, \pi]$. It can be shown [Lindsey73] that the optimum noncoherent receiver consists of a matched filter followed by a square-law detector, as illustrated in Figure 2.27. The corresponding decision variables are:

$$z_{nk} = |y_{nk}|^2 = \left|\int_{nT_S}^{(n+1)T_S} R(t)s_k^*(t)dt\right|^2, \quad k = 1, 2, \ldots \tag{2.167}$$

Because information is embedded in phase for both QAM and M-PSK modulating schemes, M-FSK is the only modulation technique that can be demodulated in a noncoherent fashion.

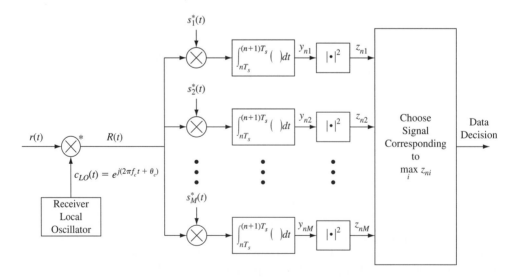

Figure 2.27 Optimum non-coherent receiver [Simon05]

Hence the output of the matched filter for noncoherent M-FSK detection is [Simon05]:

$$y_{nk} = A_C \int_{nT_S}^{(n+1)T_S} \exp\left[j2\pi(f_n - \nu_k)(t - nT_S) + j\theta_C\right]dt + N_{nk}, \quad k = 1, 2, \ldots, M \quad (2.168)$$

$$N_{nk} = \int_{nT_S}^{(n+1)T_S} \exp\left[-j2\pi\nu_k(t - nT_S)\right]N(t)dt, \quad N(t) = n(t)\exp\left[j2\pi f_C t\right] \quad (2.169)$$

For orthogonal signals ($\Delta f_{\min} = 1/T_S$), the symbol error probability is [Simon05], [Proakis02]:

$$P_{SER}(E) = \sum_{m=1}^{M-1} (-1)^{m+1} \binom{M-1}{m} \frac{1}{m+1} \exp\left[-\frac{m}{m+1}\left(\frac{E_S}{N_0}\right)\right] \quad (2.170)$$

For non-orthogonal M-FSK we get the following error probabilities [Simon05]:

$$P_{BER}(E) = Q_1(\sqrt{a}, \sqrt{b}) - \frac{1}{2}\exp\left[(a+b)/2\right]I_0(\sqrt{ab}) \quad (2.171)$$

$$a = \frac{E_b}{2N_0}\left(1 - \sqrt{1 - \rho_{12}^2}\right), \quad b = \frac{E_b}{2N_0}\left(1 + \sqrt{1 - \rho_{12}^2}\right) \quad (2.172)$$

where ρ_{12} is the correlation between the two signalling waveforms.

2.1.15 Differentially Coherent Detection (M-DPSK)

As presented in sections 2.1.7–2.1.10, although ideal coherent detection entails perfect knowledge of the carrier frequency and phase, in practice, carrier tracking loops are used to estimate the frequency and phase of the transmitted carrier. In this case, depending on the tracking mechanism, we always have a phase tracking error $\theta_e = \theta_C - \hat{\theta}_C$. When the statistical properties of θ_e are taken into account to obtain the optimum detector, detection is partially or differentially coherent. It is partially coherent when the phase tracking error θ_e is taken into account. In contrast, when only the detected phase of the received carrier θ_C (and not the tracking error) is taken into account for detection, we have differentially coherent detection.

In addition, it must be noted that maximum likelihood detection can be based on more than one observation intervals. We define the number of symbol periods used for the ML detection as N_S. We can have one-symbol ($N_S = 1$), two-symbol ($N_S = 2$) and multiple-symbol detection ($N_S > 2$). Furthermore it is shown in [Simon05], [Viterbi65] that the optimum receiver structure employs a combination of coherent and square-law detector following the matched filters.

2.1.15.1 Partially Coherent Detection

Let us consider partially coherent detection of an M-PSK modulated signal. In this case, the decision variables for $N_S \geq 2$ obtained at the output of the detection device are [Viterbi65], [Simon05]:

$$z_{nk} = \left|\sum_{i=0}^{N_S-1} \frac{1}{N_0}y_{ki}(n-i)\right|^2 + \rho_C\left(\text{Re}\left\{\sum_{i=0}^{N_S-1} \frac{1}{N_0}y_{ki}(n-i)\right\}\right), \quad ki = 1, 2, \ldots, M \quad (2.173)$$

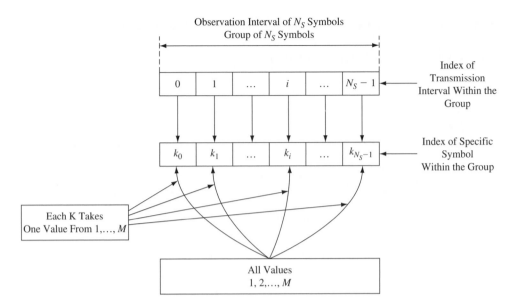

Figure 2.28   Indexing in partially coherent modulation

with $y_{ki}(n-i)$ being:

$$y_{ki}(n-i) = \int_{(n-i)T_S}^{(n-i+1)T_S} R(t)s_{ki}^*(t)\,\mathrm{d}t, \quad ki = 1, 2, \ldots, M, \quad i = 0, 1, \ldots, N_S - 1 \quad (2.174)$$

$$s_{ki}(t) = A_C e^{j\psi_{ki}} = A_C \exp\left[j\,(2ki-1)\,\pi/M\right], \quad ki = 1, 2, \ldots, M \quad (2.175)$$

where, as mentioned before, ρ_C represents the carrier recovery loop SNR, $s_{ki}(t)$ represent the baseband transmitted signal and $y_{ki}(n-i)$ represents matched filter outputs which can take M^{N_S} different values.

The way indexing is used for detection based on multiple symbols ($N_S \geq 2$) is illustrated in Figure 2.28, where it is shown that first an index k_i is assigned to a specific transmission interval i within the overall observation interval of N_S symbols. The last stage is the assignment of a specific value to each k_i from a 'pot' of M integers ($1, 2, \ldots, M$) as illustrated in the lower part of the figure. The only difference between this assignment and the one in classical M-PSK is that in the latter there is no index i at the symbol index k because there is no need to specify an observation interval of more than one symbol.

Example   The indexing mechanism and symbol assignment for $M = 4$ and $N_S = 2$ is as follows: we consider observation intervals of two successive symbols and consequently we have only two symbol indices, k_0 and k_1, which can take any of the following four values: $\pi/4$, $3\pi/4$, $5\pi/4$ and $7\pi/4$. Therefore the combination k_0, k_1 can take 4^2 different values. Therefore there are 16 different matched filter outputs $y_{ki}(n-i)$ due to the different combinations $s_{k0}(t)$, $s_{k1}(t)$.

The receiver subsystem for the simple case of $N_S = 1$ is illustrated in Figure 2.29 and consists of a combination of $\mathrm{Re}\{\bullet\}$ and $|\bullet|^2$ devices fed by the output of the matched filter. In the more

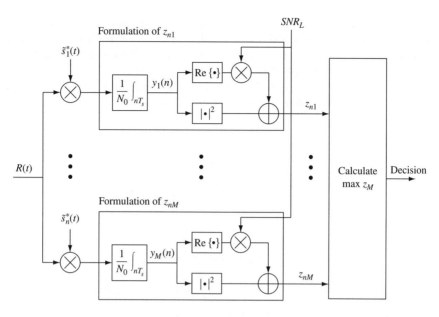

Figure 2.29   Optimum partially coherent receiver

general case $N_S \geq 2$, each matched filter block is substituted by N_S blocks, the outputs of which are summed and the summation output is fed to the Re$\{\bullet\}$ and $|\bullet|^2$ blocks. In this case we have M^{N_S} branches generating M^{N_S} decision variables $z_{n\mathbf{k}}$. This is why the $\mathbf{k}$ subscript has the form of a vector.

Because of the form of $s_{ki}(t)$ in each of the two terms in Equation (2.175), an exponential term of the form $\exp(-j\theta_i)$ is included. In each transmission interval, θ_i takes a constant value, that of the transmitted phase information in M-PSK.

The first term of Equation (2.173) corresponds to a noncoherent decision metric where the phase component introduced by the channel does not have any effect on the metric. To put it another way, if we add an arbitrary but fixed-valued phase θ_F to all phases θ_{ki}, there will be no change in the decision resulting from the first term (squared absolute value). Consequently, in the case of $\text{SNR}_L = 0$ (only the first term remains in the equation), differential encoding must be employed because otherwise there can be no unique decision on the transmitted symbol. In this case we have differentially coherent detection and we will expand on this further below.

The second term corresponds to coherent detection because it is affected by the phase component in the received signal. By employing both terms in the detection process we have partially coherent detection.

2.1.15.2 Differentially Coherent Detection

In this approach, we consider initially a noncoherent detector (squarer operation), as discussed in Section 2.1.14. Furthermore, instead of one, we use N_S observation intervals and detect the transmitted signal using the ML principle by averaging the conditional likelihood function with respect to the received signal phase θ_C and not the estimated phase error θ_e. According to considerations in the sections concerning noncoherent and partially coherent detection, this approach corresponds to letting $SNR_L = 0$. Setting $SNR_L = 0$ in the Tikhonov distribution leads to p.d.f. $p(\varphi) = 1/2\pi$ corresponding to the uniform distribution and the formulation of the decision variable contains only the first squaring operation term.

From the above discussion we can obtain the decision variable for an observation interval of two symbols ($N_S = 2$) by letting $SNR_L = 0$ in Equation (2.173):

$$
\begin{aligned}
z_{nk} &= \frac{|y_{k0}(n) + y_{k1}(n-1)|^2}{N_0^2} \\
&= \left(\frac{A_C}{N_0}\right)^2 \left| \int_{nT_S}^{(n+1)T_S} r_{bb}(t)\exp(-j\psi_{k0})\mathrm{d}t + \int_{(n-1)T_S}^{nT_S} R(t)\exp(-j\psi_{k1})\,\mathrm{d}t \right|^2,
\end{aligned}
$$
$$
k0, k1 = 1, 2, \ldots, M \tag{2.176}
$$

Alternatively, the maximization of a posteriori probability gives analogous expressions [Divsalar90].

ψ_{ki} is one of the possible values of the transmitted phase and is given by:

$$
\psi_{ki} = \frac{(2ki - 1)\pi}{M}, \quad ki = 1, 2, \ldots, M \tag{2.177}
$$

ψ_{k0} and ψ_{ki} correspond to the transmitted phases during the nth and $(n-1)^{\text{th}}$ symbol intervals, respectively. As discussed above, to resolve the phase ambiguity resulting from the nature of differential coherent PSK, differential encoding is employed:

$$
\theta_n = \theta_{n-1} + \Delta\theta_n \tag{2.178}
$$

where θ_n represents the specific differentially encoded transmitted phases at the nth interval and is part of the received signal expression. $\Delta\theta_n$ is the phase corresponding to the input data during the nth transmission interval. Since θ_n and θ_{n-1} take values from ψ_{ki} as given above, then $\Delta\theta_n$ will also take values from $\Delta\psi_{ki}$:

$$
\Delta\psi_{ki} = \frac{k\pi}{M}, \quad k = 0, 1, \ldots, M-1 \tag{2.179}
$$

Mathematical manipulations give the following expression for the decision variables:

$$
z_{nk} = \mathrm{Re}\left\{ \exp(-j\Delta\psi_k) \left[\int_{nT_S}^{(n+1)T_S} R(t)\,\mathrm{d}t \right]^* \left[\int_{(n-1)T_S}^{nT_S} R(t)\,\mathrm{d}t \right] \right\}, \quad k = 0, 1, \ldots, M-1 \tag{2.180}
$$

The decision rule is to find the highest value of z_{nk} ($\max_k\{z_{nk}\}$) while k varies form 0 to $M-1$. Figure 2.30 gives the receiver structure for differentially coherent detection of M-PSK modulated symbols when two successive symbols are considered. In DPSK modulation, the receiver contains only one branch which calculates the $\mathrm{Re}\{\bullet\}$ at the output of the multiplier following the matched filter. This then goes through a $\mathrm{sign}\{\bullet\}$ type of decision device.

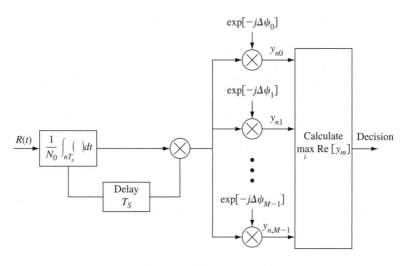

Figure 2.30 Differentially coherent receiver for M-PSK

The probability of symbol error for two-symbol observation receiver can be shown to be [Pawula82], [Simon05]:

$$P_{SER}(E) = \frac{\sqrt{g_{PSK}}}{2\pi} \int_{-\pi/2}^{\pi/2} \frac{\exp\left[-\frac{E_S}{N_0}(1 - \sqrt{1 - g_{PSK}}\cos\theta)\right]}{1 - \sqrt{1 - g_{PSK}}\cos\theta} d\theta \qquad (2.181)$$

For M-PSK and DPSK g becomes:

$$g_{PSK} \equiv \sin^2(\pi/M), \quad g_{DPSK} = 1 \qquad (2.182)$$

The resulting error probabilities for $M = 2, 4, 8, 16$ are [Simon05]:

$$P_{BER}(E) = \frac{1}{2}\exp\left(-\frac{E_b}{N_0}\right) \qquad (2.183a)$$

$$P_{BER}(E) = F\left(\frac{5\pi}{4}\right) - F\left(\frac{\pi}{4}\right), \quad M = 4 \qquad (2.183b)$$

$$P_{BER}(E) = \frac{2}{3}\left[F\left(\frac{13\pi}{8}\right) - F\left(\frac{\pi}{8}\right)\right], \quad M = 8 \qquad (2.183c)$$

$$P_{BER}(E) = \frac{1}{4}\sum_{n=7}^{8}F\left[\frac{2n-1}{M}\right] - \frac{1}{4}\sum_{n=5}^{6}F\left[\frac{2n-1}{M}\right] - \frac{1}{2}\sum_{n=1}^{2}F\left[\frac{2n-1}{M}\right], \quad M = 16 \quad (2.183d)$$

$$F(\lambda) = -\frac{\sin\lambda}{4\pi}\int_{-\pi/2}^{\pi/2}\frac{\exp\left[-\left(\frac{E_b}{N_0}\log_2 M\right)(1 - \cos\lambda\cos t)\right]}{1 - \cos\lambda\cos t}dt \qquad (2.184)$$

2.2 Digital Transmission in Fading Channels

In contrast to AWGN channels, the received signal undergoes both envelope distortion and phase distortion. The envelope $\alpha_C(t)$ is a random variable with a p.d.f. depending on the type and characteristics of the radio channel. As mentioned in Chapter 1, for microcellular and picocellular indoor and outdoor mobile communications the main channel types are Rayleigh, Rice and the more general Nakagami–m channel. Bearing in mind that the instantaneous received signal power is α_C^2 with mean square value $\overline{\alpha_C^2}$, we can define the corresponding instantaneous and average SNR per bit:

$$\text{SNR}_{\text{inst}} = \gamma = \alpha_C^2 \frac{E_b}{N_0}, \quad \text{SNR}_{\text{av}} \hat{=} \overline{\gamma} = \overline{\alpha_C^2} \frac{E_b}{N_0} \tag{2.185}$$

Consequently, to obtain BER performance the E_b/N_0 term is substituted by γ in the general P_{BER} expressions:

$$P_{\text{BER}}(E) = \int_0^\infty P_b(E; \gamma) p_\gamma(\gamma) d\gamma \tag{2.186a}$$

where $P_{\text{BER}}(E; \gamma)$ and $p_\gamma(\gamma)$ represent the conditional probability of error and the p.d.f. of γ. For the average symbol error probability we get by substituting E_S/N_0 with $\gamma \log_2 M$:

$$P_{\text{SER}}(E) = \int_0^\infty P_{\text{SER}}(E; \gamma_S) p_\gamma(\gamma) d\gamma \tag{2.186b}$$

For M-PAM, by replacing E_s/N_0 with $\gamma \log_2 M$ and using $\overline{\gamma}_S = \overline{\gamma} \log_2 M$ in Equation (2.41) and using the expression for $I(\alpha, \overline{\gamma}, \theta)$ from Table A.1 in Appendix A, Equation (2.186a) gives for Rayleigh distribution:

$$P_{\text{SER}}(E) = \left(\frac{M-1}{M}\right) \left[1 - \sqrt{\frac{3\overline{\gamma}_S}{M^2 - 1 + 3\overline{\gamma}_S}}\right] \tag{2.187}$$

2.2.1 Quadrature Amplitude Modulation

We give the error probabilities only for Rayleigh distribution. Because the P_S expression (2.108) for QAM includes $Q^2(x)$ terms we use [Simon05]:

$$I = \int_0^\infty Q^2(a\sqrt{\gamma}) p_\gamma(\gamma) d\gamma \hat{=} I_{R,Q^2}(a, \gamma) = \frac{1}{\pi} \int_0^{\pi/4} \left[1 + \frac{a^2\gamma}{2\sin^2\theta}\right]^{-1} d\theta$$

$$= \frac{1}{4}\left[1 - \sqrt{\frac{1}{x}}\left(\frac{4}{\pi}\tan^{-1}\left(\sqrt{x}\right)\right)\right] \tag{2.188}$$

where

$$x = \frac{1 + a^2\overline{\gamma}/2}{a^2\overline{\gamma}/2}.$$

Using this and integral $I(\alpha, \overline{\gamma}, \theta)$ from Appendix A as before we get:

$$
P_{\text{SER}} = 2\left(\frac{\sqrt{M}-1}{\sqrt{M}}\right)\left[1 - \sqrt{\frac{3\overline{\gamma}_S/2}{M-1+3\overline{\gamma}_S/2}}\right]
$$

$$
-\left(\frac{\sqrt{M}-1}{\sqrt{M}}\right)^2\left[1 - \sqrt{\frac{3\overline{\gamma}_S/2}{M-1+3\overline{\gamma}_S/2}}\left(\frac{4}{\pi}\tan^{-1}\sqrt{\frac{M-1+3\overline{\gamma}_S/2}{3\overline{\gamma}_S/2}}\right)\right]
$$

$$(2.189)$$

The average bit error probability for Rayleigh distribution can be only approximately evaluated [Simon05]:

$$
P_{\text{BER}}(E) \cong 2\left(\frac{\sqrt{M}-1}{\sqrt{M}}\right)\frac{1}{\log_2 M}\sum_{k=1}^{\sqrt{M}/2}\left[1 - \sqrt{\frac{(2k-1)^2\,(3\overline{\gamma}/2)\log_2 M}{M-1+(2k-1)^2\,(3\overline{\gamma}/2)\log_2 M}}\right]\tag{2.190}
$$

2.2.2 M-PSK Modulation

The special form of the integral (2.86) needs the evaluation of the following integral K:

$$
K = \frac{1}{\pi}\int_0^{(M-1)\pi/M} M_\gamma\left(-\frac{a^2}{2\sin^2\theta}\right)d\theta \tag{2.191}
$$

Rayleigh   For $M_\gamma(s) = 1/(1+s\overline{\gamma})$ we get:

$$
P_{\text{SER}} = \frac{1}{\pi}\left(\frac{M-1}{M}\right)\left\{1 - \sqrt{\frac{g_{PSK}\overline{\gamma}_S}{1+g_{PSK}\overline{\gamma}_S}}\left(\frac{M}{(M-1)\pi}\right)\right.
$$

$$
\left.\times\left[\frac{\pi}{2}+\tan^{-1}\left(\sqrt{\frac{g_{PSK}\overline{\gamma}_S}{1+g_{PSK}\overline{\gamma}_S}}\cot\frac{\pi}{M}\right)\right]\right\}
$$

$$(2.192)$$

Rice   For $M_\gamma(s)$ as in Table A.1, Appendix A, using Equation (2.191) we get:

$$
P_{\text{SER}} = \frac{1}{\pi}\int_0^{(M-1)\pi/M}\frac{(1+K)\sin^2\theta}{(1+K)\sin^2\theta + g_{PSK}\overline{\gamma}_S}\times\exp\left[-\frac{Kg_{PSK}\overline{\gamma}_S}{(1+K)\sin^2\theta + g_{PSK}\overline{\gamma}_S}\right]d\theta
$$

$$(2.193)$$

The evaluation of BER expressions is more involved but the interested reader is directed to [Simon05] for such formulas.

2.2.3 M-FSK Modulation

For orthogonal binary frequency shift keying (BFSK) it is relatively easy to employ the integrals in Appendix A. Another way is to use the results for BPSK by substituting $\overline{\gamma}$ with $\overline{\gamma}/2$. For the

Rayleigh channel we have [Simon05]:

$$P_{\text{BER}}(E) = \frac{1}{2}\left(1 - \sqrt{\frac{\overline{\gamma}/2}{1 + \overline{\gamma}/2}}\right) \tag{2.194}$$

For orthogonal 3-ary and 4-ary FSK Dong evaluated symbol error probabilities as follows [Dong99]:

$$P_{\text{SER}}(E) = \frac{2}{3} - \sqrt{\frac{\overline{\gamma}}{2 + \overline{\gamma}}} + \frac{1}{\pi}\sqrt{\frac{\overline{\gamma}}{2 + \overline{\gamma}}}\tan^{-1}\left[\sqrt{\frac{3(2 + \overline{\gamma})}{\overline{\gamma}}}\right] \tag{2.195}$$

$$P_{\text{SER}}(E) = \frac{3}{4} - \frac{3}{2}\sqrt{\frac{\overline{\gamma}}{2 + \overline{\gamma}}} + \frac{3}{2\pi}\sqrt{\frac{\overline{\gamma}}{2 + \overline{\gamma}}}\left[\tan^{-1}\left(\sqrt{\frac{3(2 + \overline{\gamma})}{\overline{\gamma}}}\right) + \tan^{-1}\left(\sqrt{\frac{2 + \overline{\gamma}}{4 + 3\overline{\gamma}}}\right)\right] \tag{2.196}$$

For the general M-ary orthogonal FSK exact calculation of symbol error rate is complicated but [Hughes92] produced, for the AWGN channel, a simple expression for an upper bound tighter than the union bound:

$$P_{\text{SER}}(E) \leq 1 - \left[1 - Q\left(\sqrt{\frac{E_S}{N_0}}\right)\right]^{M-1} \tag{2.197}$$

By substituting E_S/N_0 with γ_S we can calculate the error probability by averaging over the p.d.f. of γ_S. A simplification of this evaluation is as follows:

$$P_{\text{SER}}(E) \leq \sum_{k=1}^{M-1}(-1)^{k+1}\binom{M-1}{k}I_k, \text{ with } I_k \equiv \int_0^\infty Q^k(\sqrt{\gamma_S})p_{\gamma_S}d\gamma_S \tag{2.198}$$

A closed-form expression for I_k (for $M = 4$) is given in [Simon05].

2.2.4 Coherent Reception with Nonideal Carrier Synchronization

In this case by denoting with $A_n = A_{In} + jA_{Qn}$ the complex fading introduced by the channel we can have closed form expressions for BPSK in Rician fading [Simon05]. To do this, we assume that A_{In}, A_{Qn} have a nonzero mean, in which case the received signal at the nth time interval, at the output of the kth matched filter y_{nk} and the LO signal c_{LO} at the receiver can be expressed as follows:

$$y_{nk} = \left(\begin{array}{c}\text{specular}\\\text{component}\end{array}\right) + \left(\begin{array}{c}\text{random}\\\text{component}\end{array}\right) + \text{noise}$$

$$= s_k(\overline{A}_{in} + j\overline{A}_{Qn})\exp(j\theta_C) + s_k(\zeta_{In} + j\zeta_{Qn})\exp(j\theta_C) + n_{nk} \tag{2.199}$$

$$c_{LO} = \left(\begin{array}{c}\text{specular}\\\text{component}\end{array}\right) + \left(\begin{array}{c}\text{random}\\\text{component}\end{array}\right) + \text{noise}$$

$$= A_r\sqrt{G}\left(\overline{A}_{in} + j\overline{A}_{Qn}\right)\exp(j\theta_C) + A_r\sqrt{G}\left(\zeta_{In} + j\zeta_{Qn}\right)\exp(j\theta_C) + n_r \tag{2.200}$$

where

$$\zeta_{In} = A_{In} - \overline{A}_{in}, \quad \zeta_{Qn} = A_{Qn} - \overline{A}_{Qn} \tag{2.200a}$$

and G represents a gain factor of the reference signal c_{LO}, the same for both the specular and random components. This is directly related to the technique used for synchronization. For example, for a PLL synchronizer, [Fitz92] gives $G = 1/(B_L T)$, where B_L is the PLL equivalent bandwidth.

In addition, the different noise term in the LO signal c_{LO} indicates the different way of introducing the phase error $\hat{\theta}_C$ resulting from the carrier recovery system. In other words the recovered carrier can be expressed in two equivalent forms:

$$c_{LO} = \exp(j\hat{\theta}_C) \text{ or } c_{LO} = A_r \sqrt{G} \exp(j\theta_C) + n_r \tag{2.201}$$

Furthermore, for the Rician K factor and the power of complex fading we have:

$$K = \frac{\text{power of specular component}}{\text{power of random component}} = \frac{(\overline{a}_I)^2 + (\overline{a}_Q)^2}{\text{Var}[a_I] + \text{Var}[a_Q]} = \frac{m_I^2 + m_Q^2}{2\sigma^2} \tag{2.202}$$

$$E[|a(t_n)|^2] \equiv P_T = m_I^2 + m_Q^2 + 2\sigma^2 = (1+K)[2\sigma^2] \tag{2.203}$$

For BPSK the transmitted signal and its energy are given by $s_{kn} = A_C T_b a_n$, and $E_b = A_C^2 T_b$, where a_n is the transmitted information at time interval n. For simplicity we also assume $A_C = A_r$.

In view of the fact that the decision variable has the form $\{y_{nk} c_{LO}^*\}$, the complex cross correlation expression between the matched filter output and the reference signal is given below:

$$\rho_R = \rho_{IR} + j\rho_{QR} = \frac{1}{2\sqrt{N_{1R} N_{2R}}} \overline{(c_{LO} - \overline{c}_{LO})^* (y_{nk} - \overline{y}_{nk})}$$

$$= \frac{\frac{G}{1+K} \frac{\overline{|A_n|^2} E_b}{N_0}}{\sqrt{\left(\frac{G}{1+K} \frac{\overline{|A_n|^2} E_b}{N_0} + 1\right) \left(\frac{1}{1+K} \frac{\overline{|A_n|^2} E_b}{N_0} + 1\right)}} \tag{2.204}$$

where

$$N_{1R} = \frac{1}{2}|c_{LO} - \overline{c}_{LO}|^2, \text{ and } N_{2R} = \frac{1}{2}|y_{nk} - \overline{y}_{nk}|^2 \tag{2.205}$$

After some mathematical manipulations, the resulting expression for the average BER is [Stein64]:

$$P_{BER}(E) = \frac{1}{2} \left[1 - Q_1(\sqrt{b}, \sqrt{a}) + Q_1(\sqrt{a}, \sqrt{b}) \right]$$

$$- \frac{\overline{\gamma} \left[\sqrt{G}/(1+K) \right]}{\sqrt{\left(1 + \overline{\gamma}\frac{G}{1+K}\right) \left(1 + \overline{\gamma}\frac{1}{1+K}\right)}} \exp[-(a+b)/2] I_0(\sqrt{ab}) \tag{2.206}$$

where

$$\left\{ \begin{matrix} a \\ b \end{matrix} \right\} = \frac{1}{2} \left[\sqrt{\frac{\overline{\gamma} G \frac{K}{1+K}}{1 + \overline{\gamma} G \frac{1}{1+K}}} \mp \sqrt{\frac{\overline{\gamma} \frac{K}{1+K}}{1 + \overline{\gamma} \frac{1}{1+K}}} \right]^2 \qquad (2.207)$$

and

$$A = \frac{\frac{\sqrt{G}}{1+K} \overline{\gamma}}{\sqrt{\left(\frac{G}{1+K} \overline{\gamma} + 1 \right) \left(\frac{1}{1+K} \overline{\gamma} + 1 \right)}} \qquad (2.208)$$

Substituting $K=0$ in the above formulas we get the error probability for Rayleigh fading channel.

Employing other techniques [Simon01] based on the expansion of conditional error probability into Maclaurin series, expressions of considerable accuracy can be produced. However, an important benefit from this approach is that it can give expressions for the SNR loss, L due to sycnhronization error. For example for Rayleigh channel and BPSK modulation we have:

$$L = 20\overline{\gamma} \frac{\overline{\gamma}+1}{\overline{\gamma}} \left(\sqrt{\frac{\overline{\gamma}+1}{\overline{\gamma}}} - 1 \right) \log_{10} \left[1 + \frac{1}{G_L \sqrt{\overline{\gamma}} [\sqrt{1+\overline{\gamma}} - \sqrt{\overline{\gamma}}]} \right] \qquad (2.209)$$

For QPSK we get for the SNR loss:

$$L = 20\overline{\gamma} \frac{\overline{\gamma}+1}{\overline{\gamma}} \left(\sqrt{\frac{\overline{\gamma}+1}{\overline{\gamma}}} - 1 \right) \log_{10} \left[1 + \frac{1 + 2\overline{\gamma}}{G_L \sqrt{\overline{\gamma}} [\sqrt{1+\overline{\gamma}} - \sqrt{\overline{\gamma}}](1 + \overline{\gamma})} \right] \qquad (2.210)$$

2.2.5 Noncoherent M-FSK Detection

Owing to the nature of error probability for noncoherent M-FSK [summation of exponential terms in Equation (2.170)] in AWGN, the technique based on MGFs is used [Simon05]:

$$P_{SER}(E) = \sum_{m=1}^{M-1} (-1)^{m+1} \binom{M-1}{m} \frac{1}{m+1} M_{\gamma_S} \left(-\frac{m}{m+1} \right) \qquad (2.211)$$

where, in the expressions for M_γ in Table A.1 of Appendix A, we substitute $\overline{\gamma}$ with $\overline{\gamma}_S$ to obtain M_{γ_S}.

Appling the above, we have for the Rayleigh fading channel:

$$P_{SER}(E) = \sum_{m=1}^{M-1} (-1)^{m+1} \binom{M-1}{m} \frac{1}{1 + m(1 + \overline{\gamma}_S)} \qquad (2.212)$$

For the Ricean channel, and orthogonal M-FSK using (Table 2-A.1) we get:

$$P_{\text{SER}}(E) = \sum_{m=1}^{M-1} (-1)^{m+1} \binom{M-1}{m} \frac{1+K}{1+K+m(1+K+\overline{\gamma}_S)}$$

$$\times \exp\left[\frac{mK\,\overline{\gamma}_S}{1+K+m(1+K+\overline{\gamma}_S)}\right] \qquad (2.213)$$

For binary non-orthogonal FSK employing Equation (2.172), the error probability for fading channel in terms of the Marcum-Q function becomes:

$$P_{\text{BER}}(E) = \frac{1}{2}\left[1 - Q_1\left(\sqrt{b}, \sqrt{a}\right) + Q_1\left(\sqrt{a}, \sqrt{b}\right)\right] \qquad (2.214)$$

Elaborating on that, we obtain an expression as function of the MGF [Simon05]:

$$P_{\text{BER}}(E) = \frac{1}{4\pi}\int_{-\pi}^{\pi} \frac{1-\zeta^2}{1+2\zeta\sin\theta+\zeta^2} M_\gamma\left[-\frac{1}{4}\left(1+\sqrt{1-\rho_{12}^2}\right)\left(1+2\zeta\sin\theta+\zeta^2\right)\right]d\theta \qquad (2.215)$$

where ρ_{12} is the correlation of the two signalling waveforms of binary FSK and ζ is given as

$$\zeta \equiv \sqrt{\frac{1-\sqrt{1-\rho_{12}^2}}{1+\sqrt{1-\rho_{12}^2}}} \qquad (2.216)$$

2.3 Transmission Through Band-limited Channels

2.3.1 Introduction

In this section we give the formulation of the transmitted and received signals (continuous and sampled) in communication through a bandlimited channel. We introduce the concept of intersymbol interference and give the power spectral density of a PAM transmitted signal. The system designer should always take into account the PSD in order to ensure a good 'match' between the channel frequency response and the spectrum of the transmitted signal.

The magnitude and phase of a typical band-limited channel is shown in Figure 2.31, with channel bandwidth B_C. Mathematically, the channel is assumed to be a linear filter with frequency response $C(f)$ and impulse response $c(t)$:

$$C(f) = \begin{cases} \int_{-\infty}^{\infty} c(t)\exp\left(-j2\pi ft\right)dt & \text{for } |f| \le B_C \\ 0, & \text{for } |f| > B_C \end{cases} \qquad (2.217)$$

Figure 2.32(a) illustrates a system model for the overall baseband communication system. If the transmitted waveform through the bandlimited channel is $g_T(t)$ then the response of the channel in frequency and time are:

$$V(f) = C(f)G_T(f)$$

$$v(t) = \int_{-\infty}^{\infty} c(\tau)g_T(t-\tau)d\tau = c(t) * g_T(t) \qquad (2.218)$$

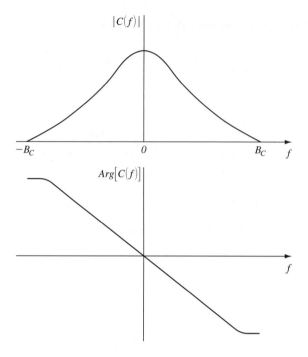

Figure 2.31 Typical band-limited channel response (magnitude and phase)

where $v(t)$ is the response of the channel to the transmitted waveform $g_T(t)$. According to Section 2.1, to maximize SNR, the received signal $r(t)$ passes through a matched filter with frequency response $G_R(f)$ matching $V(f)$:

$$r(t) = v(t) + n(t)$$
$$G_R(f) = V^*(f) \tag{2.219}$$

Figure 2.32(b) illustrates a typical smooth response of the composite system $G_T(f)C(f)G_R(f)$. If the output of the matched filter is sampled at time $t = t_0$ such that the maximum energy is obtained, we have:

$$y(t) = y_0(t) + n_0(t) \tag{2.220}$$

The signal component $y_0(t)$ at the output of the receiver matched filter at $t = t_0$, is equal to the energy E_v of the useful signal $v(t)$ at the receiver:

$$y_0(t_0) = \int_{-\infty}^{\infty} |V(f)|^2 df = E_v \tag{2.221}$$

The noise $n_0(t)$ at the output of the matched filter has zero mean and PSD $S_n(f)$:

$$S_n(f) = \frac{N_0}{2}|V(f)|^2 \tag{2.222}$$

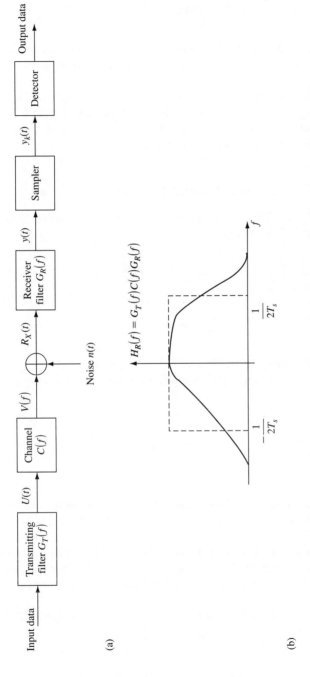

Figure 2.32 (a) Model of a baseband communication system; (b) composite system response

119

It is easy to find now the SNR at the output of the matched filter [Proakis02]:

$$\text{SNR}_0 = \frac{2E_v}{N_0} \qquad (2.223)$$

Hence SNR_0 is proportional to the received signal energy E_v. In contrast (see Section 2.1.2), when there is no bandlimited channel but only noise, SNR_0 is proportional to the transmitted signal energy E_S.

From the above considerations we can conclude that, in order to obtain the maximum SNR at the output of the matched filter, we must use a signal waveform that 'matches' in frequency the channel frequency response. With appropriate choice of $s(t)$, most of the received signal energy is confined within the channel bandwidth.

2.3.2 Baseband Transmission Through Bandlimited Channels

If we consider an M-ary PAM-modulated system, the transmitted signal, received signal and the signal at the output of the receiver filter are:

$$U(t) = \sum_{n=-\infty}^{\infty} a_n g_T(t - nT), \quad R(t) = \sum_{n=-\infty}^{\infty} a_n v(t - nT) + n(t) \qquad (2.224)$$

$$y(t) = \sum_{n=-\infty}^{\infty} a_n h_R(t - nT) + n_O(t) \qquad (2.225)$$

where $v(t) = c(t) * g_T(t)$, $h_R(t) = v(t) * g_R(t)$ and $n_O(t) = n(t) * g_R(t)$.

The output of the receiver filter $y(t)$ is sampled with sampling frequency $1/T$ to obtain the sampled version:

$$y(mT) = \sum_{n=-\infty}^{\infty} a_n h_R[(m-n)T] + n_O(mT) = h_R(0)a_m + \sum_{n \neq m} a_n h_R[(m-n)T] + n_O(mT)$$
$$(2.226)$$

Three terms appear on the right-hand side of this equation: the transmitted symbol a_m multiplied by a constant $h_R(0)$, a second term representing intersymbol interference and a third term which is the filtered noise.

The ISI term results from nonzero values of $h_R[(m-n)T]$ when $m \neq n$ and is undesirable because it changes the value of the detected sample $y(mT)$ which, in zero ISI conditions, would only contain the desired term plus noise. From Equation (2.221) it can be easily deduced that $h_R(0) = E_h$.

It can be easily shown [Proakis02] that the same hold for QAM and M-PSK modulations for the transmitted signal is now given by:

$$U(t) = \sum_{n=-\infty}^{\infty} a_n g_T(t - nT) \qquad (2.227)$$

where the information sequence a_n for QAM and PSK modulations can be expressed as follows:

QAM
$$a_n = a_{In} + ja_{Qn} \tag{2.228}$$

M-PSK
$$a_n = \exp\left[j\frac{(2i-1)\pi}{M}\right], \quad i = 1, 2, \ldots, M \tag{2.229}$$

and the corresponding bandpass signal is:

$$U_{BP}(t) = \mathrm{Re}\left[U(t)\exp\left(j2\pi f_c t\right)\right] \tag{2.230}$$

To obtain the PSD of the transmitted signal similar steps with those needed for the calculation of power spectrum of M-PSK (Section 2.2.6) are taken. First, the mean value and autocorrelation are calculated:

$$E[U(t)] = \sum_{n=-\infty}^{\infty} E[a_n]g_T(t-nT) = M_a \sum_{n=-\infty}^{\infty} g_T(t-nT) \tag{2.231}$$

$$R_U(t+\tau,t) = \sum_{n=-\infty}^{\infty}\sum_{l=-\infty}^{\infty} E[a_n a_l]g_T(t-nT)g_T(t+\tau-lT) \tag{2.232}$$

After some mathematical manipulations the PSD is given as [Proakis02]:

$$S_U(f) = \frac{1}{T}S_a(f)|G_T(f)|^2, \quad S_a(f) = \sum_{k=-\infty}^{\infty} R_a(k)\exp\left[-jk\left(2\pi f T\right)\right] \tag{2.233}$$

$S_a(f)$ and $R_a(\tau)$ represent the spectrum and autocorrelation of the data sequence whereas $G_T(f)$ is the PSD of the transmitter filter. It can be further shown that:

$$S_a(f) = \sigma_a^2 + M_a^2 \sum_{k=-\infty}^{\infty} \exp\left[-jk\left(2\pi f T\right)\right] = \sigma_a^2 + \frac{M_a^2}{T}\sum_{l=-\infty}^{\infty}\delta\left(f-\frac{k}{T}\right) \tag{2.234}$$

Consequently the PSD of the transmitted signal is as follows:

$$S_U(f) = \frac{\sigma_a^2}{T}|G_T(f)|^2 + \frac{M_a^2}{T}\sum_{k=-\infty}^{\infty}\left|G_T\left(\frac{k}{T}\right)\right|^2\delta\left(f-\frac{k}{T}\right) \tag{2.235}$$

The above equation indicates that the power spectrum of the transmitted signal consists of two terms. The first one has the shape of the frequency response of the transmitted waveform whereas the second consists of discrete lines located at multiples of $1/T$ with power following the 'envelope' of $|G_T(f)|^2$ at points k/T in frequency. The second term is frequently called a 'line spectrum'. To avoid it the data sequence can be conditioned to give a zero mean ($M_a = 0$). Another way to eliminate the line spectrum is to design $|G_T(f)|$ such that it is equal to zero for $f = k/T, k =$ integer. For example, when $g_T(t)$ is the rectangular unit pulse, $|G_T(f)|$ is equal to zero for $f = k/T$ except for the value at $k = 0$.

2.3.3 Bandlimited Signals for Zero ISI

As explained above, in order to eliminate the ISI completely the condition is:

$$h_R(nT) = \begin{cases} 1, & n = 0 \\ 0, & n \neq 0 \end{cases} \qquad (2.236)$$

It can be shown that to satisfy Equation (2.236) the necessary and sufficient condition is that the Fourier transform $H_R(f)$ of $h_R(t)$ obeys the following equation [Proakis02]:

$$\sum_{m=-\infty}^{\infty} H_R\left(f + \frac{m}{T}\right) = T \qquad (2.237)$$

Recalling that $H_R(f)$ represents the composite frequency response of transmitter filter–channel–receiver filter we have:

$$H_R(f) = G_T(f)C(f)G_R(f) \qquad (2.238)$$

Hence we must find a pulse shape for $H_R(f)$ and a value for the sampling period T_S such that Equation (2.237) is fulfilled.

$H_R(f)$ must be bandlimited [$H_R(f) = 0$, for $|f| > B_C$] and must in general have a smooth shape like that in Figure 2.32(b).
Let

$$Z(f) = \sum_{m=-\infty}^{\infty} H_R\left(f + \frac{m}{T}\right) \qquad (2.239)$$

Then, depending on the relation between symbol interval T_S and channel bandwidth B_C, we have the following possible arrangements:

(1) For $T_S < 1/2B_C$ the summation $Z(f)$ consists of separated pulses as in Figure 2.33(a). In this case, it is obvious that Equation (2.237) can never be satisfied. Hence, for such values of T_S, there is no way to eliminate ISI.
(2) When $T_S = 1/2B_C$, the sample rate is equal to the Nyquist rate and the spectrum of the composite signal $Z(f)$ is such that successive pulses just touch at their edges [see Figure 2.33(b)]. If the frequency response $H_R(f)$ at the two sides of the spectrum gradually converges to zero [as indicated by the dashed lines in Figure 2.33(b)], it is not possible to satisfy Equation (2.237). The only way Equation (2.237) can be fulfilled in this case is if the pulses are perfectly rectangular unit pulses:

$$H_R(f) = \begin{cases} T, & |f| < B_C \\ 0, & \text{otherwise} \end{cases} \qquad (2.240)$$

However, as the $h_R(t)$ in this case is a *sinc* function, it has two drawbacks – it is noncausal and the rate of convergence to zero is very slow, indicating that a timing misalignment will produce an infinite number of ISI terms of considerable amplitude.
(3) The third possibility is to have $T_S > 1/2B_C$. In this case the pulses $H_R(f)$ are overlapping [as illustrated in Figure 2.33(c)] and condition (2.237) is satisfied for many different shapes of $H_R(f)$.

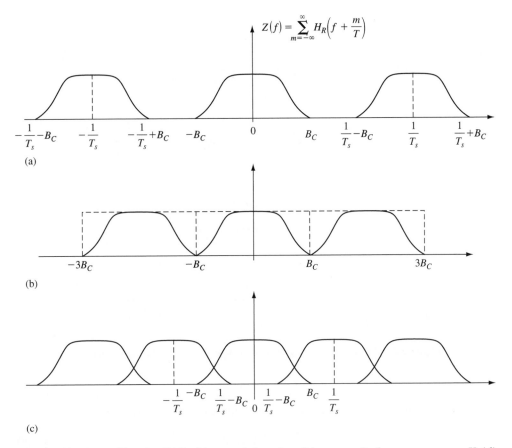

$$Z(f) = \sum_{m=-\infty}^{\infty} H_R\!\left(f + \frac{m}{T}\right)$$

(a)

(b)

(c)

Figure 2.33 Form of function $Z(f)$ of the sampled version of the composite frequency response $H_R(f)$ depending on the value of $1/T_S - B_C$ [Proakis01]

The raised cosine spectrum is one of the most frequently used filters; it has a smooth shape and fulfills condition (2.237). Its frequency response and the time domain (impulse) response are as follows:

$$H_R(f) = \begin{cases} T, & 0 \le |f| \le (1-a)/2T \\ \dfrac{T}{2}\left[1 + \cos\dfrac{\pi T}{a}\left(|f| - \dfrac{1-a}{2T}\right)\right], & \dfrac{1-a}{2T} \le |f| \le \dfrac{1+a}{2T} \\ 0, & |f| \ge \dfrac{1+a}{2T} \end{cases} \qquad (2.241)$$

$$h(t) = \frac{\sin\left(\dfrac{\pi t}{T}\right)}{\dfrac{\pi t}{T}} \frac{\cos\left(\dfrac{\pi a t}{T}\right)}{1 - 4\left(\dfrac{at}{T}\right)^2} \qquad (2.242)$$

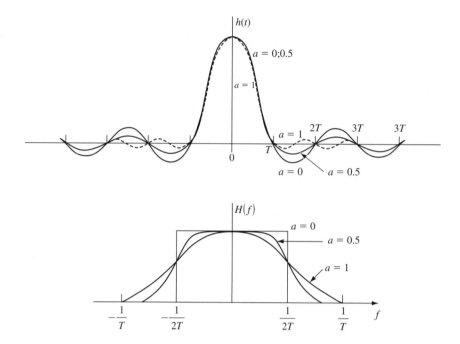

Figure 2.34 Impulse response and frequency response of a raised cosine pulse

The roll-off factor a is associated with the Nyquist rate ($B_{Nq} = 1/2T_S$) and the –6 dB bandwidth corresponding to the half-amplitude bandwidth of the raised cosine response:

$$a = \frac{B_C - B_{Nq}}{B_{Nq}} \tag{2.243}$$

When B_{Nq} is given, the value of a corresponds to the steepness of the frequency response. The lower the value of a, the steeper the response. Figure 2.34 shows the frequency response and the impulse response of the raised cosine pulse.

2.3.3.1 Implementation Aspects

The implementation of such filters for shaping the transmitted pulse is usually done in the digital domain using an FIR (finite impulse response) filter structure for two reasons. FIR filters, unlike IIR, do not have a feedback structure and this makes them suitable for data transmission applications. Secondly, FIR filters inherently exhibit linearity in phase and consequently there is no phase distortion or group delay distortion associated with them.

2.3.3.2 Eye Diagram

By plotting multiple synchronized received signal pulses within the time interval (0, T), a diagram is formed which resembles a human eye as illustrated in Figure 2.35. The positive and negative pulses are due to the randomness of the data, which can assume positive or negative

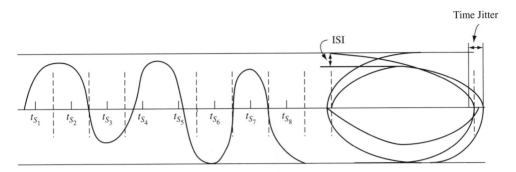

Figure 2.35   Eye diagram indicating ISI and time jitter effects

values. The optimum sampling instant is at the point where the maximum opening of the eye occurs. ISI distorts the pulses and changes their shapes and the position of zero crossings. This in turn results in eye-closure. Vertical distortion D_A is mainly due to ISI and reduces the noise margin M_N that the system can afford. Horizontal distortion T_J reduces the system resistance to timing error.

2.3.4 System Design in Band-limited Channels of Predetermined Frequency Response

In the previous section we presented the form and sampling requirements for the composite frequency response of transmit filter–channel–receive filter $H_R(f)$ in order to have zero ISI. Modifying Equation (2.238) in order to account for the ability for realizable filters we have:

$$G_T(f)C(f)G_R(f) = H_R(f)\exp[-j\pi f t_0] \tag{2.244}$$

The above equation guides us to distinguish two cases: one in which the channel $C(f)$ is perfectly known and the second where the channel response is unknown to the designer.

Let the channel response be of the form:

$$C(f) = |C(f)|\exp\left[j\theta_C\left(f\right)\right] \tag{2.245}$$

We have amplitude distortion when $|C(f)| \neq$ constant within the frequency region $|f| \leq B_C$. We have phase distortion when $\theta_C(f) \neq$ linear in frequency or when the envelope delay $\tau_{\text{env}}(f)$ is:

$$\tau_{\text{env}}(f) = -\frac{1}{2\pi}\frac{d\theta_C(f)}{df} \neq \text{constant} \tag{2.246}$$

Amplitude or phase distortion of the channel introduces ISI at the sampling instants of the received signal. An example of that is shown in Figure 2.36, where the phase expression has a quadratic form.

In this section we assume a known channel and examine transmitter–receiver filters for zero ISI and maximum SNR at the detector output. When the channel response is unknown, an equalizer filter must be designed to equalize channel distortion. Equalization techniques will be presented in some detail in a separate section further below.

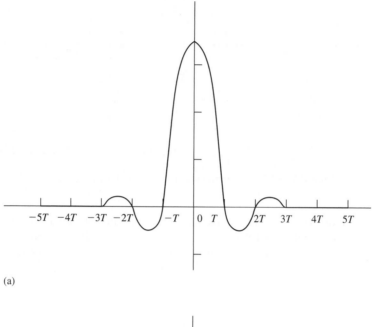

(a)

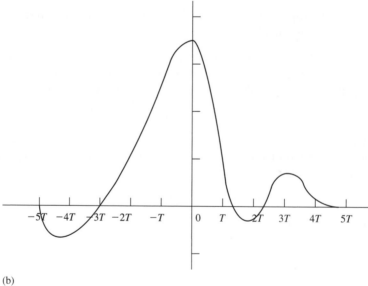

(b)

Figure 2.36 (a) Raised cosine pulse at the transmitter output, (b) resulting pulse after passing through channel with distortion

(1) *Ideal channel.* In this case the frequency response $C(f)$ has a rectangular shape:

$$C(f) = \text{Rec}\left(\frac{f}{2B_C}\right) \tag{2.247}$$

and therefore $H_R(f)$ is:

$$H_R(f) = G_T(f)G_R(f) \tag{2.248}$$

By imposing the same frequency response to both the receiver and transmitter filters (matching condition between the two filters) we get $H_R(f) = |G_T(f)|^2$ and consequently:

$$G_T(f) = \sqrt{|H_R(f)|} \exp\left[-j2\pi f t_0\right] \tag{2.249}$$

The exponential term expresses the delay and ensures realization for the filter. This approach divides equally the filtering operations between transmitter and receiver.

(2) *Nonideal channel response.* In this case we assume that the channel response is nonideal but known and the objective is to design the transmitter and receiver filters in order to maximize SNR and eliminate ISI.

If n_O represents the noise at the output of the receiver filter, its PSD is:

$$S_{no}(f) = S_n(f)|G_R(f)|^2 \tag{2.250}$$

Taking into account Equation (2.226) and the fact that ISI equals to zero, the sampled signal at the output of the receiver filter is:

$$y(mT) = h_R(0)a_m + n_O(mT) = a_m + n_{Om} \tag{2.251}$$

It is assumed that $h_R(0)$ is equal to one or (in case it is not) it is normalized to unity.

Let us use as an example binary PAM modulation for the calculation of SNR at the output of the receiver filter under channel distortion. This will reveal the impact of channel distortion on the system performance. For binary PAM we have $a_m = \pm d$.

The probability of error is given by:

$$P_{2\text{–PAM}} = Q\left(\sqrt{\frac{d^2}{\sigma_{no}^2}}\right) \tag{2.252}$$

In order to calculate SNR ($\text{SNR} = d^2/\sigma_{n_0}^2$), both numerator and denominator must be determined.

The transmitter filter response is selected as:

$$G_T(f) = \frac{\sqrt{H_R(f)}}{C(f)} \exp\left[-j2\pi f t_0\right] \tag{2.253}$$

and the receiver filter to be matched to the transmitter pulse is:

$$G_R(f) = \sqrt{H_R(f)} \exp\left[-j2\pi f t_r\right] \tag{2.254}$$

where t_r is appropriate time delay. The noise variance is given by:

$$\sigma_{no}^2 = \int_{-\infty}^{\infty} S_n(f)|G_R(f)|^2 df = \frac{N_0}{2}\int_{-\infty}^{\infty}|G_R(f)|^2 df = \frac{N_0}{2}\int_{-B_C}^{B_C} H_R(f) df = \frac{N_0}{2} \tag{2.255}$$

It can be easily shown that [Proakis02]:

$$d^2 = P_{av}T\left[\int_{-B_C}^{B_C} \frac{H_R(f)}{|C(f)|^2} df\right]^{-1} \tag{2.256}$$

Consequently the SNR at the receiver filter output is:

$$\text{SNR} = \frac{d^2}{\sigma_{no}^2} = \frac{2P_{av}T}{N_0}\left[\int_{-B_C}^{B_C} \frac{H_R(f)}{|C(f)|^2} df\right]^{-1} \tag{2.257}$$

Taking into account that the term:

$$\int_{-B_C}^{B_C} \frac{H_R(f)}{|C(f)|^2} df$$

is equal to one for $|C(f)| = 1$ (no channel distortion), it can be easily deduced that the above term represents the SNR deterioration due to the presence of channel distortion.

2.4 Equalization

2.4.1 Introduction

In Section 2.3 we considered transmission through band-limited channels and examined transmitter and receiver filters for zero ISI. It was noted that, when the transmit and receive filters $G_T(f)$ and $G_R(f)$ are designed so as to incorporate the channel response, the term:

$$\int_{-B_C}^{B_C} \frac{H_R(f)}{|C(f)|^2} df$$

represents the SNR deterioration due to channel distortion.

To eliminate channel distortion a slightly different approach is to design $G_T(f)$ and $G_R(f)$ such as they give a composite raised cosine response and in addition insert an equalizer with transfer function $G_{Eq}(f)$ to eliminate distortion due to channel response. In this case the factors in Equation (2.238) become:

$$|G_T(f)||G_R(f)| = H_R(f) \tag{2.258}$$

$$|G_{eq}(f)| = \frac{1}{|C(f)|} \tag{2.259}$$

This is shown in Figure 2.37. In this case, the noise variance at the input of the detection device is given as [Proakis02]:

$$\sigma_n^2 = \frac{N_0}{2} \int_{-B_C}^{B_C} \frac{|H_R(f)|}{|C(f)|} df \tag{2.260}$$

Comparing this with $\sigma_{n_O}^2$ in Equation (2.255), it can be deduced that we have reduction in SNR (compared with the no-equalizer system) equal to [Proakis02]:

$$\int_{-B_C}^{B_C} \frac{|H_R(f)|}{|C(f)|^2} df \tag{2.261}$$

in frequency regions where the channel attenuates the received signal ($|C(f)| \leq 1$).

This is due to the fact that the equalizer implements the above mentioned frequency response $|G_{eq}(f)| = 1/|C(f)|$. As expected, this result is the same with the one before in which the elimination of the channel response is incorporated in the transmit and receive filters.

Consequently, the price paid to reduce ISI is increased SNR. Therefore, in designing an equalizer device, one must achieve both reduction of ISI and preservation of SNR at acceptable levels for subsequent detection.

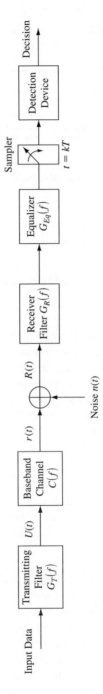

Figure 2.37 Model of baseband system including equalizer

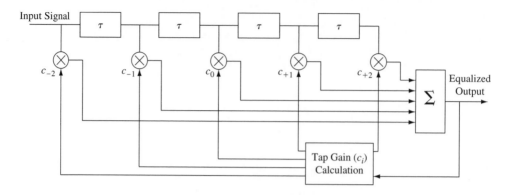

Figure 2.38   Structure of a linear equalizer

Having in mind that the impact of ISI terms on the received signal y_m is limited to a finite number of terms, Equation (2.226) can be written as:

$$y(mT) = h_R(0)a_m + \sum_{\substack{n=-L \\ n \neq m}}^{n=L} a_n h_R[(m-n)T] + n_0(mT) \qquad (2.262)$$

This helps to visualize the channel as an FIR filter in which the input symbol a_n is delayed by T seconds for a length $2L$. Each delayed branch is multiplied by the $h_R[(m-n)T]$ samples which represent the composite impulse response of the transmit filter–channel–receive filter. This was also demonstrated in Chapter 1. The above considerations help to identify the equalizer structure in the time domain, which will also have the general form of an FIR filter. Figure 2.38 illustrates the implementation of a linear equalizer with impulse response:

$$g_{eq}(t) = \sum_{n=-N}^{N} c(n)\delta(t-nD_t) \qquad (2.263)$$

where D_t represents the unit time delay. When the symbol interval is $D_t = T_S$, the equalizer is a symbol-spaced equalizer (SSE) whereas for $D_t < T_S$ it is called a fractionally spaced equalizer (FSE). FSE is useful in when $1/T < 2B_C$ (i.e. the symbol rate is less than the double channel bandwidth) to avoid aliasing distortion.

   The main categories of linear and nonlinear equalizers are shown in Figure 2.39. The design criterion for zero forcing equalizers (ZFE) is to eliminate ISI in the combined response $h(t)$. Minimum mean square error (MMSE) equalizers are linear equalizers designed to minimize MSE. On the other hand, when we have a feedback filter using the symbol decisions to produce an output which is fed back to the system, we have decision feedback equalization (DFE), which is a nonlinear equalizer. Finally, when a sequence of decisions is taken into account to obtain the maximum likelihood estimation of the transmitted symbol, we have maximum likelihood sequence estimation (MLSE), which replaces in the receiver structure the equalizer block and the decision device [Goldsmith05]. This is also a nonlinear equalizer. In this section we will examine in some detail the above types of equalizers.

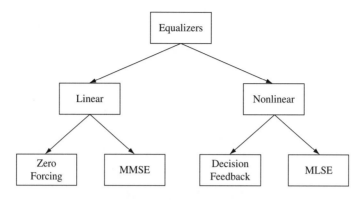

Figure 2.39 Equalization approaches

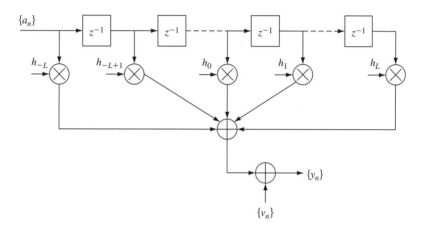

Figure 2.40 A discrete-time model for the propagation channel

2.4.2 Sampled-time Channel Model with ISI and Whitening Filter

In a practical sampled-time receiver, the continuous-time received signal is sampled in time instants $t_n = nT_S$. Consequently, it is necessary to use sampled-time processing at the receiver.

A discrete-time model for the channel with ISI is illustrated in Figure 2.40. The z^{-1} corresponds to delay equal to T_S. Because the noise at the output of the receiver matched filter is not white (nonzero correlation), it is convenient to transform the noise sequence into white by employing a whitening filter. In what follows we briefly elaborate on these concepts.

For a transmit filter with response $g_T(t)$ and a channel with impulse response $c(t)$, we employ a receiver filter $g_R(t)$ matched to the composite response at the output of the channel:

$$v(t) = g_T(t) * c(t), \quad g_R(t) = v^*(-t) \tag{2.264}$$

The overall system model is depicted in Figure 2.41, which illustrates a slightly different implementation of the receiver in Figure 2.37, in which the equalizer operates on the discrete samples y_n at the output of the matched filter.

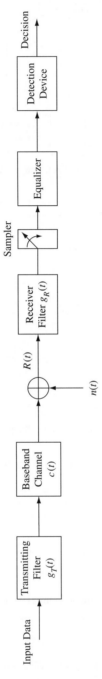

Figure 2.41 Model of baseband system with equalizer following the sampling device

The received signal $R(t)$ at the input of the matched filter and the sampled output y_n at its output are [Equations (2.224) and (2.226)]:

$$R(t) = \sum_{n=-\infty}^{\infty} a_n v(t - nT) + n(t) \tag{2.265}$$

$$y_n = y(nT) = \int_{-\infty}^{\infty} R(t) v^*(t - nT) dt \tag{2.266}$$

resulting in:

$$y_k = \sum_{n=0}^{\infty} a_n h_R(k - n) + n_0(k) \tag{2.267}$$

where $h_R(t)$ is the overall response of the combined transmit filter–channel–receive filter chain:

$$h_R(t) = v(t) * g_R(t) = g_T(t) * c(t) * g_R(t) \tag{2.268}$$

The sampled $h(nT)$ can also be expressed as:

$$h_n = h_R(nT) = \int_{-\infty}^{\infty} v^*(t) v(t + nT) dt \tag{2.269}$$

which represents the response of the receiver matched filter to the pulse $v(t)$. The above equation indicates that $h(nT)$ can also be perceived as the autocorrelation of $v(t)$.

$n_0(t)$ is an additive noise term at the output of the receiver filter corresponding to the response on $n(t)$:

$$n_{0k}(t) = n_0(k) = \int_{-\infty}^{\infty} n(t) v(t - kT) dt \tag{2.270}$$

It can be shown [Proakis04] that $n_0(k)$ is a correlated noise process. However, in order to calculate error rates using the standard techniques, it is necessary to have a white noise sequence at the output of the receiver filter. This is done by employing a whitening filter $F(z)$, which transforms $n_0(k)$ into a white sequence and is determined by the following equation:

$$H(z) = F(z) F^*(z^{-1}) \tag{2.271}$$

where $H(z)$ is the z-transform of the autocorrelation sequence h_k:

$$H(z) = \sum_{k=-L}^{L} h_k z^{-k} \tag{2.272}$$

The physical explanation of the existence of the noise whitening filter lies in the fact that AWGN is added to the system after the channel response and is only altered by the receiver filter $g_R(t)$.

The noise after passing through the receiver filter becomes coloured and has a spectral density:

$$S_v(f) = N_0 |G_R(1/z^*)|^2 \tag{2.273}$$

Consequently it is easy to see that the suitable whitening filter is:

$$F_{NW}(z) = 1/G_R(z^{-1}) \qquad (2.274)$$

To put it more generally, if the combined transmit filter–channel–receive filter response $H(f)$ can be factorized as in Equation (2.271) above, the overall response can be expressed as the output of transversal filter in which the added noise is white despite the fact that this sequence gives the discrete time received signal at the output of the receiver filter. This signal is given as:

$$U_k = \sum_{n=0}^{L} f_n a_{k-n} + n_{Wk} \qquad (2.275)$$

The set of coefficients f_k correspond to a transversal filter with transfer function $F(z)$. U_k is actually the output of the filter $1/F^*(z^{-1})$ when its input is y_k. Finally, we must note that n_{Wk} represent the samples of a white Gaussian noise process.

2.4.3 Linear Equalizers

To distinguish the ISI term in Equation (2.267) we write it as follows:

$$y_m = h_0 a_m + \sum_{\substack{n=-\infty \\ n \neq m}}^{+\infty} a_n h_{m-n} + n_{0m} \qquad (2.276)$$

Using z-transform notation, the output of the matched filter can be written as

$$Y(z) = A(z)H(z) + N_W(z) \qquad (2.277)$$

In that case, the matched filter response is expressed as: $G_R(z) = G^*(1/z^*)$.

When the equalizer is expressed as a discrete-time linear filter we have a linear equalizer. The general form of it is illustrated in Figure 2.38. However, because we have a sampled-time system, the unit delay is $D_t = z^{-1}$. Given that the length of the FIR equalizer is $(2L_E + 1)$, its impulse response and frequency response $g_E(t)$ and $G_E(f)$ are given as follows:

$$g_E(t) = \sum c_n \delta(t - nT), \quad G_E(f) = \sum_{n=-L_E}^{L_E} c_n \exp[-j2\pi fn\tau] \qquad (2.278)$$

where $\{c_n\}$ represents the set of equalizer coefficients. The overall system consists of two sub-systems. The first is the composite block 'transmit filter–channel–receive filter' with impulse response $h(t)$ and the second is the equalizer block.

The output of the overall system is the equalized signal $Eq(mT)$:

$$Eq(mT) = \sum_{n=-\infty}^{+\infty} c_n h(mT - n\tau) \qquad (2.279)$$

Then the discrete-time output signal in response to $h(t)$ is:

$$\hat{a}(kT) = \sum c_n h(mT - n\tau) = Eq(0)a(kT) + \sum_{n \neq k} a(nT)Eq(kT - nT) + \sum c_j n_{W,k-j} \qquad (2.280)$$

The zero-forcing equalizer forces all ISI terms to become zero:

$$Eq(mT) = \sum c_n h(mT - n\tau) = \begin{cases} 1, & m = 0 \\ 0, & m = \pm 1, \pm 2, \ldots, \pm N \end{cases} \tag{2.281}$$

In the z-transform domain we have:

$$Eq(z) = C(z)H(z) = 1 \Rightarrow C(z) = \frac{1}{H(z)} \tag{2.282}$$

Figure 2.42(a) shows the system model incorporating the zero-forcing equalizer in the detection process.

Alternatively, due to the fact that $H(z)$ can be expressed as a product $F(z) \cdot F^*(z^{-1})$, the overall equalizer can be seen as two successive blocks: a noise whitening filter $1/F^*(z^{-1})$ and a zero-forcing equalizer $1/F(z)$.

$$C'(z) = \frac{1}{F(z)F^*(z^{-1})} = \frac{1}{H(z)} \tag{2.283}$$

In contrast to the actual system that we analysed let us assume we have a hypothetical system represented by a channel with impulse response $\{f_n\}$ [frequency response $F(z)$] corrupted at its output by white Gaussian noise n_{Wk}. Then, the output of the equalizer, the estimated symbol at instant k and equalizer function in the z-transform domain would be respectively [Proakis04]:

$$q_n = \sum_{l=-\infty}^{\infty} c_l f_{n-l} \tag{2.284}$$

$$\hat{I}_k = q_0 I_k + \sum_{n \neq k} I_n q_{k-n} + \sum_{l=-\infty}^{\infty} c_l n_{k-l} \tag{2.285}$$

$$Q(z) = C(z)F(z) = 1 \tag{2.286}$$

In this case the equalizer z-transform function would be:

$$C(z) = \frac{1}{F(z)} \tag{2.287}$$

The system is depicted in Figure 2.42(b).

Comparing the above systems [actual, $H(z)$, and hypothetical, $F(z)$], we conclude that the ZF equalizer is always the same [$1/F(z)$]. However, in the actual system we insert the noise-whitening filter which will produce data U_k with a white Gaussian noise component. Consequently, the ZF equalizer applied to data U_k will be exactly the same as the ZF equalizer [$1/F(z)$] used for the hypothetical system corrupted only by white Gaussian noise. This formulation with whitening filter is very helpful [see Equation (2.275)] for purposes of further analysis.

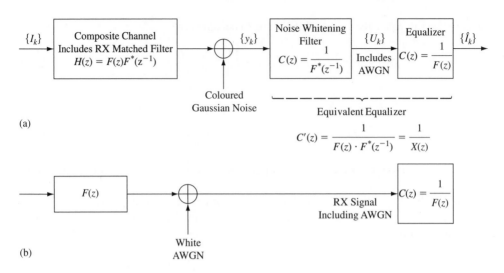

(a)

(b)

Figure 2.42 (a) System model with zero-forcing equalizer in detection stage, (b) hypothetical system with channel frequency response $F(z)$

2.4.4 Minimum Mean Square Error Equalizer

Since the zero-forcing equalizer implements the function:

$$G_{\text{ZFE}}(f) = \frac{1}{F(f)} \tag{2.288}$$

it is clear that, in the frequency region in which $C(f)$ takes low values, the ZFE provides high gain, amplifying the noise as well.

Because of the noise enhancement effect of the ZF equalizer, another criterion can be used to design a linear equalizer. This entails minimizing the undesired components in the receiver, in the mean square error sense. These components are filtered noise and remaining ISI.

Using again the concept of noise whitening filter, we denote the output of the equalizer as $Eq(kT)$, while the input is the output of the whitening filter $u(t)$:

$$Eq(kT) = \sum_{n=-N}^{N} c_n u(kT - n\tau) \tag{2.289}$$

The output of the equalizer is the estimated symbol corresponding to the transmitted data symbol a_m. The mean-square error at the output of the equalizer is the difference between $Eq(kT)$ and the transmitted symbol a_m and can be shown to be [Proakis02]:

$$E_{\text{MSE}} = E[Eq(kT) - a_m]^2 = \sum \sum c_n c_k R_{uu}(n-k) - 2 \sum c_k R_{au}(k) + E(a_m^2) \tag{2.290}$$

The above autocorrelation functions are given as:

$$R_{UU}(n-k) = E[U(mT - n\tau)U(mT - \kappa\tau)]$$

$$R_{aU}(k) = E[a_m U(mT - \kappa\tau)] \tag{2.291}$$

To determine the coefficients c_n of the equalizer, we differentiate Equation (2.290) with respect to c_n and obtain the following condition [Proakis02]:

$$\sum_{n=-N}^{N} c_n R_{UU}(n - k) = R_{aU}(k), \quad k = 0, \pm 1, \pm 2, \ldots, \pm N \tag{2.292}$$

The above equation shows that evaluation of equalizer coefficients c_n depends both on autocorrelation of noise and ISI. Both factors are present in autocorrelation $R_{UU}(n)$.

Estimates of the correlation functions $R_{UU}(n)$ and $R_{aU}(n)$ are calculated by transmitting pilot signals through the channel and obtaining time averages. By inserting Equation (2.275) from above in Equation (2.291), we obtain [Proakis04]:

$$E[U(mT - n\tau)U(mT - k\tau)] = \sum_{l=0}^{L} f_l^* f_{l-n+k} + N_0 \delta_{nk} \tag{2.293}$$

Using Equation (2.293), the resulting equations in the transform domain show clearly the main difference between the ZFE and MMSE. When the whitening filter is considered separately, we have:

$$C(z) = \frac{F^*(z^{-1})}{F(z)F^*(z^{-1}) + N_0} \tag{2.294}$$

For the overall system (including the whitening filter) we get an overall equalizer $C_T(z)$:

$$C_T(z) = \frac{1}{F(z)F^*(z^{-1}) + N_0} = \frac{1}{H(z) + N_0} \tag{2.295}$$

This is exactly the same with the ZFE $C'(z)$ in Equation (2.283) of the previous section, the only difference being the noise spectral density at the denominator.

2.4.5 Detection by Maximum Likelihood Sequence Estimation

This is a completely different approach targeting the detection of the transmitted data sequence through MLSE technique. In this method, there is no distinct equalizer block but, because MLSE implements maximum likelihood detection in the presence of ISI, it can be categorized as an alternative to equalization techniques. Following [Goldsmith05], MLSE estimates the symbol sequence a_m which maximizes the likelihood function:

$$p(r_L(t)/a_m^L, h(t)) = \prod \left[-\frac{1}{N_0} \left| r_L(n) - \sum_{k=0}^{L} a_k h_{nk} \right|^2 \right] \tag{2.296}$$

where $r_L(t)$ is the baseband received signal. Equation (2.296) gives:

$$\hat{a}^L = \arg \left\{ \max \left[2\text{Re}\left(\sum_k a_k^* y[k] \right) - \sum_k \sum_m a_k a_m^* f[k - m] \right] \right\} \tag{2.297}$$

where $y[n]$ and $f[n]$ are given as:

$$y[n] = \int_{-\infty}^{\infty} r_L(\tau) h^*(\tau - mT_S) d\tau = \sum_{n=1}^{N} r_L(n) h_{nk}^* \tag{2.298}$$

$$f[k-m] = \int_{-\infty}^{\infty} h(\tau - kT_S)h^*(\tau - mT_S)\mathrm{d}\tau = \sum_{n=1}^{N} h_{nk}h_{nm}^* \qquad (2.299)$$

To achieve this, the received signal $r_L(t)$ within a time interval $(0, LT_S)$ is expressed using orthonormal basis functions $\{\psi_n(t)\}$:

$$r_L(t) = \sum_{n=1}^{N} r_L(n)\psi_n(t) \qquad (2.300)$$

$$r_L(n) = \sum_{k=-\infty}^{\infty} \alpha_k h_{nk} + n_{0n} = \sum_{k=0}^{L} \alpha_k h_{nk} + n_{0n} \qquad (2.301)$$

$$h_{nk} = \int_{0}^{LT_S} h(t - kT_S)\psi_n^*(t)\mathrm{d}t \qquad (2.302)$$

$$n_{0n} = \int_{0}^{LT_S} n(t)\psi_n^*(t)\mathrm{d}t \qquad (2.303)$$

Because there is no equalizer filter in the MLSE detector structure, no deterioration due to noise is noticed. However, the complexity of this algorithm is very high. The Viterbi algorithm can be used to reduce its complexity, which still increases exponentially as a function of the 'width' L of ISI [Proakis04]. Consequently, MLSE is practically useful only for very small values of L.

2.4.6 Decision Feedback Equalizer

This type of equalizer, illustrated in Figure 2.43, consists of three parts. The first is a forward filter which implements the same mechanism as a linear equalizer, discussed in the previous sections. The second is a decision circuit the output of which feeds the third part of the equalizer, which is a feedback filter. The feedback filter estimates the channel response and subtracts the effect of ISI from the output of the feedforward filter.

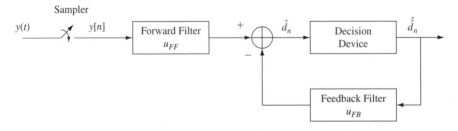

Figure 2.43   Decision-feedback equalizer

Assuming that a_n is the input data and that $u_{FF}(i)$ and $u_{FB}(i)$ represent the forward filter (FF) and feedback filter (FB) coefficients, respectively, we have:

$$\hat{a}(k) = \sum_{i=-N1}^{0} u_{FF}(i) \cdot y(k-i) - \sum_{i=1}^{N2} u_{FB}(i) \cdot Eq[(k-i)] \tag{2.304}$$

$Eq[(k-i)]$ are previously estimated data [in the form of $\hat{\hat{a}}(k)$] used as inputs to the feedback filter to produce an output which will be subtracted from the output of the feedforward filter.

The way filter coefficients $u_{FF}(i)$ and $u_{FB}(i)$ are determined depends on the optimization criterion. The zero-forcing criterion can be chosen (eliminate ISI) or the criterion for mini-mization of the MSE can also be employed. We comment briefly on the MMSE since it is more frequently used. The MMSE requirement is to minimize the error E_{MSE}:

$$E_{MSE} = E[\hat{a}(k) - a(k)]^2 \tag{2.305}$$

This results in the following expression for calculation of feedforward and feedback filter coefficients [Goldsmith05], [Proakis04]:

$$\sum \psi(l-i)u_{FF}(i) = f^*(-l) \tag{2.306}$$

with:

$$\psi(l-i) = \sum_{m=0}^{-l} f_m^*(m)f(m+l-i) + N_0\delta(l-i) \tag{2.307}$$

where $f(i)$ represents the composite transmitter filter–channel–receiver filter response.

From the above, the coefficients $u_{FF}(i)$ are calculated. Using these, coefficients $u_{FB}(k)$ can be determined using the following equation:

$$u_{FB}(k) = -\sum u_{FF}(i)f(k-i) \tag{2.308}$$

For total removal of ISI there must be no errors at the output of the decision circuit $\{Eq[(k-i)] = a_{k-i}\}$. This is expected since this output serves as input to the feedback filter of the DFE. It can be shown [Proakis04], [Goldsmith05] that the minimum error is given by:

$$E_{MSE\,min} = \exp\left[T_S \int_{-1/2T_S}^{1/2T_S} \ln\left[\frac{N_0}{H_{sum}(f) + N_0}\right]df\right],$$

$$H_{sum}(f) = \frac{1}{T_S}\sum_{n=-\infty}^{\infty} \left|H\left(2\pi f + \frac{2\pi n}{T_S}\right)\right| \tag{2.309}$$

In general, the DFE is superior to linear equalizers if there are no decision errors. In the opposite case, decision errors worsen the performance by a few decibels [Proakis04] because they are amplified by the feedback filter.

2.4.7 Practical Considerations

Three factors are the most important in the behaviour and performance of various types of equalizers presented in the previous sections. The first is ISI and how effectively can it be reduced by equalization. The second is the impact of noise and the third is the overall BER reduction and performance in the various types of equalizers.

Regarding ISI reduction in conjunction with noise contribution, linear equalizers are inferior to decision feedback equalizers. Consider a channel with severe ISI, which means spectral nulls in the frequency response. To compensate for this it will require a linear equalizer with large gains. Consequently, we will have higher noise levels in linear equalizers designed to combat severe ISI.

On the other hand, DFE uses the feedback mechanism to detect ISI at the feedback filter output and subtract it from the output of the forward filter. Spectral nulls in the channel frequency response do not affect it as in linear equalizer [Goldsmith05], which estimates the inverse of the channel response [Equations (2.283) and (2.288)]. Because of this, the latter has high sensitivity to spectral nulls.

2.4.8 Adaptive Equalization

Equations (2.281) and (2.292) represent the ZF and MMSE equalizers respectively and can be expressed in a unified way using the following matrix equation:

$$\mathbf{A} \cdot \overline{\mathbf{c}} = \overline{\mathbf{C}}_{\text{ch}} \tag{2.310}$$

where $\mathbf{A}$ is a $(2N + 1) \times (2N + 1)$ matrix containing samples of the input signal u_k, $\overline{\mathbf{c}}$ is a vector of length $(2N + 1)$ representing the equalizer coefficients and $\overline{\mathbf{C}}_{\text{ch}}$ is a $(2N + 1)$ vector associated with channel coefficients. The objective is to determine in the best possible way the equalizer coefficient vector $\overline{c}$ without having to calculate the inverse matrix $\mathbf{A}^{-1}$. A simple and efficient way to do that is to begin with an arbitrarily chosen vector $\overline{\mathbf{c}}(0)$ and use iterations following the steepest descent method to calculate successive values of the gradient vector $\mathbf{g}_k$ [Proakis04].

Initially, a starting value $\mathbf{g}_0$ is chosen from the MSE surface $J = E|I_k - I_k|^2$ by taking partial derivatives with respect to each element of vector $\overline{\mathbf{c}}$:

$$\mathbf{g}_0 = \frac{1}{2} \cdot \frac{\partial J}{\partial \overline{\mathbf{c}}(0)} \tag{2.311}$$

$$\mathbf{g}_k = \frac{1}{2} \cdot \frac{\partial J}{\partial \overline{\mathbf{c}}(k)} = \mathbf{A}\overline{\mathbf{c}}(k) - \overline{\mathbf{C}}_{ch} \tag{2.312}$$

The estimated gradient vectors $\mathbf{g}_k$ can then be used to update vector $\overline{\mathbf{c}}$ as follows:

$$\overline{\mathbf{c}}(k + 1) = \overline{\mathbf{c}}(k) - \mu \mathbf{g_k} \tag{2.313}$$

where μ is a parameter which determines the size of the step in the updating procedure and takes small positive values.

Using this method it takes hundreds of iterations for the algorithm to converge to a coefficient vector $\overline{c}$ reasonably close to $\overline{\mathbf{c}}_{\text{opt}}$ [Proakis02]. Since equalizers have the form of a tap delay line, it takes T_S seconds to complete one iteration and consequently the steepest descent method needs several hundreds of T_S time-units to obtain a satisfactory estimation for equalizer coefficients vector $\overline{\mathbf{c}}$.

In applications like radio transmission, where channel characteristics change frequently, adaptive equalization is necessary and therefore the gradient and the channel coefficient vectors are substituted with their corresponding estimates in Equation (2.313):

$$\hat{\bar{\mathbf{c}}}(k+1) = \hat{\bar{\mathbf{c}}}(k) - \mu\hat{\mathbf{g}}_{\mathbf{k}} \qquad (2.314)$$

Using the MSE criterion, the kth estimate of the gradient vector can be expressed as:

$$\hat{\mathbf{g}}_{\mathbf{k}} = -e_k\mathbf{r}_k \qquad (2.315)$$

where e_k represents the error between the transmitted symbol a_k and the equalizer output $Eq(kT)$:

$$e_k = a(kT) - Eq(kT) \qquad (2.316)$$

$\mathbf{r}_k$ is the input signal vector at the equalizer structure at the kth time instant. When there is a whitening filter before the equalizer, the input vector is $\mathbf{u}_k$, corresponding to the time waveform $u(t)$. Recursive Equation (2.314) is the well-known LMS adaptation algorithm [Widrow85].

The step-size parameter μ is inversely proportional to the number of equalizer taps and the receiver power [Proakis04].

2.5 Coding Techniques for Reliable Communication

2.5.1 Introduction

Coding is used to enhance overall system performance either by reducing the error rate or by reducing the transmitted power. The penalty paid for that is redundancy and complexity. Redundancy entails a small increase in bandwidth. As we shall see, to increase performance without coding, the bandwidth should be increased exponentially.

We will examine coding in AWGN channels and in fading channels. Although coding techniques for AWGN channels cannot be directly applied in fading channels, they can be used provided they are combined with interleaving.

We must note that, owing to recent advances in silicon technology and processing capabilities, complexity and implementation are not major issues any more when the goal is increased performance. As a result, coding techniques have developed significantly during the last decade. There are two types of channel coding, waveform coding and digital sequence coding (the latter is also defined as structured sequences by [Sklar01].

In coded transmission, digital sequence coding is concerned with transforming the digital information sequence into a new one with some form of redundancy, which will help to reduce the symbol or the bit error rate of the wireless system. This is usually done by transforming an information sequence of k bits into the corresponding coded sequence of n bits $(n > k)$. To do that and still preserve the information rate (transmitted information bits/s) constant, more bits must be sent per unit time and the gross bit rate must be increased by n/k. Consequently, the corresponding bandwidth will be increased by the same factor. Inversely, if we preserve the gross bit rate constant to a transmission rate Rt, then, the information rate is decreased by k/n. Figure 2.44 illustrates two BER performance curves for uncoded and coded transmission.

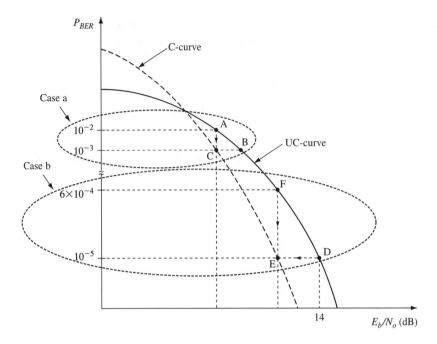

Figure 2.44 Coded and uncoded error rate performance indicating benefits of coding

In uncoded transmission, from the previous sections we can notice that the probability of error is always a function of euclidean distance or signal energy. For example in M-ary PSK transmission we have seen previously that:

$$P_{\mathrm{BER}}(\mathrm{MPSK}) = \frac{2(M-1)}{M} Q\left(\frac{d_{\mathrm{nb}}}{\sqrt{2N_0}}\right) \tag{2.39}$$

where $d_{\mathrm{nb}} = 2\sqrt{E_g}$ is the distance between neighbouring constellation points. In binary transmission this is $d_{\mathrm{nb}} = 2\sqrt{E_g}$.

To increase the signal energy E_g (or E_b), the transmitted power must be increased but this is an impractical approach due to electronic equipment limitations. The other way to increase the signal energy is to increase the symbol duration T.

To express the above concept quantitatively we can consider the symbol error probability bound which for orthogonal signals are given as [Proakis02]:

$$P_{\mathrm{SER}(E)} < \begin{cases} 2\exp\left[-\dfrac{RT}{2}(E_b/N_0 - 2\ln 2)\right], & 0 \le \ln M \le E_S/4N_0 \\[2ex] 2\exp\left[-RT\left(\sqrt{E_b/N_0} - \sqrt{\ln 2}\right)^2\right], & E_S/4N_0 \le \ln M \le E_S/N_0 \end{cases} \tag{2.317}$$

From this it can be seen that, to reduce P_{SER}, the product RT must become very large ($RT \to \infty$). If T must be kept constant, the bit duration must be reduced by $T/2^{RT}$, requiring an exponential increase in bandwidth.

Comparing the bandwidth requirements for coded and uncoded transmission, we conclude that to preserve performance in coded systems a small increase in bandwidth is necessary, whereas in uncoded transmission, the bandwidth has to be increased in exponential fashion.

2.5.2 Benefits of Coded Systems

Figure 2.44 (case a) can provide us with some more detailed explanations on the trade-offs of coding in transmission systems.

(1) *Case a* – the BER performance of the uncoded system is not satisfactory. To reduce BER, we either move on the uncoded (UC) curve increasing the transmitted power or we move to the corresponding coded curve (C-curve), which will give the required BER performance without any increase in signal power. For example instead of moving from the unsatisfactory performance at point A to point B, we move to the C-curve at point C.

(2) *Case b* – where the error rate is satisfactory but the required E_b/N_0 is too high, coding will reduce it considerably. An example from Figure 2.44 (case b) is moving from point D of the UC-curve to point E of the C curve. Note that the gain obtained from the signal power reduction in this case is considerable because, owing to the fact that the BER is very low, the operating point D is located low on the BER performance curve. In this region the gain in E_b/N_0 is higher than that at higher P_{BER}.

(3) *Case c* – when the performance and signal power are acceptable but we need to increase the transmission rate, the result will be reduction of E_b/N_0, as noted from the equation [Sklar01]:

$$E_b/N_0 = P_r/(N_0 R) \qquad (2.318)$$

This will result in increase of error rate. By employing coding will restore both the required error rate and transmitted power. This scenario can be seen in Figure 2.44 (case b) where from point D of the UC-curve the rate increase moves the system operating point up to point F. By employing coding we move from point F to point E where, for the reduced E_b/N_0, we can still have the required error performance.

The above considerations led to the definition of the following quantities:

- *Coding gain* – defined as the reduction of the necessary E_b/N_0 in coded transmission while retaining the same error rate:

$$CG|_{dB} = \frac{E_b}{N_0}(\text{uncoded})|_{dB} - \frac{E_b}{N_0}(\text{coded})|_{dB} \qquad (2.319)$$

- *Code rate* – represents the amount of redundancy in the form of a ratio introduced due to coding. For example, in a block coding scheme, if a block of k bits is transformed into a codeword containing n bits $(n > k)$, then the code rate is k/n. The inverse of code rate is called the bandwidth expansion factor $B_C = k/n$ and indicates the required bandwidth increase when the resulting codeword of length n must be transmitted in the same time interval as the uncoded sequence of length k. The reduction of the bit duration results in bandwidth expansion.

2.5.3 Linear Block Codes

A k-bit information sequence $\mathbf{a}_m$, through a mapping, can be transformed to an n-bit encoded word (called codeword) $\mathbf{c}_m$:

$$\mathbf{a}_m = [a_{m1}, a_{m2}, \ldots, a_{mk}], \quad \mathbf{c}_m = [c_{m1}, c_{m2}, \ldots, c_{mn}] \qquad (2.320)$$

Definition: when the mapping of the k information bits to the n codeword bits in a (n, k) block code is a linear mapping, it is called a linear block code.

Codeword symbols can be bits (we call them codeword bits) or multilevel symbols as in Reed–Solomon coding (we call them codeword symbols).

Definition: if the 2^k codewords of length n in an (n, k) block code form a subspace of B_n then it is a linear block code.

Definition: a subset S_C is a *subspace* of B_n if the following properties are true:

(1) the vector containing all zeros belongs to S_C; and
(2) if $S_i \in S_C \wedge S_j \in S_C$ then $S_i + S_j \in S_C$.

Definition: Hamming distance $d_{ij}^{\mathrm{H}} = d^{\mathrm{H}}(\mathbf{c}_i, \mathbf{c}_j)$ between codewords $\mathbf{c}_i$ and $\mathbf{c}_j$ is an integer equal to the number of different bits in them.

Definition: *minimum distance* is the smallest possible distance between any two codewords in a code. It can be easily shown that this is equal to the minimum distance $d_{\min}$ of all codewords with respect to the all-zero codeword:

$$d_{\min} = d_{0i} \text{ when } d_{0i} < d_{0j} \ \forall j \neq i \tag{2.321}$$

The encoded words can be generated as:

$$c_{mj} = a_{m1}g_{1j} + a_{m2}g_{2j} + \cdots + a_{mk}g_{kj} = \sum_{i=1}^{k} a_{mi}g_{ij}, \quad j = 1, \ldots, n \tag{2.322}$$

For this reason, codewords of linear block codes are generated as linear combinations of the row vectors $\mathbf{g}_i$ of the $(k \times n)$ generator matrix $\mathbf{G}$:

$$\mathbf{c}_i = \mathbf{a}_i \mathbf{G}, \quad \mathbf{G} = \begin{bmatrix} g_{11} & g_{12} & \cdots & g_{1n} \\ g_{21} & g_{22} & \cdots & g_{2n} \\ \vdots & \vdots & & \vdots \\ g_{k1} & g_{k2} & & g_{kn} \end{bmatrix} \tag{2.323a}$$

The systematic form of the generator matrix given below helps to define the corresponding systematic linear block code:

$$\mathbf{G} = [\mathbf{I_k}|\mathbf{P}] = \begin{bmatrix} 1 & 0 & \cdots & 0 & p_{11} & p_{12} & \cdots & p_{1(n-k)} \\ 0 & 1 & \cdots & 0 & p_{21} & p_{22} & \cdots & p_{2(n-k)} \\ \vdots & \vdots & \vdots & \vdots & \vdots & \vdots & \vdots & \vdots \\ 0 & 0 & \cdots & 1 & p_{k1} & p_{k2} & \cdots & p_{k(n-k)} \end{bmatrix} \tag{2.323b}$$

where $\mathbf{I_k}$ is the identity $(k \times k)$ matrix and $\mathbf{P}$ gives the parity check bits.
The produced codeword in the form of a vector is:

$$\mathbf{c}_i = \mathbf{a}_i \mathbf{G} = [a_{i1}, a_{i2}, \ldots a_{ik}, p_1, p_2, \ldots, p_{n-k}] \tag{2.324}$$

It is clear that the first k bits are the information bits whereas the last $(n - k)$ bits represent the parity bits derived from the following equation:

$$p_j = a_{i1}p_{1j} + \ldots + a_{ik}p_{kj}, \quad j = 1, \ldots, n - k \tag{2.325}$$

The parity check matrix $\mathbf{H}$ can be defined as:

$$\mathbf{H} = [\mathbf{P}^{\mathsf{T}}|\mathbf{I}_{\mathbf{n-k}}] \tag{2.326}$$

Because $\mathbf{c}_i\mathbf{H}^{\mathsf{T}} = \mathbf{0}_{n-k}$, it is straightforward to decide whether the received vector $\mathbf{r}_C$ is a valid codeword or it has been corrupted by noise and altered. The test is called syndrome testing and is based on checking the value of syndrome vector defined as [Goldsmith05]:

$$\mathbf{S} = \mathbf{r}_C\mathbf{H}^{\mathsf{T}} \tag{2.327}$$

where $\mathbf{r}_C$ is a codeword corrupted by noise $\mathbf{e}$:

$$\mathbf{r}_C = \mathbf{c} + \mathbf{e}, \quad \mathbf{e} = [e_1, e_2, \dots, e_n] \tag{2.328}$$

Where $\mathbf{r}_C$ is a valid codeword ($\mathbf{r}_C = \mathbf{c}_i$) then:

$$\mathbf{S} = \mathbf{r}_C\mathbf{H}^{\mathsf{T}} = \mathbf{c}_i\mathbf{H}^{\mathsf{T}} = \mathbf{0}_{n-k} \tag{2.329}$$

When $\mathbf{S}$ is other than zero, there are errors in the received codeword $\mathbf{r}_C$. These errors can be detected because the syndrome $\mathbf{S}$ is a function of the error vector only.

2.5.4 Cyclic Codes

Cyclic codes are the most frequently used subclass of linear block codes. If every codeword is a cyclic shift of the elements of some other codeword, then we have a cyclic code. To put it more formally, the shifting by i elements (codeword bits) of a codeword will give a valid codeword:

$$\mathbf{c}_{-i} = (c_{n-i}, c_{n-i+1}, \dots, c_{n-1}, c_0, c_1, \dots c_{n-i-1}) \tag{2.330}$$

Using polynomial representations we can express the message bit sequence $a(x)$ of length k, the generator polynomial $g(x)$ and the resulting codeword sequence $c(x)$ of length n as follows:

$$a(x) = a_0 + a_1x + \cdots a_{k-1}x^{k-1}, \quad g(x) = g_0 + g_1x + g_2x^2 + \cdots + g_{n-k}x^{n-k},$$
$$g_0 = g_{n-k} = 1 \tag{2.331}$$

$$c(x) = a(x)g(x) = c_0 + c_1x + c_2x^2 + \cdots + c_{n-1}x^{n-1} \tag{2.332}$$

Inversely, the basic principle for a codeword c to be a valid codeword is for $g(x)$ to divide $c(x)$ with no remainder. This is necessary and sufficient condition:

$$\frac{c(x)}{g(x)} = q(x) \tag{2.333}$$

Finally, we must note that a cyclic code can also be arranged in systematic form with the first k codeword bits corresponding to the information sequence whereas the last $n - k$ correspond to the parity bits. The following formula shows us the way to do that:

$$\frac{x^{n-k}a(x)}{g(x)} = q(x) + \frac{p(x)}{g(x)} \tag{2.334}$$

We first multiply the message bit polynomial by x^{n-k} and then divide by $g(x)$. If $p(x)$ is the remainder polynomial of degree less or equal to $n - k - 1$:

$$p(x) = p_0 + p_1x + p_2x^2 + \cdots + p_{n-k-1}x^{n-k-1} \tag{2.335}$$

It can be shown that the systematic form of the code is as follows [Goldsmith05]:

$$[a_0, a_1, \ldots a_k, p_0, p_1, \ldots, p_{n-k-1}] \qquad (2.336)$$

2.5.4.1 Some Useful Cyclic Codes

Hamming Codes
These are derived from the corresponding linear block Hamming code, which are characterized by the following expression:

$$(n, k) = (2^m - 1, 2^m - 1 - m) \qquad (2.337)$$

where m is a positive integer. Because they have a minimum distance $d_{min} = 3$, they can only correct only one error ($t = 1$).

Golay Codes
They are linear codes with $(n, k) = (23, 12)$ and $d_{min} = 7$. The respective generator polynomial is:

$$g(p) = p^{11} + p^9 + p^7 + p^6 + p^5 + p + 1 \qquad (2.338)$$

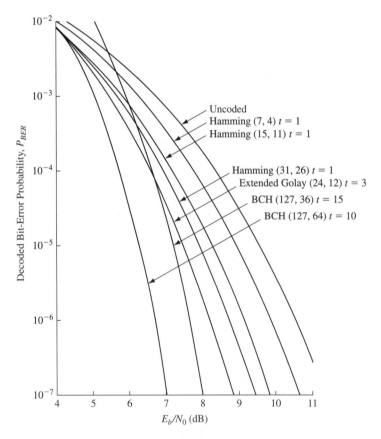

Figure 2.45 BER performance of coded systems (reprinted from B. Sklar, 'Digital Communication Fundamentals and Applications', 2nd ed, copyright © 2001 by Prentice Hall)

Maximum Length Shift Register (MLSR) Codes

These codes are generated by an m-stage shift register with feedback when m bits of information are loaded to the shift register. All code-words produced by the shift register can be generated by cyclic shifting of a single word. MLSR are characterized by $(n,k) = (2^m - 1, m)$.

Finally, it must be mentioned that the sequence produced at the output is periodic with period $n = 2^m - 1$.

Bose–Chandhuri–Hoquenghem (BCH) Codes

BCH codes comprise a class of block codes. They are cyclic codes and they have clear advantages with respect to other cyclic codes, especially for block lengths of hundreds of symbols. In addition, at high code rates they are superior to other block codes. Another feature is that BCH codes may have negative gain at low SNRs due to the fact that they have small energy per symbol compared with other block codes. Figure 2.45 shows the performance of several cyclic codes for coherent BPSK demodulation. Note the high coding gains at high SNR and the negative ones at low SNR.

2.6 Decoding and Probability of Error

2.6.1 Introduction

In this section we employ BPSK modulation in order to present in a simple way the underlying concepts. The modulated data go through an encoder and at the receiver side, depending on the transmitted bit, the received symbol r_j, can be:
When the jth transmitted codeword bit is 0:

$$r_j = \sqrt{E_C} + n_j \tag{2.339a}$$

When the jth transmitted codeword bit is 1:

$$r_j = -\sqrt{E_C} + n_j \tag{2.339b}$$

where E_C is the energy per codeword symbol and n_j represents AWGN with zero mean and variance $N_0/2$.

Because of the noise component, the received symbol r_j at the output of the matched filter can take many values around $\sqrt{E_C}$. If we decide about the transmitted symbol based on the fact that r_j is closer to $\sqrt{E_C}$ or $-\sqrt{E_C}$ (see Figure 2.46) we have hard decision decoding (HDD) and the distance between the transmitted and received symbols is not taken into account. When this distance is accounted for, a correlation metric is formed which is used to decide upon the transmitted codeword. In this case we have soft decision decoding (SDD). Below we examine briefly both decoding approaches and their respective performance.

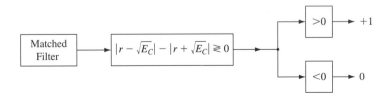

Figure 2.46 Hard decision decoding

2.6.1.1 Hard Decision Decoding and Error Performance in AWGN

As mentioned above, in BPSK-modulated systems, HDD makes a decision that a 1 was transmitted if the received symbol is closer to $\sqrt{E_b}$. On the contrary, when r_j is closer to $-\sqrt{E_b}$ the decision on the transmitted symbol is 0. HDD employs the minimum-distance criterion, deciding that the transmitted codeword is the one that has the minimum Hamming distance from the received one:

$\quad$ $\mathbf{c}_j$ is chosen as the transmitted codeword if:

$$d(\mathbf{c}_j, \mathbf{r}_C) \le d(\mathbf{c}_i, \mathbf{r}_C) \quad \forall i \ne j \tag{2.340}$$

In AWGN channel, the minimum Hamming distance criterion is equivalent to the maximum likelihood criterion for decoding given below [Goldsmith05]:

$$\text{Choose } \mathbf{c}_j : \mathbf{c}_j = \arg\max p(\mathbf{r}_C|\mathbf{c}_i), \quad i = 1, \dots, 2^k \tag{2.341}$$

To determine the probability of error, we must first note that, owing to errors, the decoded codeword may be different than any of the valid codewords. In that case, the received codeword can still be decoded correctly if the number of detected errors in this word is less than $d_{\min}/2$ [Proakis04]. Equivalently, there will be an error decision if the codeword contains more than t errors:

$$t = \left\lfloor \frac{1}{2}(d_{\min} - 1) \right\rfloor, \quad \lfloor x \rfloor = \text{the integer part of } x \tag{2.342}$$

Pictorially this is shown in Figure 2.47, where valid codewords correspond to the centres of spheres of radius t. Spheres represent the decision regions within which a received codeword is decoded as the one corresponding to the centre of the sphere. Figure 2.47 shows an example of four valid codewords. Codewords c_1, c_4 are the ones located at the minimum distance $d_{\min}$. Received codewords c'_i, c'_j will be decoded correctly as c_1, c_4 respectively, whereas a decision error will occur for c'_x.

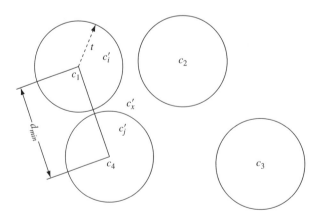

Figure 2.47 Spherical decision regions around codewords C_1, C_2, C_3, C_4

This means that there exists another valid codeword (not the transmitted one) closer to the decoded word. Using the above considerations, the probability P_e of decoding a received codeword in error is:

$$P_e < \sum_{j=t+1}^{n} \binom{n}{j} p^j (1-p)^{n-j} \qquad (2.343)$$

where p represents the uncoded probability of error P_{BER} for the modulation used for transmission. For BPSK, this probability is given in Section 2.1.7.3 by Equation (2.85) and the fact that $Q(\infty) = 0$:

$$p_{\text{BPSK}} = Q\left(\sqrt{\frac{2E_C}{N_0}}\right) \qquad (2.344)$$

The energy per codeword symbol is $E_C = kE_b/n$ where E_b is the energy per bit. From this one can see that if an (n, k) code is used with high redundancy (large n) the energy per codeword E_C is reduced with respect to uncoded bit energy E_b (low k/n). This would result in coding loss. However, because these codes have strong error correction properties, at high SNR, there is an overall coding gain. This is the case for BCH codes discussed previously.

At high SNRs, a lower bound can be estimated based on the fact that the received codeword can be only mistaken for one of its nearest neighbours. The upper bound is a Union bound assuming that all other valid codewords (there are $2^k - 1$ codewords) have a distance $d_{\min}$ from the transmitted codeword. These considerations give the following expression:

$$\sum_{j=t+1}^{d_{\min}} \binom{d_{\min}}{j} p^j (1-p)^{d_{\min}-j} \leq P_e \leq (2^k - 1) \sum_{j=t+1}^{d_{\min}} \binom{d_{\min}}{j} p^j (1-p)^{d_{\min}-j} \qquad (2.345)$$

By applying the Chernoff bound a tighter bound can be obtained:

$$P_e \leq (2^k - 1)[4p(1-p)]^{d_{\min}/2} \qquad (2.346)$$

Because the bound of P_e is a function of p and in turn p is a function of euclidean distance it should be seriously considered that the best codes should maximize the euclidean distance between all possible codewords [Goldsmith05]. However, the euclidean distance is a feature of modulated waveforms. Hence joint design of channel code and modulation scheme can achieve maximization of euclidean distances [Goldsmith05]. This approach is examined in Section 2.6.8. A good approximation of the P_e is [Sklar02]:

$$P_e \approx \frac{d_{\min}}{n} \sum_{j=t+1}^{n} j \binom{n}{j} p^j (1-p)^{n-j} \qquad (2.347)$$

2.6.1.2 Soft Decision Decoding and Error Performance

As explained above, in hard decision decoding and BPSK transmission, the decoder decides whether a *0* or a *1* has been transmitted based only on whether the received codeword is closer to $\sqrt{E_b}$ or $-\sqrt{E_b}$, respectively. However, the distance itself from $\sqrt{E_b}$ or $-\sqrt{E_b}$, as a quantity, is not used. On the other hand, when the decoder bases its decision on a correlation metric formulated between the received codeword $\mathbf{r}_C = [r_1, \ldots, r_n]$ and each of the valid codewords $\mathbf{c}_i$, we have soft decision decoding. The metric is the following [Proakis04]:

$$C(\mathbf{r}_C, \mathbf{c}_i) = \sum_{j=1}^{n} [2c_i(j) - 1]r_j, \quad (2c_i(j) - 1) = \begin{cases} +1 & if\ c_i(j) = 1 \\ -1 & if\ c_i(j) = 0 \end{cases} \tag{2.348}$$

The decoder decides that the codeword c_i with the highest correlation metric has been transmitted. Hence, when the received codeword $\mathbf{r}_C$ corresponds to a specific c_i, then most of the products (from $j = 1$ to $j = n$) within the summation are close to $\sqrt{E_C} \pm n_j$:

$$(2c_i(j) - 1)r_j = \begin{cases} (+1)\left(\sqrt{E_C} + n_j\right) = \sqrt{E_C} + n_j & if\ c_i(j) = 1\ AND\ r_j\ bit\ corresponds\ to\ 1 \\ (-1)\left(-\sqrt{E_C} + n_j\right) = \sqrt{E_C} - n_j & if\ c_i(j) = 0\ AND\ r_j\ bit\ corresponds\ to\ 0 \end{cases}$$

$$\tag{2.349}$$

Thus the correlation of the received codeword $\mathbf{r}_C$ to $(2\mathbf{c}_i - 1)$ corresponding to the actually transmitted codeword $\mathbf{c}_t$ will give a correlation metric with average value close to $n\sqrt{E_C}$:

$$\overline{C(\mathbf{r}_C, \mathbf{c}_i)} = \begin{cases} n\sqrt{E_C}, & when\ c_i = c_t \\ \ll n\sqrt{E_C}, & when\ c_i \neq c_t \end{cases} \tag{2.350}$$

To proceed with approximate expressions for the probability of error, we give the following properties [Proakis04] without proof.

(1) In an AWGN channel P_e is the same for any transmitted codeword of the linear code.
(2) Given that the transmitted codeword is $\mathbf{c}_1$, the correlation metric $C(\mathbf{r}_C, \mathbf{c}_i)$ has Gaussian distribution with mean $n\sqrt{E_C}(1 - \rho_i)$ and variance $nN_0/2$. ρ_i is the cross-correlation coefficient between the transmitted codeword $\mathbf{c}_1$ and all other codewords ($\rho_i = 1 - 2w_i/n$, $i = 2, 3, \ldots,$), and w_i is the weight of the ith word.
(3) The probability of error for a given codeword $\mathbf{c}_i$ $P_e(\mathbf{c}_i)$ is given as follows:

$$P_e(\mathbf{c}_i) = p[C(\mathbf{r}_C, \mathbf{c}_1) < C(\mathbf{r}_C, \mathbf{c}_i)] = P(x < -2w_i\sqrt{E_C}) \tag{2.351}$$

where x is a Gaussian r.v. with variance equal to $2w_iN_0$.

Thus we have [Proakis04]:

$$P_e(\mathbf{c}_i) = Q\left(\frac{2w_i\sqrt{E_C}}{\sqrt{2w_iN_0}}\right) = Q\left(\sqrt{2w_i\gamma_bR_C}\right) \tag{2.352}$$

where $\gamma_b = E_b/N_0$ (SNR pet bit) and R_C is the code rate.

To evaluate the average probability of a code error $\overline{P}_e$ we use the union bound:

$$\overline{P}_e \leq \sum P_e(c_i) = \sum_{i=2}^{2^k} Q\left(\sqrt{2w_i\gamma_bR_C}\right) \tag{2.353}$$

The above bound needs knowledge of the distribution for w_i which can be eliminated and substituted by $d_{\min}$ because $w_i \geq d_{\min}$. In this case, the above bound becomes looser:

$$P_e \leq (2^k - 1)Q\left(\sqrt{2\gamma_bR_Cd_{\min}}\right) < (2^k - 1)\exp\left(-\gamma_bR_Cd_{\min}\right) < 2^k \exp\left(-\gamma_bR_Cd_{\min}\right)$$

$$\tag{2.354}$$

The coding gain in dB can be determined by comparison of the above quantity with the uncoded BER for BPSK, $P_{\mathrm{BER}} = Q(\sqrt{2\gamma_b}) < \exp[-\gamma_b]$:

$$G_C = 10 \log_{10}\left[(\gamma_b R_C d_{\min} - k \ln 2)/\gamma_b\right] \tag{2.355}$$

2.6.2 Convolutional Codes

A convolutional coder uses a shift register of K stages to produce an n-bit coded sequence from a k-bit information sequence. Equivalently, the encoder can be partitioned in K shift registers, each of which has a length of k bits. This is illustrated in Figure 2.48. At each time instant, the n coded bits are produced from the linear combinations of a number of bits coming from a number of the $K \cdot k$ overall bits of the structure. At time instant t_i, k information bits enter the shift register and n bits are produced at the output as the coded sequence. At instant t_{i+1} the next k bits enter the encoder and so on. Owing to this process, the values of the n bits of the coded sequence depend both on the corresponding k bits that have entered the encoder and on the $(K - 1)k$ bits located in the encoder before the arrival of the recent k bits. For this reason, convolutional codes are characterized by memory whereas the code rate is $R_C = k/n$.

Furthermore, the state of the convolutional code is directly associated with its memory and is defined as the content of the $(K - 1)k$ previous bits of the overall encoder. The quantity K is defined as the constraint length of the convolutional encoder. Figure 2.49 illustrates an example of a convolutional encoder with $K = 3, k = 1, n = 3$.

The encoding procedure and the generated sequences can be represented in the form of a state diagram and a trellis diagram.

2.6.2.1 State Diagram

This diagram consists of a number of boxes corresponding to all possible states of the convolutional code. The states are connected with lines indicating the transitions from one state to

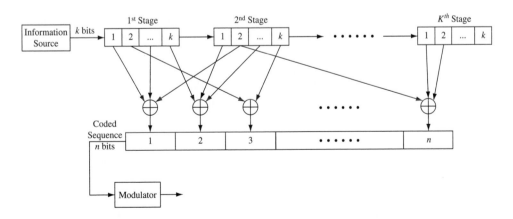

Figure 2.48   Structure of convolutional encoder with K stages

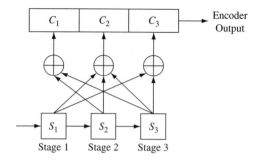

Figure 2.49 Convolutional encoder with $K = 3$, $k = 1$, $n = 3$

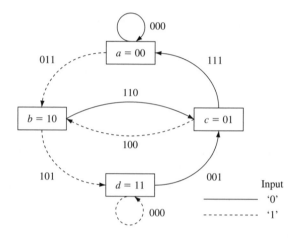

Figure 2.50 State diagram of convolutional encoder example

another depending on the inputs to the encoder. There are 2^k lines going into each state which correspond to the possible inputs to the encoder. Similarly, there are 2^k lines leaving each state which represent the number of states from which a particular state can be reached. Figure 2.50 is the state diagram for the specific example. The boxes represent the four different states of the encoder; the arrows give the transitions from one state to the next. Dashed lines correspond to a '1' input whereas solid lines represent the transitions due to a '0' input. On the top of the transition branches is the resulting branch word.

The state diagram is a condensed way of representing the convolutional encoder. However, it is important to have a diagram showing the state transitions of the convolutional coder as a function of time. Tree diagrams and Trellis diagrams are two ways to see the coder transitions in time.

2.6.2.2 Trellis Diagram

This type of diagram shows the evolution of the output of the encoder in time when the input is provided with 0 or 1 bits. It is a representation of the operation of the convolutional encoder in the time domain while at the same time providing information on the transition between

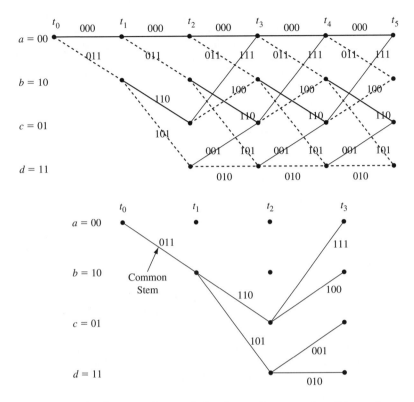

Figure 2.51 (a) Trellis diagram of convolutional encoder example; (b) survivor paths and common stem

the encoder states depending on its input. The transition information is contained in the state diagram but the time dimension results in a very effective representation of the encoding procedure. We use Figure 2.51 to explain the trellis diagram operation and the representation of information in it. The nodes correspond to the possible states of the encoder. Lines correspond to the transitions from one state to another due to a specific input. Dashed lines give the transitions when the input bit of the encoder is 1, whereas solid lines represent the transitions when the input bit is 0. Starting from the 000 state, the trellis diagram develops in time and changes states depending on the input bit. After $K - 1$ stages (in our case $K = 3$), each node has 2^k outgoing paths and 2^k incoming paths, which is in agreement with our discussion of the state diagram. At that time the steady state of the trellis is reached. In addition, the binary number written on each branch represents the output of the encoder when the corresponding input bit is applied. From Figure 2.51 it is easy to realize that the trajectory corresponding to the input sequence 110110 is:

$$a(t_0) \rightarrow b(t_1) \rightarrow d(t_2) \rightarrow c(t_3) \rightarrow b(t_4)$$

where letters correspond to nodes and indexed time t_i corresponds to the specific time instant.

It is important to note that each codeword sequence **C** corresponds to a specific path through the trellis diagram. This will be useful when the decoding procedure is discussed later.

Table 2.1 Trajectories through the trellis and bit errors for the example of Figure 2.51

Complete path	Successive paths	Hamming distances	Input bit sequence	Bit errors
Path 1	00–10–01–00	$2+2+3$	10000	1
Path 2	00–10–01–00	$2+2+3$	10000	1
Path 3	00–10–11–01–00	$2+2+1+3$	11000	2
Path 4	00–10–01–10–01–00	$2+2+1+2+3$	10100	2

2.6.2.3 Distances Through Trellis Diagrams and Minimum Distance Paths

Apart from the above properties, the trellis diagram can be used to calculate the distances between sequences of codewords. This is very useful in determining the capability of error correction of convolutional codes, because error probabilities are directly related to distances between specific paths in the trellis. Calculation of the distances through the trellis between various codeword sequences can be done by choosing a particular state (for easiness of graphical representation the all-zero state is the best choice) and then finding all paths diverging from this state and getting back to it again. For reasons of convenience we call the paths starting from the all-zero state and merging to it again, complete paths. In addition we call all the paths showing the transition of one state to the next 'successive paths'. A complete path consists of a number of successive paths. Each of these complete paths corresponds to a particular input bit sequence. Furthermore, each successive path corresponds to a Hamming distance between it and the all-zero state. By adding the Hamming distances of all successive paths constituting a complete path, we obtain the Hamming distance of the complete path. It is now clear that the complete path having the smaller Hamming distance corresponds to the input sequence having the lower number of errors (different bits between the input sequence and the all-zero bit sequence).

Definition: minimum free distance d_f for a convolutional code is the minimum Hamming distance of all possible trellis paths which start and finish at the all-zero state.

As an example consider Figure 2.51(a). There are four complete paths diverging from state 00 and merging back to it again. Each of them is labelled by the corresponding Hamming distance from the all-zero state. Table 2.1 gives the trajectory of each complete path in terms of successive paths, the Hamming distances, the input bit sequences corresponding to the particular complete path and the bit errors with respect to the all-zero sequence.

2.6.3 Maximum Likelihood Decoding

In contrast to block codes, ML detection for convolutional codes is achieved by finding the most likely codeword sequence $\mathbf{C}^{opt}$ under the assumption that sequence $\mathbf{R}$ has been received:

$$p(\mathbf{R}|\mathbf{C}^{opt}) \geq p(\mathbf{R}|\mathbf{C}) \quad \forall \mathbf{C} \tag{2.356}$$

As noted in Section 2.6.2.2 a codeword sequence corresponds to a specific path in the trellis diagram and consequently ML decoding entails finding the ML path through the trellis. Using

C_i and R_i to represent the ith part of codeword sequence $\mathbf{C}$ and $\mathbf{R}$, respectively (and the ith branch of the trellis) the likelihood function of the above equation will give [Goldsmith05]:

$$p(\mathbf{R}|\mathbf{C}) = \prod_{i=0}^{\infty} p(R_i|C_i) = \prod_{i=0}^{\infty} \prod_{j=1}^{n} p(R_{ij}|C_{ij}) \tag{2.357}$$

R_{ij} and C_{ij} represent the specific jth symbol (or bit) of the corresponding parts R_i and C_i of the codeword sequences. The products in the above equation reflect the fact that symbols are affected independently by AWGN.

Because $\log(x)$ is a monotonically increasing function, we can define a log-likelihood function from the above, which transforms products to summations and introduces the concept of branch metric:

$$\Lambda_L = \log\left[p(\mathbf{R}|\mathbf{C}^l)\right] = \sum_{i=0}^{B_l} \log\left[p(R_i|C_i^l)\right] = \sum_{i=0}^{B_l} \sum_{j=1}^{n} \log\left[p(R_{ij}|C_{ij})\right] \tag{2.358}$$

where $\mathbf{C}^l$ represents the codeword sequence and B_l the number of branches of the lth path.
From the above, the branch metric is defined as:

$$B_i = \sum_{j=1}^{n} \log\left[p(R_{ij}|C_{ij})\right] \tag{2.359}$$

2.6.3.1 ML Hard Decision Decoding

We will show that the Hamming distance can be used as an equivalent metric in hard decision decoding. If $\mathbf{R}$ and $\mathbf{C}$ have a length of L bits and a Hamming distance d (differ in d bits) then we have:

$$p(\mathbf{R}|\mathbf{C}) = p^d \cdot (1-p)^{L-d} \tag{2.360}$$

where p represents the probability of error for hard decision decoding which is equal to the uncoded P_{BEP} of the corresponding modulation scheme. In addition it should be remembered that R_{ij} takes the values of 0 or 1 only. The log-likelihood is:

$$\Lambda_L = -d \log\left(\frac{1-p}{p}\right) + L \log(1-p) \tag{2.361}$$

It is easy to see that, since $p < 0.5$, minimization of Λ_L is equivalent to minimizing the Hamming distance d. Consequently, ML decoding in convolutional codes is equivalent to finding the coded sequence $\mathbf{C}$ with the minimum Hamming distance.

Example According to the notation of Section 2.4.3 we consider two complete paths starting and finishing in the all-zero state and the corresponding decoder outputs:

$$(s_0, s_1, s_2, s_3) = (00 - 00 - 00 - 00), \quad C_0 = 000000000$$

$$(s_0, s_1, s_2, s_3) = (00 - 10 - 01 - 00), \quad C_1 = 011110111$$

As noted above, s_i corresponds to the state (or node) on the trellis diagram at time instant t_i.

Assuming that the received sequence is $R = 011000110$, the Hamming distance of it from C_0 and C_1 is 4 and 3 respectively. Let p be the P_{BEP} for the corresponding modulation. Using Equations (2.358) and (2.360) and setting d equal to 4 and 3, respectively, the two path metrics are:

$$M(C_0) = \sum_{i=0}^{2}\sum_{j=1}^{3} \log[p(R_{ij}|C_{ij})] = 4\log p + 5\log(1-p), \quad M(C_1) = 3\log p + 6\log(1-p)$$

(2.362)

If we assume that $p = 0.4$ then $M(C_0) = -3.0528$ and $M(C_1) = -2.3484$. Consequently, the second metric has a higher value and leads to the decision that C_1 was the ML sequence in the interval $t_0 - t_3$.

2.6.3.2 ML Soft Decision Decoding

In contrast to HDD, in SDD the actual decimal values of the received sequence R_{ij} are used instead of binary numbers (0,1). Furthermore, taking into account the generic expression for the correlation metric in Section 2.6.1.2, in a similar way, the received code sequence elements and their likelihood can be expressed as:

$$R_{ij} = \sqrt{E_{\text{SC}}}(2C_{ij} - 1) + n_{ij}$$

(2.363)

$$p(R_{ij}|C_{ij}) = \frac{1}{\sqrt{2\pi}\sigma} \exp\left[-\frac{\left(R_{ij} - \sqrt{E_C}\left(2C_{ij} - 1\right)\right)^2}{2\sigma^2}\right]$$

(2.364)

where

$$E_C = kE_b/n$$

(2.365a)

and n_{ij} represents zero-mean Gaussian noise with variance:

$$\sigma^2 = N_0/2$$

(2.365b)

Finally, an equivalent branch metric to Equation (2.359) is:

$$M_i = \sum_{j=1}^{n} R_{ij}(2C_{ij} - 1)$$

(2.366)

where scaling factors have been removed. Similarly terms which are the same for any branch digits C_{ij} are not accounted for in the above summation.

2.6.4 The Viterbi Algorithm for Decoding

The Viterbi algorithm was developed by A. Viterbi in 1967 and is an effective approach to reducing the complexity of maximum likelihood decoding. To do this, it operates iteratively through the trellis and discards paths which cannot give the maximum possible path metric. In this way, it retains a survivor path while it advances through the trellis.

To explain the basic principle of Viterbi decoder we use superscripts to denote the codeword sequence up to specific points in time (or trellis branches, or nodes). For example, $\mathbf{C}^j$ denotes the codeword sequence up to trellis point at time t_j (assuming that it all starts at t_0), or trellis branch j. If we denote Hamming distances at time t_j by $D_H(\mathbf{R}^j, \mathbf{C}^j)$ it can be shown that the following relationship holds for minimization of Hamming distance at a particular point in time i (or a particular node in the trellis) [Gitlin92]:

$$\min_{s_1, s_2, \ldots s_{i-1} | s_i} \left\{ D_H\left(\mathbf{R}^{i+1}, \mathbf{C}^{i+1}\right) \right\} = \min_{s_{i-1} | s_i} \left\{ d(\mathbf{r}_i, \mathbf{c}_{i-1}) + \min_{s_1, s_2, \ldots s_{i-2} | s_{i-1}} \left[D_H\left(\mathbf{R}^{i-2}, \mathbf{C}^{i-2}\right) \right] \right\}$$

$$(2.367)$$

Finding the path with the highest metric corresponds to finding the path with the minimum Hamming distance as expressed by the above equation. To understand the Viterbi algorithm we assume that the optimum path goes through the state (or node) s_i. The above equation indicates that the minimum distance to state s_i consists of the optimum partial path up to state s_{i-2} plus the single-path Hamming distance $d(\mathbf{r}_i, \mathbf{c}_{i-1})$ from state s_{i-1}. Hence, under the assumption that the highest metric passes through node s_i, the partial path beyond that state corresponding to the minimum distance (highest metric) will include the optimum partial path before state s_i. An equivalent scheme of the above minimum distance path finding technique can be expressed in terms of partial branch metrics [Goldsmith05]:

$$\max\left(P^L\right) = \max\left(P^i\right) + \max\left[\sum_{k=i}^{L} B_l^k \right] \qquad (2.368)$$

where $P_l^i = \sum_{k=0}^{i} B_l^k$ represents the partial branch metrics up to state i of the l possible paths.

Using the above procedure the Viterbi algorithm can now be implemented by deleting all paths that converge to a given node s_i (or state) except the one which has the highest metric. This is a partial path because it represents the code trajectory up to node s_i. The only remaining partial path is called the survivor path. Going backwards, to keep a specific branch as a part of the overall survivor path, it must be a common path to all survivor partial paths of subsequent trellis branches. This path is often called a common stem [Goldsmith05] and an example is given in Figure 2.51(b), where one can see that path C_i is the common stem of all the survivor paths between time instants t_i and t_{i+3}.

2.6.5 Transfer Function for Convolutional Codes

We present briefly the concept of transfer functions because they are very useful in determining probabilities of error. To determine the transfer function of an encoder the state diagram is used. The branches of the state diagram are labelled with D^j, where the exponent j represents the Hamming distance between the corresponding codeword for this branch and the all-zero codeword.

The next step is to use the state equations corresponding to the specific state diagram. The last step is to produce, from the above, the transfer function for the code implementing a specific complete path. As an example we use the encoder of Figure 2.50 to first formulate the state equations corresponding to the four states of the state diagram:

$$X_c = D^3 X_a + D X_b, \quad X_b = D X_c + D X_d$$

$$X_d = D^2 X_c + D^2 X_d, \quad X_e = D^2 X_b \qquad (2.369a)$$

where $X_a, \ldots, X_e$, are dummy variables and the transfer function using the above equation is finally given by [Goldsmith05]:

$$T(D) = \frac{X_e}{X_a} = \sum_{d=d_f}^{\infty} a_d D^d = \frac{D^6}{1 - 2D^2} = D^6 + 2D^8 + 4D^{10} + \cdots \qquad (2.369b)$$

The exponents j and coefficients of the D^j factors in the above equation represent the Hamming distance and the number of paths with the particular Hamming distance. For example, the factor $2D^8$ means that there are two paths with Hamming distance equal to 8.

By introducing two more variables a more detailed transfer function can be produced which can be used to obtain P_{BEP} expressions [Proakis04], [Goldsmith05].

2.6.6 Error Performance in Convolutional Codes

To calculate the probability of error for both hard and soft decision decoding, we assume that the transmitted sequence is the all-zero vector. The probability of error is then the probability that the decoder will decide on another sequence. Choosing other than the all-zero as the transmitted sequence will give the same results. The principle formulas for deriving P_{BEP} expressions are similar to those for linear codes. Here the Hamming distance is involved instead of the Hamming weight. In addition, bounds for the error rate can be expressed with the help of transfer functions.

2.6.6.1 Hard Decision Decoding

If d is the distance of the path of the state s_i at some point in time t_i, then if d is odd, the decision will be erroneous if there are more than $(d+1)/2$ errors [Proakis04]:

$$P_2(d) = \sum_{k=(d+1)/2}^{d} \binom{d}{k} p^k (1-p)^{d-k} \qquad (2.370)$$

If d is even, the lower limit of the summation in the above formula is substituted by $d/2$.

In convolutional coding the probability of error can be determined from the paths which converge back to the all-zero path and their corresponding metrics. The first step is to define the first-event probability, as the probability that the metric of a path converging to the all-zero path is higher than the metric of the all zero path for the first time [Proakis04]. We call this path i_{ft}. If this path has d different bits compared with the all-zero path, we define the pairwise comparison error probability $P_2(d)$, which is expressed as:

$$P_2(d) = P\left[CM^{(i_{ft})} \geq CM^{(0)} \right] \qquad (2.371)$$

where $CM^{(i)}$ is the path metrics for path i.

If a_d gives the number of paths with distance d from the all-zero path, an error bound can be given as [Proakis04]:

$$P_e < \sum_{d=d_{free}}^{\infty} a_d P_2(d) < \sum_{d=d_{free}}^{\infty} a_d [4p(1-p)]^{d/2} < T(D)\big|_{D=\sqrt{4p(1-p)}} \qquad (2.372)$$

The last inequality comes from the fact that the transfer function $T(D)$ provides detailed information about all paths converging to the all-zero path and the corresponding distances.

2.6.6.2 Soft Decision Decoding

For soft decision decoding, the pairwise error probability $P_2(d)$ is:

$$P_2(d) = Q\left(\sqrt{\frac{2E_C}{N_0}}d\right) = Q\left(\sqrt{2\gamma_b R_C d}\right) \tag{2.373}$$

Making use of the properties of the transfer function and the union bound, an upper bound of the error probability can be expressed as [Goldsmith05]:

$$P_e \leq \sum_{d=d_f}^{\infty} a_d Q\left(\sqrt{2\gamma_b R_C d}\right) \tag{2.374}$$

The property of Q function to be bounded by the respective exponential function

$$Q\left(\sqrt{2\gamma_b R_C d}\right) \leq \exp\left[-\gamma_b R_C d\right] \tag{2.375}$$

helps to express P_e using the transfer function:

$$P_e < T(D)|_{D=\exp[-\gamma_b R_C]} \tag{2.376}$$

The bit error probability can be found with the help of the transfer function again [Goldsmith05]:

$$P_b \leq \sum_{d=d_f}^{\infty} a_{df}(d) Q\left(\sqrt{2\gamma_b R_C d}\right) \tag{2.377}$$

Figure 2.52 shows the BER performance of hard-decision Viterbi decoding for BPSK modulation, binary symmetric channel and a trellis memory of 32 bits.

2.6.7 Turbo Codes

The turbo encoder consists of two recursive systematic convolutional (RSC) coders arranged in parallel, as shown in Figure 2.53(a). The top branch represents the information sequence $\mathbf{a}_m$, which goes through the first RSC encoder. The main characteristic of the Turbo encoder is the interleaver through which the information sequence passes before entering the second encoder. The three output branches are combined to produce the composite encoded stream.

Turbo coding was introduced in 1993 by Berrou *et al.* [Berrou96] and is considered as one of the most important breakthroughs since it can achieve performance close to the capacity limit for satellite channels. One important feature of a Turbo coder is that it is linear because its constituent components (two RSCs, one interleaver) are linear. This is useful because the all-zero sequence can be used to determine its performance. A second aspect is the produced code rate, which is given as:

$$\frac{1}{R} = \frac{1}{R_1} + \frac{1}{R_2} - 1 \tag{2.378}$$

where R_1 and R_2 represent the code rates of the two RSC coders.

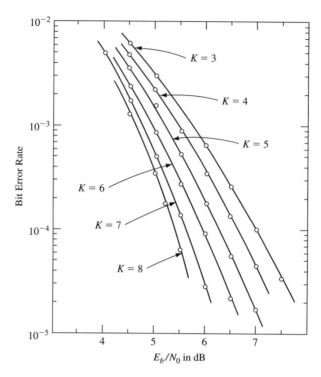

Figure 2.52 Error performance for convolutional codes with $K = 3$, 4, 5, 6, 7 and 8 (reprinted from J. Heller, I. Jacobs, 'Viterbi decoding for satellite and space communication', IEEE Trans. Commun. Technology, vol. COM-19, October 1971, pp. 835–848, © IEEE 1971)

A third characteristic is the role of interleaving which on the one hand produces an equivalent code of considerable length and on the other serves as a decorrelating mechanism. The latter is useful when we consider the iterative nature of decoding as presented immediately below.

It can be shown that if we use an iterative procedure we can have a suboptimum turbo decoder, the basic nature of which is depicted in Figure 2.53(b). Let r_a, r_{c1} and r_{c2} represent the received sequences corresponding to the information sequence $\mathbf{a}_m$ and the encoded sequences c_1 and c_2 from the two RSC encoders respectively. As illustrated in Figure 2.53(b) there are two decoders with three inputs each; r_a and r_{c1} are two of the inputs of decoder 1. The de-interleaved output sequence L_2 from decoder 2 is fed back as an input to decoder 1. This sequence is produced on the basis of a log-likelihood function L_{21} [Ryan98]. Similarly, from a log-likelihood function L_{12}, an output L_1 is produced from decoder 1 which goes through an interleaver to produce a better estimate of the received sequence, which is applied to one of the inputs of the second decoder. The second input of decoder 2 is the sequence r_a after passing through an interleaver. The third input is the received sequence r_{c2}. After some iterations the two decoders produce output sequences very close to the previous ones. At this point, decoding is terminated and the decoded sequence is obtained after passing L_2 through the de-interleaver at the output of decoder 2. The two log-likelihood functions L_{12} and L_{21} are calculated taking into account the trellis trajectory of the code. The decoding process is based on the calculation of a log-likelihood function $L(u_k)$, which is computed from the log-likelihood functions L_{12} and L_{21} at

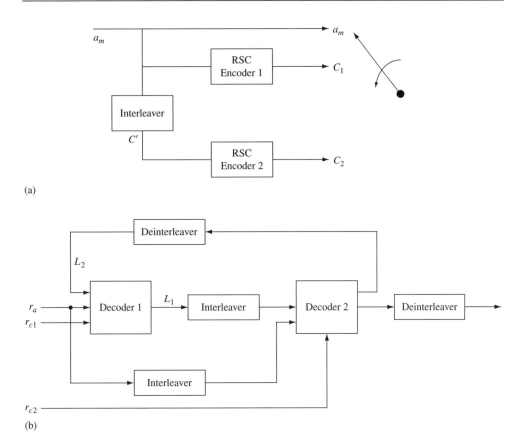

Figure 2.53 (a) Structure of turbo encoder; (b) structure of turbo decoder

the outputs of decoders 1 and 2, respectively [Ryan98]. The decision on the input encoded bit u_k is expressed as [Ryan98]:

$$\tilde{u}_k = \text{sgn}[L(u_k)] = \text{sgn}\left[\log\left(\frac{P(u_k = +1|\mathbf{r}_a)}{P(u_k = -1|\mathbf{r}_a)}\right)\right] \qquad (2.379)$$

[Berrou96] suggests a technique to stop the iterative procedure and produce the decoded output based on the variance of the bit estimate.

Figure 2.54 gives the performance of a turbo decoder for 1, 2, 3, 6 and 18 iterations. The rate of the encoder is $R = 1/2$ and its memory is $M = 4$. The interleaving depth is $d = 2^8 \times 2^8$. One can notice its remarkable performance at 18 iterations, which is very close to the Shannon limit. This is due to two factors. These are the structure of the interleaver, which increases the equivalent length of the code and the iterative structure of the decoder. However, this kind of turbo code can exhibit an error floor at BER $= 10^{-6}$ [Goldsmith05]. One way to overcome this problem is by increasing the constraint length of the code [Goldsmith05].

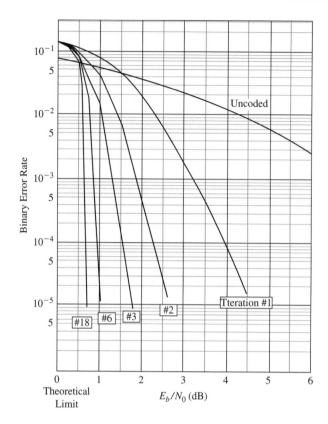

Figure 2.54 BER performance of turbo decoder for 1, 2, 3, 6 and 18 iterations (reprinted with permission from C. Berrou, A. Glavieux, 'Near optomum error correcting coding and decoding: turbo codes', IEEE Trans. Commun., vol. 44, October 1996, pp. 1261–1271)

2.6.8 Coded Modulation

From Equation (2.355) it was demonstrated that coding can result in gain in system performance with the consequence of bandwidth increase, since it is necessary to include redundancy due to coding. However, if there are bandwidth constraints in the target application, a way has to be devised to employ coding for robust reception without having to increase the bandwidth. This can be done by combining modulation and coding in a technique called coded modulation. The main principle is to increase the number of signals used for transmission. This can compensate for losses due to redundancy from coding.

The straightforward way to do this is to increase spectral efficiency by employing more complex constellation formats. For example, 16-QAM is more spectrally efficient than QPSK. However, more complex modulations will also introduce deterioration of system performance, as is clear from previous sections.

Hence, the key point is to consider modulation and coding being interdependent, such that the introduction of more signals for transmission is done, while at the same time the designed scheme should increase the minimum euclidean distance between the coded signals. This can be done by mapping the coded bit sequences into constellation points in such a way that we

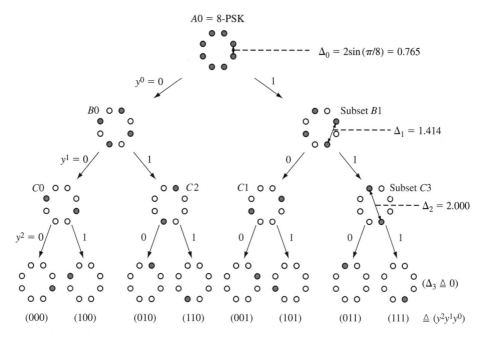

Figure 2.55 Set partitioning for 8-PSK constellation (adapted from [Ungerboek82] © IEEE 1982)

maximize the minimum euclidean distance. Ungerboeck discovered [Ungerboeck82] that this can be done by set partitioning. Figure 2.55 shows an example of a set partitioning for an 8-PSK constellation system. The minimum distance in this partitioning is increased each time, starting from d_0 with no partition and reaching d_2 when there are four partitions. The values are:

$$d_0 = \sqrt{(2 - \sqrt{2})E}, \quad d_1 = \sqrt{2E}, \quad d_2 = 2\sqrt{E} \qquad (2.380)$$

Although in this example we have three partition stages, not all of them have to be used.

Figure 2.56 illustrates the general structure of a coded modulation scheme. The coded bits at the output of the encoder are used to select one of the 2^n possible subsets. This is done in the 'subset selection block'. The uncoded bits are then used to select one of the 2^{k_2} signal points in the selected subset. This is done in the 'signal point selection block'. Finally, the 'constellation mapper' maps the resulting symbol to one of the N-dimensional signal constellation. The end result is that, although the constellation format is the same as in the uncoded case, the mapping of the resulting data symbols to the constellation points is produced by the combination process of encoding and modulation described above.

As an example we look at the 8-PSK constellation. We use an encoder of Figure 2.57(a) as in [Proakis04] to produce a coded sequence (c_1, c_2, c_3) combining two coded bits (c_1, c_2) and one uncoded (c_3). By employing the partitioning $C_i (i = 1, 2, 3, 4)$ of Figure 2.55, each of the four values of the set (c_1, c_2) corresponds to a specific member of the partitioned set C_i. If the resulting coded-modulation mapping is as in Figure 2.57(b), the set of triplets $(c_1, c_2, c_3) = (010, 011)$ corresponds to the set member C_2 in which the two possible outcomes of the uncoded bit c_3 (0 and 1) are appended.

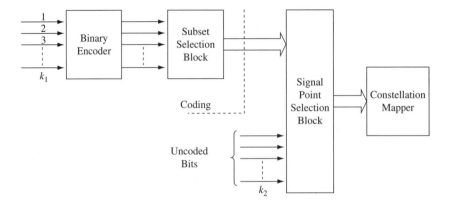

Figure 2.56 Structure for coded modulation

Two factors are important in increasing the overall gain of a coded modulation scheme. The first is to determine the suitable coding scheme, which will provide the higher possible code distances. The other refers to optimum placing of the signal constellation points.

The probability of error decreases with the square of the separation between symbols representing N-bit sequences (d_u^2). Similarly, P_{BER} decreases when the density of symbols D_u increases. In the case of a coded sequence in coded modulation schemes we seek to maximize, through the encoder, the minimum distance $d_{\min}$, while at the same time increasing the density of possible symbols to D. In this way, the channel coding gain becomes [Goldsmith05]:

$$G_C = \frac{d_{\min}^2}{d_u^2} \cdot \frac{D}{D_u} \tag{2.381}$$

2.6.9 Coding and Error Correction in Fading Channels

The coding techniques presented in this chapter have been designed mostly for AWGN channels and do not have the ability to correct for bursts of errors occurring when the channel is in a deep fade during transmission. The most effective way to solve this problem is to separate the neighbouring symbols occurring during the deep fade and spread them in some way along the transmitted sequence. In this way, errors will not appear in bursts and can be corrected by the coder–decoder mechanism, as it is done in AWGN. This can be done using interleaving, according to which the information sequence is re-arranged such that neighbouring symbols at the input of the interleaver are separated by a number of other symbols at its output. The rearrangement is done using a well-defined rule such that, at the receiver, the information symbols can assume their original position in the time sequence.

The following considerations provide the tools to determine performance in fading channels when coding and interleaving are used [Goldsmith05]:

(1) Coding and interleaving can provide enhanced performance because they implement diversity mechanisms. For this reason, the basic requirement is to enhance the diversity characteristics by suitable design of the coding structure.
(2) Channel fading characteristics are considered known to the receiver by channel estimation through special training symbols.

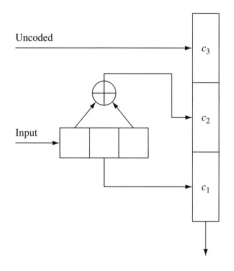

(a)

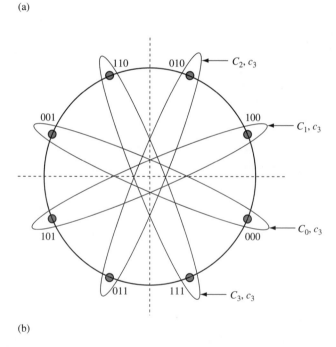

(b)

Figure 2.57 (a) Structure of encoder for coded modulation for 8-PSK; (b) coded modulation mapping for 8-PSK constellation

2.6.9.1 Block Coding and Interleaving

Figure 2.58 illustrates the structure of an interleaver–deinterleaver for block coding. The interleaver can be conceived as a matrix consisting of m rows and n columns. Each row contains an (n, k) codeword as it is stored into the interleaver. Therefore the ith codeword $\mathbf{c}_i = [c_{i1}, c_{i2}, \ldots, c_{in}]$ represents the ith row of the interleaver matrix. However, the output

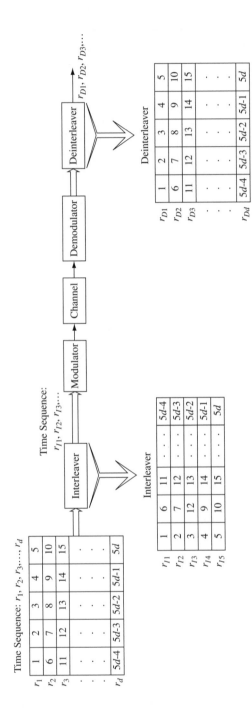

Figure 2.58 Interleaver–deinterleaver operation

from the interleaver is read by columns and sent to the modulator in this format. The opposite takes place in the de-interleaver block. Hence neighbouring codeword bits will be separated by $m - 1$ bits when they enter the modulator–channel–demodulator structure. Therefore, if there is a deep fade during the ith codeword which destroys successive bits c_{i5}, c_{i6}, c_{i7}, the way the codeword bits appear at the modulator–demodulator structure is $\ldots, c_{(i-1)5}, c_{i5}, c_{(i+1)5}, \ldots$ and c_{i6} is located $(m - 1)$ code bits away from c_{i5}. The number of rows I_r represents the interleaving depth. Symbols belonging to the same codeword before interleaving will have a distance $I_r T_S$ at the output of the interleaver. During transmission they will experience independent fading if $I_r T_S > T_C$, where T_C represents the coherence time of the multipath channel.

Using the union bound expressions can be calculated for the error probability when the interleaving depth guarantees independent fading. As an example, [Proakis04] gives the following bound for (n, k) block code in BPSK modulation and soft decision decoding:

$$P_e < 2^k \binom{2d_{\min} - 1}{d_{\min}} \left(\frac{1}{4R_C \overline{\gamma}_b}\right)^{d_{\min}} \tag{2.382}$$

2.6.9.2 Convolutional Coding and Interleaving

Because codewords in convolutional encoding are not defined in the same way as in block coding, the interleaving procedure is somewhat different. The output of the encoder is multiplexed in such a way that code bits are placed at buffers of increasing size (from 1 to $N - 1$).

Figure 2.59 shows a so-called periodic interleaver and deinterleaver. J represents the increase of the size of storage between successive register units. The first unit at the top has a storage size equal to zero and therefore transmits the encoded symbol without delay. On the receiver side, the inverse processing takes place and must be synchronized to the input interleaver.

Figure 2.60 is adopted from [Sklar01] and illustrates an example of convolutional interleaver–deinterleaver with $J = 1$ and $N = 4$. Each time, four symbols enter the interleaver while the previous ones are shifted to the right. When they are in the right-most position in the interleaver, the next time instant they appear at the input of the deinterleaver on the

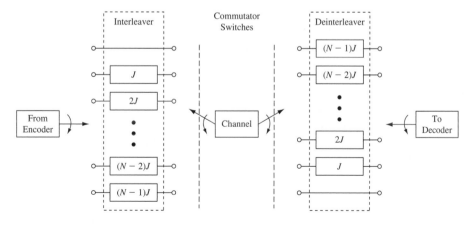

Figure 2.59   Periodic interleaver–deinterleaver structure (adapted from [Sklar01])

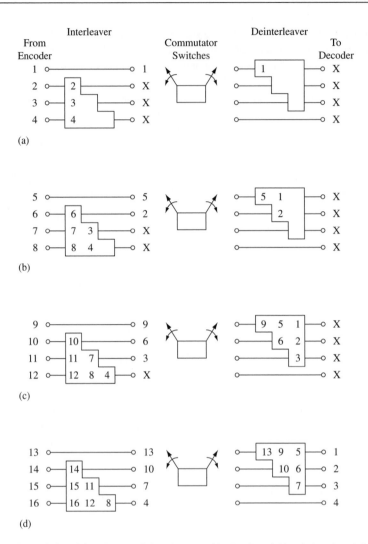

Figure 2.60 Convolutional interleaver–deinterleaver with $J = 1$ and $N = 4$ (reprinted from B. Sklar, 'Digital Communication Fundamentals and Applications', 2nd edn, copyright © 2001 by Prentice Hall)

same line. Note that the symbol appearing at the top line (which has no storage unit) appears simultaneously at the input of the deinterleaver.

It is reported that the performance of convolutional interleaving is the same as that for block interleaving. However, the storage and delay requirements are half as much compared with block interleaving [Forney71].

References

[Aulin81]: T. Aulin, C. Sundberg, 'Continuous phase modulation – Part I: full response signaling', IEEE Trans. Commun., vol. COM-29, March 1981, pp. 196–209.

[Benvenuto02]: N. Benvenuto, G. Cherubini, 'Algorithms for Communications Systems and their Applications', John Wiley & Sons Ltd, Chichester, 2002.

[Berrou96]: C. Berrou, A. Glavieux, 'Near optomum error correcting coding and decoding: turbo codes', IEEE Trans. Commun., vol. 44, October 1996, pp. 1261–1271.

[Cooper86]: G. Cooper, C. McGillem, 'Modern Communications and Spread Spectrum', McGraw Hill, New York, 1986.

[Craig91]: J. Craig, 'A new, simple and exact result for calculating the probability of error for two-dimensional signal constellations', IEEE MILCOM'91 Conf. Record, Boston, MA, 1991, pp. 25.5.1–25.5.5

[Divsalar90]: D. Divsalar and M. Simon, 'Multiple-symbol differential detection of MPSK', IEEE Trans. Commun., vol. COM-38, no. 3, March 1990, pp. 300–308.

[Dong99]: X. Dong, N. Beaulieu, 'New analytical probability of error expressions for classes of orthogonal signals in Rayleigh fading', Proc. GLOBECOM'99, Rio de Janeiro, December 1999, pp. 2528–2533.

[Fitz92]: M. Fitz, 'Further results in the unified of digital communication systems', IEEE Trans. Commun., vol. 40, no. 3, March 1992, pp. 521–532.

[Forney71]: G.D. Forney, 'Burst error correcting codes for the classic bursty channel', IEEE Trans. Commun. Technol., October 1971, pp. 772–781.

[Gitlin92]: R. Gitlin, J. Hayes, S. Weinstein, 'Data Communications Principles', Plenum Press, London, 1992.

[Goldsmith05]: A. Goldsmith, 'Wireless Commuications', Cambridge University Press, Cambridge, 2005.

[Heller71]: J. Heller, I. Jacobs, 'Viterbi decoding for satellite and space communication', IEEE Trans. Commun. Technology, vol. COM-19, October 1971, pp. 835–848.

[Hughes92]: L. Hughes, 'A simple upper bound on the error probability for orthogonal signals in white noise', IEEE Trans. Commun., vol. COM-40, no. 4, April 1992, p. 670.

[Lindsey73]: W Lindsey, M. Simon, 'Telecommunication Systems Engineering', Prentice Hall, Englewood Cliffs, NJ, 1973.

[Lu99]: J. Lu, K.B. Letaief, J.C.-I. Chuang, M.L. Liou, 'M-PSK and M-QAM BER computation using signal-space concepts', IEEE Trans. Commun., vol. 47, no. 2, February 1999, pp. 181–184.

[Pawula82]: R. Pawula, S. Rice, J. Roberts, 'Distribution of the phase angle between two vectors perturbed by Gaussian noise', IEEE Trans. Commun., vol. COM-30, no. 8, August 1982, pp. 1828–1841.

[Peebles87]: P. Peebles Jr, 'Diital Communication Systems', Prentice Hall, Englewood Cliffs, NJ, 1987.

[Proakis02]: J. Proakis, M. Salehi, 'Communication Systems Engineering', 2nd edn, Prentice Hall, Englewood Cliffs, NJ, 2002.

[Proakis04]: J. Proakis, 'Digital Communications', 4th edn, McGraw-Hill, New York, 2004.

[Rappaport02]: T. Rappaport, 'Wireless Communications, Principle and Practice', end edn, Prentice Hall, Englewood Cliffs, NJ, 2002.

[Ryan98]: W. Ryan, 'A Turbo Code Tutorial', Invited Talk, the Analytic Science Corporation, Reston, VA, 1998.

[Simon01]: M. Simon, M.-S. Alouini, 'Simplified noisy reference loss evaluation for digital communication in the presence of slow fading and carrier phase error', IEEE Trans. Veh. Technol., vol. 50,no. 2, March 2001, pp. 480–486.

[Simon05]: M. Simon, M.-S. Alouini, 'Digital Communications over Fading Channels', 2nd edn, John Wiley & Sons Inc., New York, 2005.

[Sklar01]: B. Sklar, 'Digital Communications: Fundamental and Applications', 2nd edn, Prentice Hall, Englewood Cliffs, NJ, 2001.

[Steele92]: R. Steele, 'Mobile Radio Communications', Pentech Press, London, 1992.

[Stein64]: S. Stein, 'Unified analysis of certain coherent and noncoherent binary communication systems', IEEE Trans. Inform. Theory, vol. IT-10, no. 1, January 1964, pp. 43–51.

[Ungerboeck82]: G. Ungerboeck, 'Channel coding with multilevel/phase signals', IEEE Trans. Inform. Theory, vol. 28, January 1982, pp. 55–67.

[Viterbi63]: A. Viterbi, 'Phase-locked loop dynamics in the presence of noise by Fokker–Planck techniques', Proc. IEEE, vol. 51, December 1963, pp. 1737–1753.

[Viterbi65]: A.J. Viterbi, 'Optimum detection and signal selection for partially coherent binary communication', IEEE Trans. Inform. Theory, vol. IT-11, April 1965, pp. 239–246.

[Viterbi66]: A. Viterbi, 'Principles of Coherent Communication', McGraw-Hill, New York, 1966.

[Widrow85]: B. Widrow, S. Stearns, 'Adaptive Signal Processing', Prentice Hall, Englewood Cliffs, 1985.

[Yoon02]: D. Yoon, K. Cho and J. Lee, 'Bit error probability of M-ary quadrature amplitude modulation', IEEE Trans. Commun., vol. 50, no. 7, July 2002, pp. 1074–1080.

Appendix A

We provide below the formulas needed for the evaluation of the probability or symbol and bit error for most of the modulation schemes. These expressions were extracted from [Simon05] and [Simon98], where the reader can find more details on the derivations.

One of the most frequently encountered integrals for the evaluation of average error probabilities is the following $I(a, \bar{\gamma}, \theta)$:

$$I(a, \bar{\gamma}, \theta) = \int_0^\infty Q(a\sqrt{\gamma})p_\gamma d\gamma = \frac{1}{\pi} \int_0^{\pi/2} \left[\int_0^\infty \exp\left(-\frac{a^2\gamma}{2\sin^2\theta}\right) p_\gamma d\gamma \right] d\theta$$

Using the moment generating function (MGF) of γ it is sometimes convenient to have $I(a, \bar{\gamma}, \theta)$ in the following form:

$$I(a, \bar{\gamma}, \theta) = \frac{1}{\pi} \int_0^{\pi/2} M_\gamma\left(-\frac{a^2\gamma}{2\sin^2\theta}\right) d\theta$$

where:

$$M_\gamma(s) \equiv \int_0^\infty \exp(s\gamma)p_\gamma d\gamma$$

Taking into account the above [Simon05] gives expressions for the p.d.f.s, the moment generating functions and integral $I(a, \bar{\gamma}, \theta)$ for the Rayleigh, Rice and Nakagami-m distributions as shown in table A.1.

Table A.1 Probability density functions, moment generating functions and Integral $I(a, \overline{\gamma}, \theta)$ for various distributions

	Probability density function Moment generating function	Integral $I(a, \overline{\gamma}, \theta)$
Rayleigh	$p_\gamma(\gamma) = \dfrac{1}{\overline{\gamma}} \exp\left(-\dfrac{\gamma}{\overline{\gamma}}\right)$ $M_\gamma(-s) = \dfrac{1}{1 + s\overline{\gamma}}$	$I_R(a, \overline{\gamma}) = \dfrac{1}{2}\left(1 - \sqrt{\dfrac{a^2\overline{\gamma}/2}{1 + a^2\overline{\gamma}/2}}\right)$
Rice	$p_\gamma(\gamma) = \dfrac{(1+K)e^{-K}}{\overline{\gamma}} \exp\left[-\dfrac{(1+K)\gamma}{\overline{\gamma}}\right] I_0\left(2\sqrt{\dfrac{K(1+K)\gamma}{\overline{\gamma}}}\right)$ $M_\gamma(-s) = \dfrac{1+K}{1+K+s\overline{\gamma}} \exp\left[-\dfrac{Ks\overline{\gamma}}{1+K+s\overline{\gamma}}\right]$	$I_{\text{Rice}} = \dfrac{1}{\pi}\displaystyle\int_0^{\pi/2} \left(\dfrac{(1+K)\sin^2\theta}{(1+K)\sin^2\theta + a^2\overline{\gamma}/2}\right)$ $\times \exp\left[-\dfrac{Ka^2\overline{\gamma}/2}{(1+K)\sin^2\theta + a^2\overline{\gamma}/2}\right] d\theta, \quad s \geq 0$
Nakagami-m	$p_\gamma(\gamma) = \dfrac{m^m \gamma^{m-1}}{\overline{\gamma}^m \Gamma(m)} \exp\left(-\dfrac{m\gamma}{\overline{\gamma}}\right)$ $M_\gamma(-s) = \left(1 + \dfrac{s\overline{\gamma}}{m}\right)^{-m}, \quad s > 0$	$I_m(a, m, \overline{\gamma}) = \dfrac{1}{2}\left[1 - \sqrt{\dfrac{a^2\overline{\gamma}/2}{m + a^2\overline{\gamma}/2}} \displaystyle\sum_{k=0}^{m-1}\binom{2k}{k}\left(\dfrac{1 - \dfrac{a^2\overline{\gamma}/2}{m + a^2\overline{\gamma}/2}}{4}\right)^k\right],$ $m = \text{integer}$

3

RF Transceiver Design

The approach to follow in the presentation of the RF transceiver architecture depends on a variety of factors, which stem not only from technical specifications but from implementation and commercial considerations as well. The technical specifications refer mainly to the functional requirements whereas the degree of integration refers mostly to the implementation aspects and the layout technology. In this book we concentrate mainly on transceiver systems with a considerable amount of integration. Integration issues have gained a continuously increasing importance in recent years due to their adverse impact on the transceiver architecture.

In this chapter we will present the various receiver and transmitter architectures. On the receiver side we will use the frequency domain to demonstrate the main characteristics and problems associated with the different topologies such as image reject architectures, the classical super-heterodyne receiver and the direct conversion receiver. On the transmitter side we will present the two-stage upconversion and the direct conversion transmitter architectures.

The objective of the chapter is two-fold: first, to give a systematic presentation of the approaches on transceiver architecture; and second, to demonstrate a variety of realizations of these topologies in modern technology where integration and power consumption are the most decisive factors. The goal is to bring forward the prevailing techniques and current trends in transceiver system design.

3.1 Useful and Harmful Signals at the Receiver Front-End

At the input of the antenna there are different groups of signals: the ones generated by other radios of the same system (standard) and those operating at radio frequencies located out of the band of the standard. For example, if the radio operates as a GSM phone, the signals from the users of the GSM system can be located anywhere within 890–960 MHz. Within this band there is a 20 MHz gap between the uplink and downlink frequency bands.

The interferers coming from the same system are called 'in-band interferers'. More specifically, signals that are located in the neighbourhood of the useful RF frequency f_{RF} are defined as adjacent channel interferers, whereas signals coming from radios operating outside of the useful band of the standard are called 'out-of-band interferers'.

Finally, there is a special frequency related to the receiver RF frequency that is called the 'image frequency'. This corresponds to the value which, when mixed down by the

Digital Radio System Design   Grigorios Kalivas
© 2009 John Wiley & Sons, Ltd

downconverter, will produce an IF signal at exactly the same frequency as the useful one. It is easy to see that this frequency is the symmetrical of the useful signal f_{RF} with respect to the local oscillator frequency f_{LO}. As a result, $f_{image} = f_{RF} + 2f_{IF}$.

Figure 3.1 is a representative picture of a typical received spectrum and illustrates all the above-mentioned signals. We have to note that, at the transmitter side, the useful radio signal centred at f_{RF} will be the strongest, whereas this will not be the case at the receiver side where interferers can be much stronger than the useful signal.

To briefly address the issue of useful signal extraction at the back end of the radio receiver, we will consider a conventional super-heterodyne receiver as illustrated in Figure 3.2. In this receiver architecture, the RF signal is translated to baseband by using one or more downconverting (mixing) stages.

The out-of-band signals are eliminated by band-select RF filters located just after the antenna. The image reject filter placed just after the low noise amplifier (LNA), is used to eliminate the image frequency which is located $2f_{IF}$ away from f_{RF}. For this reason, this filter must have a one-sided bandwidth of $2f_{IF}$.

In-band interference signals go through the first two filtering stages and can therefore create inter-modulation distortion if the front-end components (LNA, mixers) exhibit considerable nonlinearity. The resulting products will fall in the useful signal band. Finally, the role of IF filters is to suppress the neighbouring frequencies (in-band interferers) and consequently

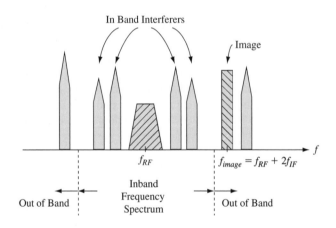

Figure 3.1 Typical receiver spectrum

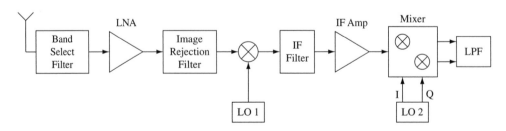

Figure 3.2 General super-heterodyne receiver with two frequency translations

characterize the ability of the receiver to 'select' the useful RF signal among the interferers. These filters define the selectivity capability of the receiver.

3.2 Frequency Downconversion and Image Reject Subsystems

The principal role of the RF front-end is to downconvert the received signal to a much lower frequency for further processing by the analogue or digital parts of the receiver chain. Downconversion is performed by multiplication (mixing) of the signal by a sinusoid produced by the local oscillator. This multiplication will translate the RF signal down to an intermediate frequency, which is equal to the difference $\omega_{RF} - \omega_{LO}$. Here, ω_{RF} is the centre frequency of the RF signal whereas ω_{LO} represents the frequency of the local oscillator. This multiplication process corresponds to a convolution in the frequency domain between all the signals passed through the front-end filter and the LO signal. However, since we deal with real signals, their Fourier transforms produce complex conjugate signals placed symmetrically around zero frequency. On the other hand, the local oscillator is transformed into two impulses positioned at f_{LO} and $-f_{LO}$ on the frequency axis. All of the above are illustrated in Figure 3.3, which verifies those mentioned in the previous section about the image frequency: The image will be downconverted and coincide with the useful IF signal. Therefore, the image frequency has to be removed before the mixing operation takes place. Another aspect to clarify is that the image problem is due to the fact that a real sinusoid in the place of LO produces two impulses at positive and negative frequencies. This means that, if the mixing of the received signal is done using a complex exponential instead of the sinusoid, the image problem could be avoided [Mirabbasi00].

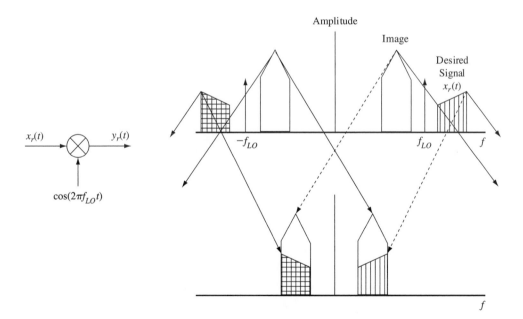

Figure 3.3 Downconversion of a real signal with a sinusoid local oscillator

To implement that kind of multiplication, two multipliers are needed that will operate on real signals and will produce in-phase and quadrature components, as shown in Figure 3.4. These mixers will operate using the cosine and sine of the local oscillator frequency f_{LO}. After some complex operations it can be proven [Tasic01] that if the received signal $R(t)$ at the input of the system is given by:

$$x_r(t) = R(t) = Si(t) + IM(t) = A \cdot \cos(\omega_{RF} t + \theta) + B \cdot \cos(\omega_{IM} t + \varphi) \qquad (3.1)$$

then the output of the downconverter will be the complex summation of the in-phase and quadrature branches:

$$y(t) = y_I(t) + jy_Q(t) = 2 \cdot (I + jQ) = [Si(t) \cdot e^{-j\omega_{IF} t}]^* + [IM(t) \cdot e^{j\omega_{IF} t}]^* \qquad (3.2)$$

where $[\bullet]^*$ denotes complex conjugation.

In Equation (3.1), $R(t)$ stands for the received signal at the input of the dowconverter whereas $Si(t)$ and $IM(t)$ represent the useful signal and the image signal respectively. As repeatedly stated in the above section, the problems associated with the image frequency stem from the

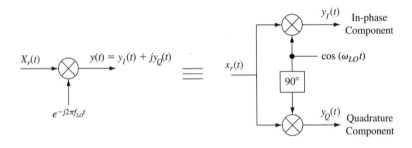

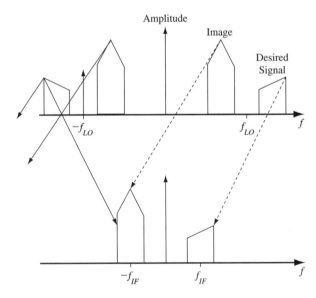

Figure 3.4 Mixing of a real signal with a complex local oscillator

fact that downconversion has to be performed using a local oscillator frequency different from the carrier frequency (heterodyne receiver). In the case where these frequencies are the same, we have direct downconversion to baseband where there is no image frequency. This technique will be presented in detail later in this chapter. Because of the problems created by the image frequency in heterodyne receivers, a variety of systems have been investigated that could be used to reject the image. The most well known in the literature are the Hartley [Hartley28] and the Weaver [Weaver56] image reject receivers. These will be presented in the rest of this section.

Both architectures achieve image rejection using single sideband demodulation (SSB). This is realized by applying phase cancellation techniques according to which the image is cancelled by producing opposite replicas of it at the output of the two branches of both Hartley and Weaver receivers.

3.2.1 Hartley Image Reject Receiver

This topology is illustrated in Figure 3.5(a). The RF input signal is split into two branches. At the lower branch the signal is mixed with the LO signal $\cos \omega_{LO} t$, whereas at the upper branch it is mixed with the quadrature LO signal. In the upper branch, after lowpass filtering to remove the high-frequency terms, it is shifted by 90°. Then, the two resulting signals are added to produce the final IF output signal. Below we give an outline of the mathematical analysis and the frequency domain interpretation [Razavi98] of the Hartley architecture.

If the RF input signal is $x(t)$ and $x_A(t)$, $x_B(t)$ and $x_C(t)$ represent the signal at the corresponding points in Figure 3.5, simple trigonometric manipulations give the following expressions:

$$x(t) = A_{RF} \cdot \cos(\omega_{RF} t) + A_{IM} \cdot \sin(\omega_{IM} t) \tag{3.3}$$

$$x_A(t) = 0.5 \cdot [-A_{RF} \cdot \sin(\omega_{RF} - \omega_{LO})t + A_{IM} \cdot \sin(\omega_{RF} - \omega_{IM})t] \tag{3.4}$$

$$x_B(t) = 0.5 \cdot [A_{RF} \cdot \cos(\omega_{LO} - \omega_{RF})t + A_{IM} \cdot \cos(\omega_{LO} - \omega_{IM})t] \tag{3.5}$$

$$x_C(t) = 0.5 \cdot [A_{RF} \cdot \cos(\omega_{RF} - \omega_{LO})t - A_{IM} \cdot \cos(\omega_{LO} - \omega_{IM})t] \tag{3.6}$$

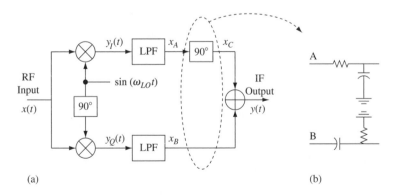

(a)   (b)

Figure 3.5   (a) Hartley receiver topology; (b) 90° phase shift implementation

As a result, the IF output signal is:

$$y(t) = x_B(t) + x_C(t) = A_{RF} \cdot \cos(\omega_{RF} - \omega_{LO})t \tag{3.7}$$

The above equation shows that, due to the cancellation of the image terms $A_{IM} \cos(\omega_{LO} - \omega_{IM})t$ in Equations (3.5) and (3.6), only the useful signal component is present at the output of the structure.

It gives more insight to show the above processing in the frequency domain graphically. This is demonstrated in Figure 3.6. The input spectrum is convolved with the Fourier transforms of sine and cosine which produce x_A and x_B. After that, x_A going through the 90° phase shifter, it is multiplied by $-j$ in the positive frequencies and $+j$ in the negative frequencies. As can be seen, x_B and x_C contain the image signal with opposite polarities, which in turn results in its cancellation at the output.

Regarding implementation issues, first we must note that the 90° phase shifter is realized using an RC–CR network as in Figure 3.5(b), which produces +45° and −45° phase shifts at its output.

The main disadvantage of the Hartley receiver comes from component mismatches, which result in reduced image rejection. In practice, the image rejection is a function of the amplitude and phase imbalances on the two paths of the receiver. More specifically, if the gains of the two paths are not the same and the respective phases are not in quadrature, the above theoretically predicted image cancellation is not infinite. In what follows we give a brief mathematical treatment of this matter. To do so we assign the error factors ε in the amplitude and θ in the phase of the local oscillator signals $X_{LOupper}$ and $X_{LOlower}$ at the two branches. ε represents the relative voltage gain mismatch given by the ratio of ΔA to A:

$$X_{LOupper}(t) = A_{LO} \cdot \sin(\omega_{LO}t) \tag{3.8}$$

$$X_{LOlower}(t) = (A_{LO} + \Delta A) \cdot \cos(\omega_{LO}t + \theta) \tag{3.9}$$

The image rejection ratio (IRR) is defined as:

$$IRR = [P_{IM}/P_{SIG}]_{out}/[P_{IM}/P_{SIG}]_{in} \tag{3.10}$$

After some trigonometric manipulation we obtain the following expression [Razavi98]:

$$IRR = \frac{1 - 2 \cdot (1 + \varepsilon)\cos\theta + (1 + \varepsilon)^2}{1 + 2 \cdot (1 + \varepsilon)\cos\theta + (1 + \varepsilon)^2} \tag{3.11}$$

When ε and θ are small ($\varepsilon \ll 1$, $\theta \ll 1$ rad) then an approximate expression for IRR is:

$$IRR = \frac{\varepsilon^2 + \theta^2}{4} \tag{3.12}$$

Although Equation (3.11) is derived using a gain and phase mismatch in the LO signal only, similar expressions can be obtained by incorporating mismatches in the mixers, the LPFs and the 90° phase shifter.

Figure 3.7 shows graphically the image rejection as a function of phase and amplitude imbalance. The amplitude error is a parameter in this diagram. It is easy to see that a phase error of 3° combined with an amplitude imbalance of 1 dB will lead to an image rejection of 25 dB. Typical receivers should provide for image suppression of 40–60 dB and for this reason very small figures for the phase and amplitude imbalances must be achieved.

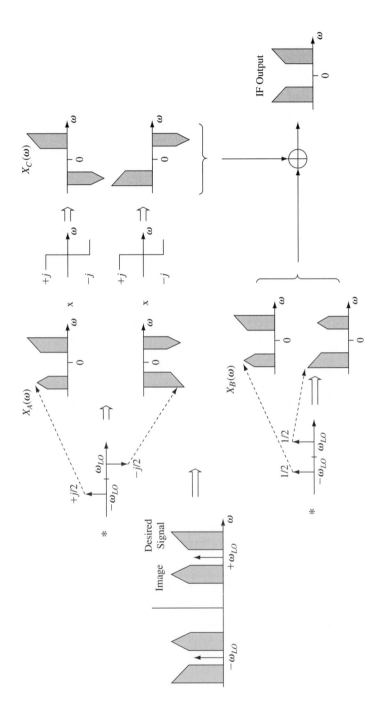

Figure 3.6 Frequency domain processing in Hartley receiver

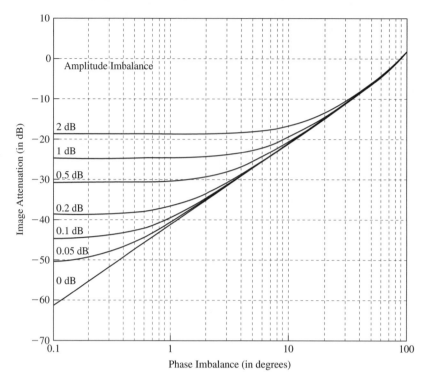

Figure 3.7   Image rejection as a function of amplitude and phase imbalances (reprinted from A. Abidi, 'Direct-conversion radio transceivers for digital communications', IEEE J. Solid-St. Circuits, December 1995, pp. 1399–1410, © 1995 IEEE)

Coming back to the RC–CR network, we must note that the phase difference at the output of the two branches of the network is always 90°, provided that there is local matching of the R and C elements and does not depend on the frequency of operation. On the contrary, the amplitude imbalance is a function of both fluctuations in the values of R and C and of the frequency of operation. More specifically, the perfect amplitude balance holds only at the centre frequency of operation, which is equal to $1/(2\pi RC)$. The overall impact of the above parameters on the amplitude imbalance is given by the following expression [Pache95]:

$$\frac{\Delta V}{V_{\text{LO}}} = V_{\text{LO}}\left(\frac{\Delta R}{R} + \frac{\Delta C}{C} + \frac{f_C - f_{\text{LO}}}{f_C}\right) \qquad (3.13)$$

From the above discussion on RC–CR networks it is easy to realize that it would be impossible to use them in high-frequency systems without modifications and design improvements to compensate for the impairments. These will be presented later in this chapter.

3.2.2 Weaver Image Reject Receiver

Figure 3.8 shows a Weaver image reject subsystem. The 90° phase shifter is replaced by a second quadrature mixing stage. In this way, gain imbalance problems associated with it are

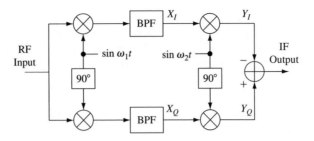

Figure 3.8 Weaver image-reject receiver

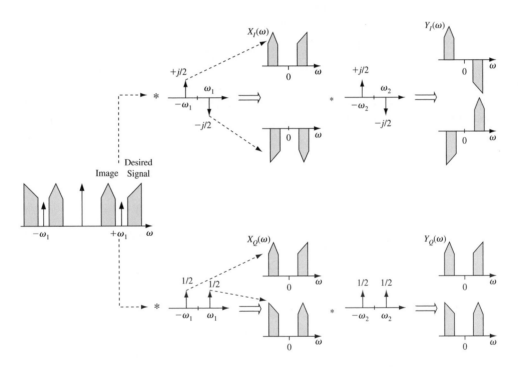

Figure 3.9 Frequency-domain processing in Weaver receiver

eliminated. In addition, the lowpass filters following the first mixing stage must be substituted by band pass filters in order to eliminate the 'secondary image' problem [Razavi98], which will be presented later in this section.

The operation of the Weaver mixer system is illustrated graphically in the frequency domain in Figure 3.9. The spectrum of the input signal is convolved with the Fourier transforms of sine and cosine functions at frequency ω_1 to produce signals x_I and x_Q. Another convolution with delta functions at frequency ω_2 (corresponding to the Fourier transform of sine and cosine at ω_2) gives the final frequency domain signals $Y_I(\omega)$ and $Y_Q(\omega)$ at the corresponding points. Subtraction of these signals completely eliminates the images.

The second mixing stage creates the problem of the 'secondary image', which is an interferer at frequency $2\omega_2 - \omega_{in} + 2\omega_1$, where ω_1 is the frequency of the first quadrature LO, ω_2 is that of the second LO and ω_{in} is the input frequency. Figure 3.10 [Razavi98] shows graphically the frequency domain operations resulting in the secondary image. We observe that after the first downconversion, the interferer has moved at frequency $2\omega_2 - \omega_{in} + \omega_1$, which is the image of the useful signal at $\omega_{in} - \omega_1$.

There are a number of ways to address the secondary image problem. The most effective of them are downconversion to baseband, downconversion to a low-IF and complex mixing, all applied at the second mixing stage.

Translation of the first IF signal to baseband represents direct downconversion and is associated with most of the problems of this technique like DC offset and $1/f$ noise (see corresponding section). However, since the first LO frequency ω_1 can be chosen to be far from the input frequency ω_{in}, DC components created by self-mixing at the first LO can be removed by the BPF that comes after the first mixing stage.

The second technique, which involves a low-IF second-mixing stage, attempts to alleviate the problems mentioned above associated with direct downconversion. Finally, complex mixing is an effective solution to eliminate the secondary image problem. As mentioned at the beginning of this section, complex mixing is equivalent to performing the mixing process using a complex exponential as the LO signal. Theoretically this operation will eliminate the image-related

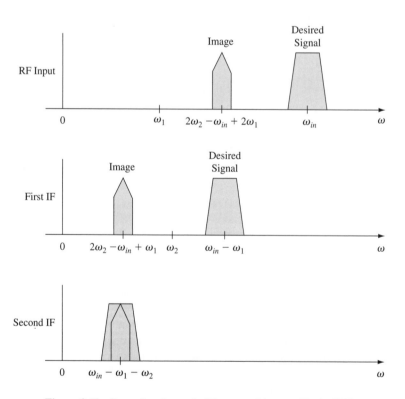

Figure 3.10   Secondary image in Weaver architecture [Razavi98]

problems. It is upon this principle that double quadrature structures, consisting of four mixers, are used to implement the second mixing stage in low-IF receiver techniques presented later in this chapter.

3.3 The Heterodyne Receiver

Having presented previously the frequency translation process along with some of its implications, we will now proceed directly in a detailed examination of the characteristics and problems associated with the heterodyne receiver. Figure 3.11 is a more detailed illustration of Figure 3.2 and represents the general structure of a typical heterodyne receiver. The band-select filter after the antenna is used to reject signals that do not belong to the useful band. After that, the in-band signals go through moderate amplification (LNA) and image rejection filtering. The purpose of the IR filter is to suppress as much as possible the image frequency. Channel selection is implemented using a filter placed just after the first downconversion. The major amount of channel selection is implemented at this stage. Subsequently a quadrature mixing operation translates the signal into a form suitable for digital processing.

Since the quadrature mixing stage translates the input signal at FF' to baseband ($\omega_{LO2} = \omega_{IF}$) we have a single-IF heterodyne receiver. The generated in-phase (I) and quadrature (Q) components are filtered and converted to digital for further processing. This is done in all-digital modulation systems because the information is mapped in the two-dimensional cartesian space. However, analogue modulation does not need I and Q downconversion to baseband.

Another IF stage can be inserted between the variable gain amplifier (VGA) and the quadrature mixer at FF' if necessary. In that case we would have a double-IF heterodyne receiver. In what follows we will examine the factors that are crucial to decide upon the number of frequency translations and to determine the values of the LO and IF frequencies in the receiver.

We use Figure 3.12 as an example of a general heterodyne architecture with two IF stages to demonstrate the important aspects and problems associated with this kind of receiver. At each point in the receiver chain, the effect of the previous component is shown in this figure. It is interesting to note the effect of the various filters along the receiver chain in order to present the associated problems. Observing the difference between A and B, we can see that the band select filter filters out the out-of-band channels and partially the image. Then, the image reject filter mostly suppresses the image (spectrum at point C). Finally, the two stages of channel select filters suppress the neighbouring in-band channels to different degrees. The first one

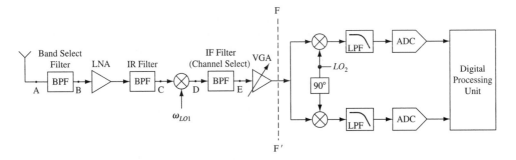

Figure 3.11 General structure of an heterodyne receiver

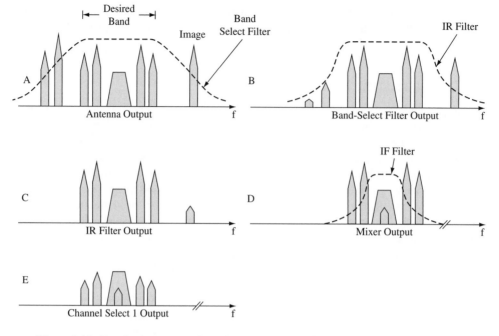

Figure 3.12 Received spectrum along the receiver chain at the output of cascaded devices

gives a moderate filtering of the in-band channels, while the second brings them to the desired low level.

Two questions arise from the above discussion: whether the image can be moved closer or further away from the useful channel and if it is possible to discard the second IF by designing the first channel select filter with the appropriate selectivity in order to suppress the in-band channels satisfactorily. The first question is directly associated with the choice of the IF frequency because IF frequency and image rejection are both present in the following empirical formula:

$$\text{Image rejection} \approx 6N \log_2\left(\frac{2f_{\text{IF}}}{BW/2}\right) \tag{3.14}$$

where BW is the system bandwidth (pass-band) whereas N indicates the number of filter poles. The value of image rejection in the above equation can be estimated by examining the parameters and locations of radio transmitters operating at the image frequency. In most cases, image rejection lies within 40–60 dB. If we fix the filter parameters (bandwidth and order N), Equation (3.14) also indicates that IR increases as f_{IF} becomes larger. However, this has an implication for the channel selection quality of the channel select filter. If f_{IF} is large, the IF filter selectivity is poor in contrast to a low IF at which the IF filter can have considerably better selectivity. The above considerations lead to the conclusion that when f_{IF} is large we have high image rejection but poor channel selectivity, whereas low f_{IF} degrades receiver sensitivity but the channel select filter have much better selectivity. These are also illustrated in Figure 3.13 where it is pictorially demonstrated that, when we have a high f_{IF},

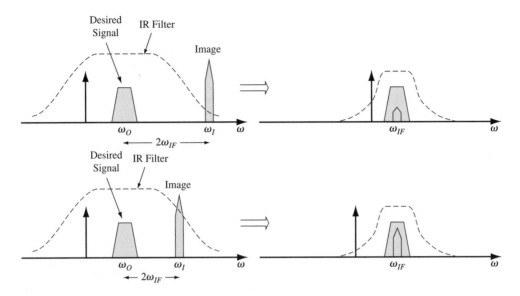

Figure 3.13   Image and Interference rejection as fuction of f_{IF}

image rejection is very drastic but adjacent channel rejection is moderate, whereas in the opposite case the inverse is true.

Another problem that can occur in heterodyne receivers is associated with a possible interferer at the frequency of $(f_{\text{in}} + f_{\text{LO}})/2$. This will translate down to $f_{\text{IF}}/2$ and the second-order distortion problem after the mixer will produce an undesirable signal falling at the useful signal frequency band. This is the half-IF problem and can be controlled by minimizing second order distortion.

3.4 The Direct Conversion Receiver

If the local oscillator frequency is set equal to that of the incoming RF signal, the IF frequency becomes equal to zero. As illustrated in Figure 3.14, the incoming signal is translated to DC with its bandwidth spreading from $-\omega_S$ to ω_S along the frequency axis, ω_S denoting half of the information bandwidth. As most of the digital modulated signals carry information on both sides of the centre frequency, a quadrature mixer must be used providing for in-phase (I) and quadrature (Q) outputs. The local oscillator input to the two mixers must have a phase difference of 90° in order to produce I and Q components. Another arrangement would be to split the RF input signal to two branches having 90° phase difference, which in turn will drive the mixers. However, implementation of such a 90° shifter would be difficult for wideband input signals whereas there is no such demand if the 90° shifter is used in the LO path. The discussion in Section 3.2 about the RC–CR phase shift network indicates clearly its narrowband operation. In addition, trade-offs in noise, power and gain [Razavi98] may result in front-end performance deterioration.

The main advantage of the direct conversion receiver is the elimination of the image reject filter due to the absence of the image frequency problem. Along with it, the LNA can drive the

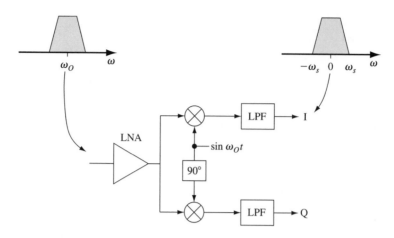

Figure 3.14 Direct conversion receiver with quadrature mixing

mixers directly (no filter in-between) and therefore there is no need for matching to the usual
50 Ω input impedance of the filter. This leads to integration improvement at the front-end.
A second advantage is that there is no need for intermediate frequency stages and IF filters,
since channel selection takes place at baseband and is implemented using lowpass filters, which
are easier to integrate and operate with tunable bandwidths for applications of multimode or
multiband terminals.

Nevertheless, the direct conversion architecture exhibits a few disadvantages as well. These
are sensitivity to I/Q mismatch, DC offset, even-order distortion and $1/f$ noise. In addition
to these, channel selection also needs more careful design. In the rest of this section we will
examine in some detail each of these issues.

3.4.1 DC Offset

DC offset is created by self-mixing of two types of signals: the LO signal feeding the mixer
and strong unwanted signals entering the RF mixer input. As shown in Figure 3.15, the strong
LO signal, due to finite LO–RF mixer isolation, passes to the RF port and is reflected upon
the amplifier output and input. This 'LO leakage' signal enters the RF mixer port and is mixed
with the LO signal, resulting in a DC component at the output of the mixer. The second type
of signal that can create DC components can be a strong interferer entering the receiver, or a
strong reflected ray from the transmitted signal. The unwanted signal enters the RF port of the
mixer and part of it 'leaks' to the LO mixer input. Then, it is reflected back to the LO input
and mixed with the RF incoming replica producing a DC component. The DC offset can also
vary with time. This variation can be produced because the strong reflected signals that are
mixed with the LO to create DC can vary with time due to the terminal movement.

From approximate calculations, it can be deduced that, if the LO–RF isolation is worse than
40 dB, the self-mixing signal at the input of the mixer will have an amplitude of a few tenths
of microvolts. In turn, this will produce a DC offset at the output of the mixer which after
amplification of 50–70 dB will give a very strong DC component [Lindquist93], [Razavi98].

There are two basic techniques to eliminate the DC offset problem. The first one uses AC
coupling in the receiver chain whereas the second relies on signal processing techniques. To

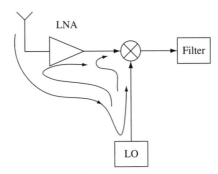

Figure 3.15   Self-mixing of LO signal and incoming interferer

implement AC coupling, a highpass filter (HPF) is employed in the receiver signal path. This HPF is used to cut off the signal components around DC such that the DC offset is eliminated as well. However, depending on the modulation type there may be substantial energy at zero frequency and around it. If there is a dip in the energy around DC, AC coupling will have almost no effect on the performance of the radio. For example, in wideband-FSK modulation used in pagers, the energy at DC is −25 dB below maximum, which takes place at ±4.5 KHz [Abidi95]. In this application, a first-order HPF with a corner frequency of 1KHz integrated along with the channel select filter results in no deterioration in the system performance. However, in some other cases, the design practice may not be so clear. An important reason for this difficulty is the time-varying nature of the DC offset level. This time variability can be due to self-mixing of reflections in a moving receiver or it can be created by blocking signals that appear at random times in the receiver input. In such cases, measurements that will give an average of the time-varying DC offset can be very helpful. In order to be effective, these measurements must take place during idle periods of the receiver. One such case is a TDMA system where the measurement can take place during the idle time between two successive receive bursts. Measurement is implemented with a capacitor that is charged during the idle period and stores the level of DC offset [Razavi98]. When the received TDMA burst is coming in, the stored value of the DC offset is subtracted. However, this method will not work when large blocking signals appear at the input of the receiver. In this case, the front-end design must accommodate for good reverse isolation in the mixer and good linearity. In addition to these, signal processing techniques can be used which can reduce the required SNR by more than 10 dB to achieve satisfactory performance [Mashhour01].

DC offset cancellation by signal processing usually involves feedback loops in the baseband section of the receiver. These methods employ DC-offset removal by estimating an average DC offset after digitization of the baseband signal. One of these methods calculates the error between the DC-compensated value and the output of the decision device and after integration it employs an updating mechanism to calculate the estimated DC offset [Sampei92]. This method operates on burst data transmission using the preamble of the packet and can be used for any kind of modulation. However, high dynamic range analog-to-digital conversion is needed to accommodate for large DC offsets. Another approach addresses the high dynamic range problem by using an adaptive delta modulator in the feedback path [Lindquist93]. For moderate DC offsets (10%) the SNR improvement achieved by the digital signal processing techniques is in the range of 2–3 dB.

3.4.2 I–Q Mismatch

The 90° phase shifter and the combination of mixers, filters and amplifiers in the quadrature downconverter (Figure 3.14) will introduce amplitude and phase imbalances at the I and Q branches of the downconverted signal. These amplitude and phase errors will result in distortion of the signal constellation at the output of the two branches, which will inevitably increase the bit error rate of the system. To show an example on the detrimental effects of I–Q mismatch we can take the case of a QPSK received signal:

$$x_r(t) = \cos \left[\omega_C t + \theta_m + \frac{\pi}{4} \right], \quad \theta_m = n\pi/2, \ n = 0, ..., 3 \tag{3.15}$$

In addition, we introduce amplitude and phase errors, $\varepsilon/2$ and $\theta/2$ respectively, in the I branch and $-\varepsilon/2$ and $-\theta/2$ in the Q branch of the local oscillator signal (as in Section 3.2). By splitting the errors into two parts, the effects of the mismatch can be observed in both I and Q axes on the constellation diagrams, as shown in the equations below. After some elementary trigonometric manipulations and lowpass filtering we get the following expressions for the baseband I and Q components [Razavi97]:

$$x_I(t) = a_I \left(1 + \frac{\varepsilon}{2}\right) \cos \frac{\theta}{2} - a_Q \left(1 + \frac{\varepsilon}{2}\right) \sin \frac{\theta}{2} \tag{3.16}$$

$$x_Q(t) = -a_I \left(1 - \frac{\varepsilon}{2}\right) \sin \frac{\theta}{2} + a_Q \left(1 - \frac{\varepsilon}{2}\right) \cos \frac{\theta}{2} \tag{3.17}$$

where a_I and a_Q represent the information sequence split in I and Q branches. The effect of the above equations on the constellation diagram is shown in Figure 3.16. To illustrate clearly the result of isolated imbalances in amplitude and in phase, their effect is shown separately. In Figure 3.16(a) we assume only amplitude error ($\theta = 0$), whereas in Figure 3.16(b) we assume only phase imbalance ($\varepsilon = 0$).

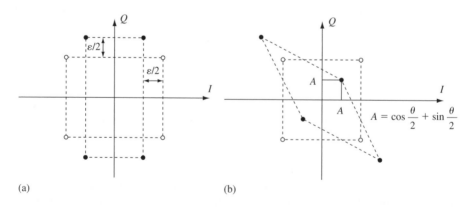

Figure 3.16 Effect of amplitude (a) and phase (b) imbalance in I, Q channels for QPSK

3.4.3 Even-Order Distortion

This type of problem is due to second-order nonlinearities of the components in the front-end subsystem. If we have a nonlinearity of the type:

$$y(t) = a_1 x_R(t) + a_2 x_R^2(t) \tag{3.18}$$

and the input to the receiver is two closely spaced sinusoids,

$$x_R(t) = V_R \cos(\omega_1 t) + V_R \cos(\omega_2 t) \tag{3.19}$$

then this will produce second order products of the form:

$$\text{Second order products} = a_2 \cdot V_R^2 \cdot \frac{1 + \cos(2\omega_1 t)}{2} + a_2 \cdot V_R^2 \cdot \frac{1 + \cos(2\omega_2 t)}{2}$$
$$+ 2 \cdot a_2 \cdot V_R^2 \cdot \left(\frac{\cos(\omega_1 + \omega_2)t + \cos(\omega_1 - \omega_2)t}{2} \right) \tag{3.20}$$

The term containing the frequency $\omega_1 - \omega_2$ in the equation above will create problems because the resulting frequency will be very low (due to the close values of ω_1 and ω_2) and will subsequently translate down to the DC along with the useful signal. Second-order nonlinearity can be created by the LNA or by the mixer or both. This kind of nonlinearity is quantified by defining the second-order intercept point (IP2). By employing suitable RF IC design techniques, IP2 can be considerably reduced.

3.4.4 1/f Noise

This type of noise is also known as flicker noise and it is very strong at low frequencies, which is evident from the slope of its spectrum. Flicker noise is mostly due to the switching operation of active devices. The rms flicker noise voltage referred to the input of a switching FET is given by the following expression [Darabi00]:

$$V_n(\text{rms}) = \sqrt{2 \times \frac{K_f}{W_{\text{eff}} \cdot L_{\text{eff}} \cdot C_{\text{ox}} \cdot f}} \tag{3.21}$$

where K_f, W_{eff}, L_{eff} and C_{ox} are constants related to the device geometry whereas f represents the frequency at which the flicker noise fits best the $1/f$ slope. This frequency is usually in the range of few kilohertz to 1 MHz depending on the device.

A similar formula is given in [Razavi97] where the flicker and the thermal noise power components are calculated and compared. It is found that the relative increase of flicker noise with respect to thermal noise is approximately 17 dB in the range 10 Hz to 200 kHz.

Because of the low-frequency nature of this noise component it will always be down-converted at DC where the useful signal is centred at zero IF receivers. To minimize the flicker noise component, the switching devices must be designed with a large area [Razavi97], [Manstretta01]. This will lead to long switching times which can be reduced by reducing the bias currents of the devices.

3.5 Current Receiver Technology

3.5.1 Image Reject Architectures

In this section we will examine in some detail receiver topologies, which belong to the general category of image rejection architectures. First, some general considerations about the technology and the implementation aspects will be given and subsequently we will present two specific techniques in more detail; wideband-IF and low-IF receiver architecture. These techniques possess some distinct characteristics, which have given them remarkable advantages in certain applications.

At first we categorize the existing image reject implementations in two classes that follow either the Hartley or the Weaver architectures. This classification distinguishes the image reject receivers in these which produce I and Q signals with 90° phase difference at the IF frequency, and those which use a second pair of mixers in order to eliminate the image, instead of the 90° phase shift network.

In the first class of receivers, the 90° phase shift is implemented either by using poly-phase networks ([Crols95], [Behbahani01], [Katzin98], [Ugajin02], [Imbornone00], [Baumberger94]) or by employing frequency division [Maligeorgos00], which inherently creates quadrature output signals.

Polyphase networks are also called Hilbert filters and reject the image by creating a notch at the place of the mirror frequency. The way in which polyphase filters are placed in the receiver's IF path is illustrated in Figure 3.17. They are usually symmetric RC networks, which can consist of two or more stages and must be matched in amplitude and phase. Component mismatch will create similar problems with I–Q mismatch in DCR (Section 3.4). One disadvantage of polyphase filters is their considerable insertion loss, which, depending on the number of stages, can vary between 5 and 25 dB [Behbahani01]. This problem can be solved by inserting amplifiers between the filter stages.

The other approach for implementing the 90° phase shifter is based on the inherent principle of frequency dividers to provide two outputs in quadrature. In [Maligeorgos00] the local oscillator phase shifter is implemented by first doubling the LO frequency and then dividing it by 2. Power consumption can be kept low by using analogue frequency division whereas image rejection can exceed 45 dB in a bandwidth of 1 GHz. Although frequency division can give broadband performance, it requires generation of frequencies twice that of the local oscillator. This may create implementation problems in systems employing frequencies beyond a few GHz.

We now take a brief look at implementations, which follow the Weaver architecture. As mentioned in Section 3.2, the main problems associated with this approach come from I–Q mismatches and the secondary image, which is situated at $2\omega_2 - \omega_{in} + 2\omega_1$ and is downconverted to the same frequency with the useful channel at the second mixing stage. By employing double quadrature downconversion at the second IF [Crols95], the phase error between the two (quadrature) paths can be diminished to fractions of a degree. As shown in Figure 3.18, a double quadrature downconversion block consists of four mixers driven in both inputs (RF and LO) by quadrature signals. This implementation will improve the image rejection of the receiver. Double quadrature downconversion combined with low second IF will be examined in more detail later in this section.

The secondary image can be eliminated by using bandpass filters in place of the lowpass filters in between the two mixing stages [Razavi98]. Similarly, zero IF at the second downconversion stage will eliminate the secondary image. Problems associated with direct conversion

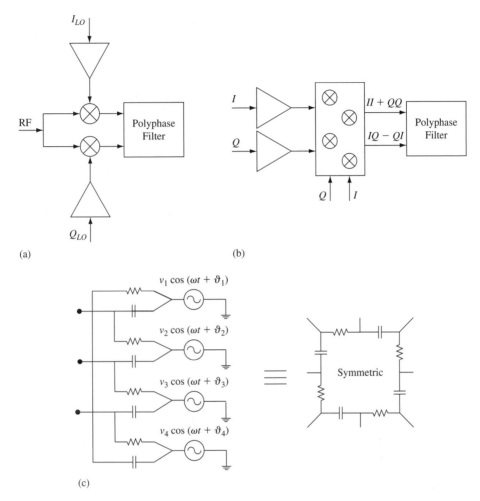

(a)   (b)

(c)

Figure 3.17 (a) Quadrature mixing followed by polyphase filtering; and (b) double quadrature mixing followed by polyphase filtering. (c) Classic RC polyphase filter in two different forms

to baseband (DC offset, $1/f$ noise, etc.), after the second mixing stage will be diminished if the frequency of the first LO is far from ω_{in} [Razavi98].

3.5.1.1 Wide-band IF Receiver Architecture

In this approach, the complete useful frequency band is translated, as a block, to a first IF by performing a first downconversion. Since the centre of the useful band and the IF frequency are both fixed, the RF local oscillator frequency is also fixed. Figure 3.19 demonstrates the principle of operation of the wide-band IF architecture. Channel selection is performed after the second mixing stage. This is done by tuning the second LO such that the desired channel is downconverted to DC. Then, by appropriate channel filtering, the neighbouring channels can be eliminated. The most distinct characteristic of this architecture is that the first LO is fixed,

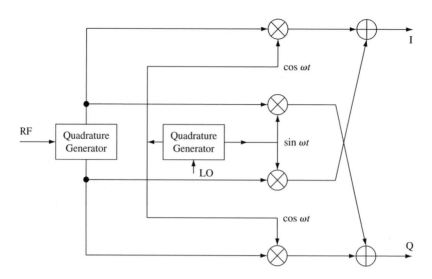

Figure 3.18   The double quadrature down-converter technique [Crols95]

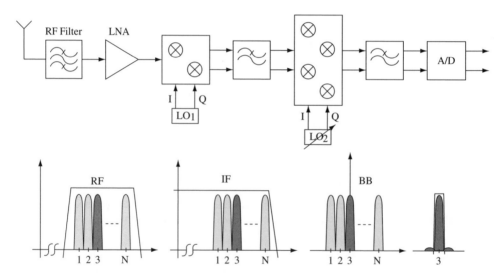

Figure 3.19   Wideband-IF with double conversion architecture (reprinted from J. Rudell *et al.*, 'A 1.9-GHz wide-band IF double conversion CMOS receiver for cordless telephone applications', IEEE J. Solid-St. Circuits, December 1997, pp. 2071–2088, © 1997 IEEE)

allowing channel selection to be performed in the second IF. The image rejection capabilities are closely linked to the form of two downconversion stages. In the classical wide-band IF architecture [Rudell97], the first downconverter is a quadrature mixer whereas the second consists of a double quadrature structure employing four mixers. As a result, the overall image rejection structure resembles the Weaver architecture. However, in this case, the second downconversion is a complex mixing stage.

Figure 3.20 illustrates the functional operation of this architecture by means of frequency domain operations. We have two real (non-complex) signals at the input of the receiver: the desired signal and the image. The desired signal (represented by sidebands 1 and 4) is located above the frequency of the first local oscillator whereas the image lies below the first LO (sidebands 2 and 3). Consider that the spectra of the local oscillator signal $\cos \omega_{LO}t$ and its quadrature $\sin \omega_{LO}t$ are two delta functions in the real and the imaginary axis respectively (Figure 3.21). The convolutions of the four sidebands (1, 2, 3 and 4) with the above-mentioned delta functions will translate the signals as illustrated in Figure 3.20. At the output of the two-stage image rejection mixer we have only the desired signals (1 and 4) whereas the image, because of the cascading structure of a quadrature and a double-quadrature mixer, is cancelled out.

From the above discussion, we can make two observations about the properties of this architecture. First, due to its fixed frequency the first local oscillator can be designed with a very low phase noise. This will result in no apparent impact of the phase noise of the RF local oscillator on the overall system performance. Second, as one can notice from Figure 3.20, any incoming signals with operating frequency above the first LO will pass without attenuation while all signals located below the first local oscillator will be rejected. Hence, the first LO defines the edge of the image rejection band and therefore, ideally, this represents a very wide-band image-reject subsystem. In addition to the above advantages, this architecture allows for the design of the frequency synthesizer (second LO) with reduced phase noise performance. This can result from reduced noise contribution from the frequency reference source, the phase detector and the frequency divider employed in the synthesizer [Rudell97].

The main disadvantage of this architecture stems from the fact that the overall useful band (all channels) is translated to the first IF and consequently channel selection is performed after the second downconversion. Therefore, the second LO must have a wider tuning range (as a percentage of the centre synthesizer frequency) in order to be able to convert all channels down to baseband. A second issue is that the required dynamic range is higher, due to all the strong neighbouring channel interferers that may be present.

Since this is an image-reject architecture, the amount of image rejection is still a function of amplitude and phase mismatches in the two mixing stages. It can be proved [Rudell97] that the image rejection (IR) is given by the following equation:

$$IR = 10 \log \frac{1 + (1 + \Delta A)^2 + 2(1 + \Delta A) \cos (\varphi_{LO1} + \varphi_{LO2})}{1 + (1 + \Delta A)^2 - 2(1 + \Delta A) \cos (\varphi_{LO1} - \varphi_{LO2})} \text{ dB} \qquad (3.22)$$

where ΔA represents the total amplitude error between the I and Q paths in the two stages and φ_{LO_1} and φ_{LO_2} are the phase deviation from $90°$ in the two branches in LO_1 and LO_2, respectively. Note the resemblance between the above equation and Equation (3.11). The difference is that in Equation (3.11) IRR represents the ratio of $(Image/Signal)_{out}$ to $(Image/Signal)_{in}$, whereas Equation (3.22) gives directly the ratio Image/Signal at baseband.

Using the wideband-IF receiver architecture, a DECT prototype receiver was designed and built [Rudell97], as shown in Figure 3.22. The first LO frequency was set equal to 1.7 GHz and the first IF band extends from 181 to 197 MHz. The second stage downconversion translates the signal to baseband and for this reason the second LO is a synthesizer tuning in the 181–197 MHz band. For purposes of noise coupling reduction, the signal path along the receiver

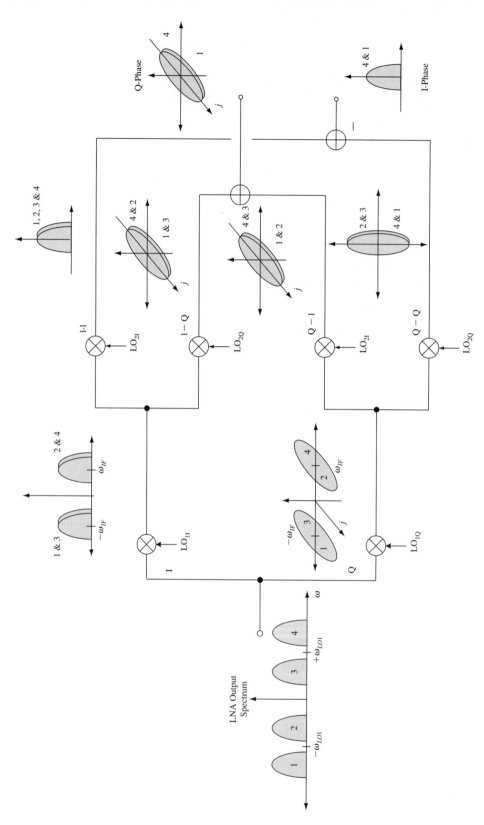

Figure 3.20 Frequency-domain processing in IR mixer fro wideband-IF architecture (reprinted from J. Rudell *et al.*, 'A 1.9-GHz wide-band IF double conversion CMOS receiver for cordless telephone applications', IEEE J. Solid-St. Circuits, December 1997, pp. 2071–2088, © 1997 IEEE)

$$\cos{(\omega t)} = \frac{e^{j\omega} + e^{-j\omega}}{2} \qquad \cos\left(\omega t + \frac{\pi}{2}\right) = -\sin{(\omega t)} = \frac{-e^{j\omega} + e^{-j\omega}}{2j}$$

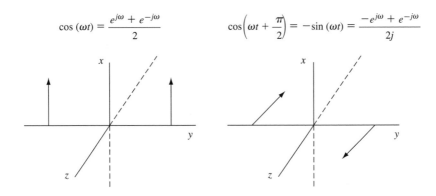

Figure 3.21 In-phase and quadrature local oscillator signals

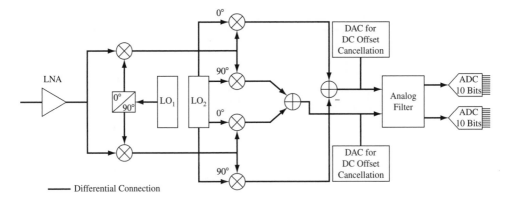

Figure 3.22 A DECT prototype receiver [Rudell97]

chain is fully differential whereas, in order to eliminate low frequency noise and DC offset, two six-bit current DACs are employed after the second downconversion. The system was designed and fabricated on a 0.6 μm CMOS process and gave a total image rejection of 55 dB, a sensitivity of −90 dBm and an input IP3 of −7 dBm.

Ahola *et al.* [Ahola04] employ a wideband-IF architecture to implement the receiver of an 802.11a/b/g wireless LAN using 0.18 μm CMOS technology. The first LO is fixed at 3840 MHz downconverting the two bands (2.4 and 5.2 GHz) at the 1300–1500 MHz IF band. The variable second LO produces the 1300–1500 MHz band in steps of 1 MHz. The receiver has a noise figure of 4.1/5.6 dB at 2.4/5.2 GHz. The sensitivity is −75 dBm at the data rate of 54 Mb/s.

3.5.1.2 Low-IF Receiver Architecture

As mentioned previously in this section, the low-IF architecture [Crols95], [Crols98] can alleviate problems associated with direct downconversion while at the same time it eliminates the image problem by complex signal operations. The principle idea that resulted in the low-IF

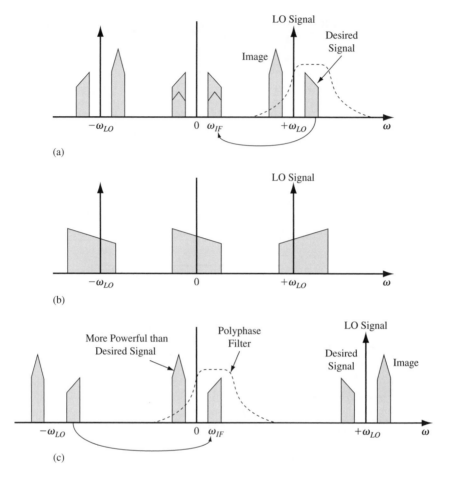

Figure 3.23 Frequency domain operations in heterodyne, direct-conversion and low-IF receiver architecture

architecture comes from the fact that image problems are created because the useful signal, the image and the LO all have two components in the frequency domain; positive and negative. Figure 3.23(a) demonstrates this situation.

Indeed, we have convolutions of the LO at negative frequency with the signal and image at positive frequencies, and convolutions of the LO at positive frequency with the signal and image at negative frequencies. This results in the superimposition of signal and image at the IF frequency. Hence, the solution would be to filter out the image frequency before down-conversion. This represents the classical super-heterodyne receiver architecture. If the LO at negative frequency is eliminated and the signal carrier frequency is identical to that of the local oscillator, there is only one convolution which translates the useful signal to baseband [Figure 3.23(b)]. This eliminates the problem of image, which is inherently associated with the difference between the LO and useful signal frequencies. To eliminate completely the local oscillator at negative frequencies, the real LO signal must be replaced by a complex signal.

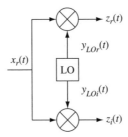

Figure 3.24 Basic receiver structure in low-IF approach

This procedure has been presented in some detail in Section 3.2. In this case, similarly to Figure 3.4, the basic receiver structure is shown in Figure 3.24, representing the direct conversion receiver architecture. In the last arrangement downconversion is performed using again only the positive LO frequency (complex mixing), which must be different from the wanted signal frequency. In this case, the image frequency will not fall on the useful signal because of the absence of the negative frequency in the LO [Figure 3.23(c)]. In addition we can have a low value for the IF. The downconverted image, located in the negative frequencies, can now be filtered out by one-sided (asymmetric) polyphase filtering. Since there is no need to use image reject filter in the front-end, it is not necessary to have a high IF value. This last arrangement leads to the low-IF architecture.

Taking an example system of this architecture as in Figure 3.25, we can see the basic points in the operation of this architecture. At the input of the receiver we have the first stage of complex mixing of a signal $x(t)$ with a complex local oscillator $y_{LO}(t)$:

$$y_{LO}(t) = y_{LOr}(t) + j \cdot y_{LOi}(t) \tag{3.23}$$

$$y_{LOr}(t) = \cos \omega_{LO} t \rightarrow Y_{LOr}(j\omega) = \delta(\omega - \omega_{LO}) + \delta(\omega + \omega_{LO})$$

$$y_{LOi}(t) = j \sin \omega_{LO} t \rightarrow Y_{LOi}(j\omega) = -j \cdot [\delta(\omega - \omega_{LO}) - \delta(\omega + \omega_{LO})] \tag{3.24}$$

The output signal $z(t)$ is:

$$z(t) = y_{LO}(t) \cdot x(t) = y_{LOr}(t) \cdot x_r(t) + j \cdot y_{LOr}(t) \cdot x_i(t) - y_{LOi}(t) \cdot x_i(t) + j \cdot y_{LOi}(t) \cdot x_r(t) \tag{3.25}$$

In the frequency domain, after some algebraic manipulations we get [Crols98]:

$$Z(j\omega) = Y_{LO}(j\omega) \otimes X(j\omega) = [Y_{LOr}(j\omega) \otimes X_r(j\omega) - Y_{LOi}(j\omega) \otimes X_i(j\omega)]$$
$$+ j \cdot [Y_{LOr}(j\omega) \otimes X_i(j\omega) + Y_{LOi}(j\omega) \otimes X_r(j\omega)] \tag{3.26}$$

It is clear that the above result is represented graphically from the combined sidebands as they appear in Figure 3.25 after the first mixing. After that, the quadrature signals go through a complex bandpass filter, which must have a frequency response as illustrated in Figure 3.26. This will eliminate the negative frequency components and, after a second quadrature mixing stage, only the useful signal is recovered at baseband for further digital detection. We note again that, in this structure, the IF frequency does not have to be high because a low chosen IF does not create implementation problems in the subsequent complex filtering.

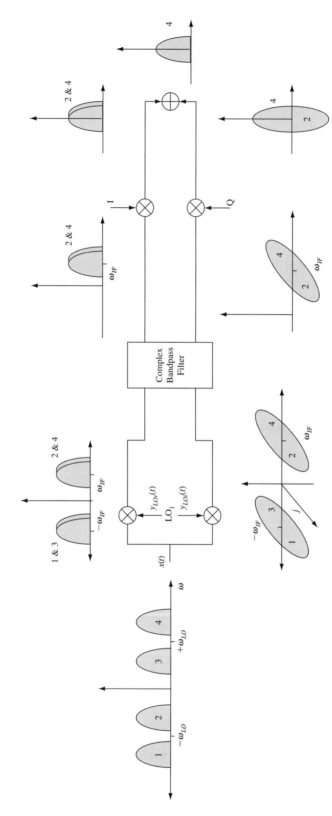

Figure 3.25 Frequency domain processing through the stages of the low-IF receiver

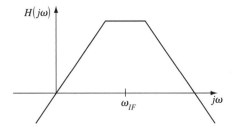

Figure 3.26 Complex bandpass filter transfer function

The most challenging aspects in the implementation of such an architecture are complex signal processing and proper selection of the IF frequency. Complex signal processing in the receiver refers to complex mixing and complex filtering. Extensive discussion was given above on complex mixing.

Let us briefly discuss some important aspects of complex filtering. The most important characteristic of such a filter is that the frequency response in positive frequencies is not related to that in negative frequencies. As a result, a bandpass filter can be designed such that it has a passband in the positive frequency domain while it eliminates the mirror negative frequency band. This is very important in low-IF receiver design because it is necessary to eliminate the translated image frequency, which is situated at the negative frequency band [Figure 3.23(c)]. These filters are usually implemented as passive or active RC-networks [Behbahani01], [Crols98].

Kluge *et al.* use a low-IF receiver architecture to design and implement a receiver for Zigbee application (corresponding to the IEEE 802.15.4 standard) [Kluge06]. A passive polyphase filter and a mixer block of four passive switching pair devices follow the LNA. Passive mixers lead to the elimination of flicker noise. To account for the gain loss, the LNA consists of two stages resulting in a differential gain of 33 dB. A two-stage polyphase filter is implemented in order to reduce the IQ signal imbalances. The overall receiver has a noise figure of 5.7 dB and a sensitivity of −101 dBm. The image rejection and IP3 are 36 dB and −16 dBm, respectively. The chip, which also includes the transmitter, was implemented using a 0.18 μm CMOS technology.

Cho *et al.* [Cho04] designed a dual-mode CMOS transceiver for Bluetooth and IEEE 802.11b. The Bluetooth receiver realizes a low-IF architecture. A common front-end is used for both Bluetooth and WLAN systems, downconverting the RF signals to 800 MHz. The complex filter connected at the outputs of the IQ downconverter implements a bandpass response for Bluetooth centred at a 2 MHz low-IF. Following the filter are limiter-amplifiers with 15 dB gain and subsequent bandpass filters to eliminate the 6 MHz harmonics produced by the limiters. The Bluetooth receiver has a sensitivity of −80 dBm at a BER of 10^{-1} and adjacent channel rejection of 30 dB at 2 MHz.

A double low-IF architecture is demonstrated in [Boccuzzi98], in which a first downconversion to 100 kHz takes place whereas the second IF is equal to 10.7 MHz. As illustrated in Figure 3.27, this system implements a Weaver architecture.

3.5.1.3 Image Rejection based on System Design

There are a few receiver designs that tackle the image rejection issue by careful consideration of the frequency planning aspect. In this way, it is possible to determine approximate values needed in image rejection.

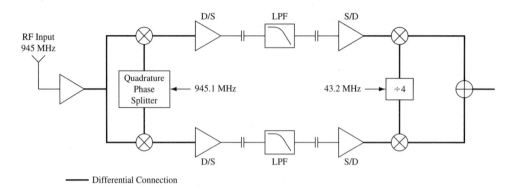

Figure 3.27 Double low-IF receiver architecture (reprinted from V. Boccuzzi, J. Glas, 'Testing the double low-IF receiver architecture', IEEE Personal, Indoor and Mobile Radio Communications (PIMRC) Conference, 1998, © 1998 IEEE)

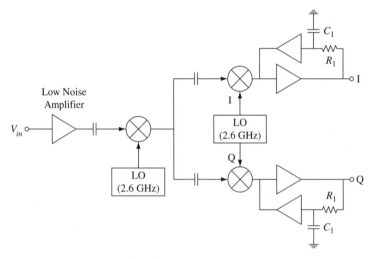

Figure 3.28 A double downconversion 5.2 GHz CMOS receiver (reprinted from B. Razavi, 'A 5.2-GHz CMOS receiver with 62–dB image rejection', IEEE J. Solid-St. Circuits, May 2001, pp. 810–815, © 2001 IEEE)

The receiver in Figure 3.28 [Razavi01] is designed to operate in the 5.2 GHz band. This architecture places the image band at DC by choosing the first IF (2.6 GHz) equal to half the RF centre frequency. The second downconversion translates the input signal to DC. In this way, the local oscillators of both mixing stages are equal to 2.6 GHz. Capacitive coupling permits for DC offset cancellation at the second frequency translation.

One advantage of this architecture comes from the fact that, owing to the placement of the image band at DC, there is high attenuation of the image at the antenna and the front-end and, as a result, there is no need for image reject filter.

Another advantage is associated with the value of the LO which is half that of the RF. This will allow for lower requirements for the VCO and the synthesizer as a system.

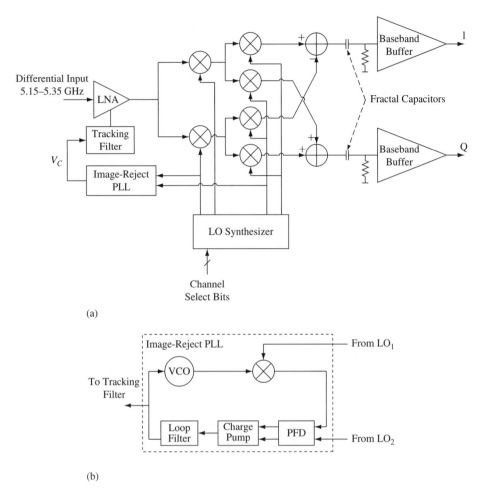

Figure 3.29 (a) Double-conversion architecture for a 5 GHz CMOS WLAN receiver, (b) image-reject PLL to tune the tracking filter

This architecture has three disadvantages, mostly associated with the high value of the local oscillators. The first one comes from the fact that low frequency noise (mostly flicker noise) situated near DC is upconverted to the passband in the RF mixer. Hence, measures must be taken to reduce the amount of generated low frequency noise in the LNA and RF mixer.

The second drawback has to do with LO–IF feedthrough in the first mixer. All this feedthrough at 2.6 GHz will fall at the centre of the 2.6 GHz first IF band. There is no way to filter this energy. To address this problem the first mixer has to be designed with very low LO–IF feedthrough. Finally, the third problem [Razavi01] stems from the difficulty of implementing channel select filters in the 2.6 GHz IF. This will transfer the design problem to the second stage of mixers, which must have very good linearity properties. Circuit design techniques in mixers and amplifiers along the receiver chain are used [Razavi01] to eliminate these problems.

Another CMOS receiver for wireless LAN at 5.2 GHz is presented in [Samavati00], [Lee02] and illustrated in Figure 3.29(a). Although the image rejection property of the receiver is based

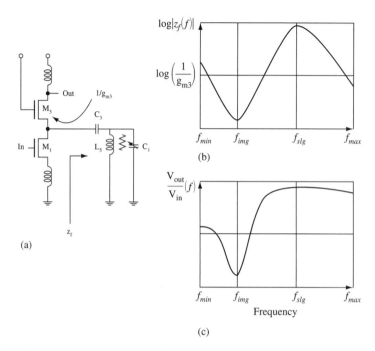

Figure 3.30 LNA with integrated notch filter (reprinted from T. Lee, H. Samavati, H. Rategh, '5-GHz CMOS wireless LANs', IEEE Trans. Microwave Theory Techniques, January 2002, pp. 268–280, © 2002 IEEE)

on the Weaver architecture, we categorize it in this class due to the peculiarity of its frequency planning. Indeed, the first local oscillator is chosen equal to 16/17 of the RF input frequency whereas the second LO is equal to 1/17 of the RF input. In this way, the image signal falls on the frequency spectrum of the downlink of a satellite system and therefore it is a weak signal. This means that the requirements for image rejection are not very high.

The heart of the image reject receiver is the double quadrature downconverter, as discussed in the low-IF receiver architecture previously (Figure 3.18). In addition, an LNA with an integrated combined notch filter is used to increase image rejection. The transfer function of a third-order filter is illustrated in Figure 3.30. One can notice that the filter is designed such that it exhibits considerable rejection at the image frequency while, at the same time, it preserves the signal level at the desired frequency. Furthermore, this notch must be tunable in frequency. This is achieved by a tracking PLL, as in Figure 3.29(b), which generates a control voltage V_C that tunes the variable capacitance (varactor) in the notch filter. Figure 3.31 illustrates the synthesizer generating the quadrature local oscillator signals for the 5 GHz WLAN receiver.

Mehta *et al.* [Mehta05], proposed an architecture for IEEE 802.11 g, based on a sliding IF frequency equal to $2/3\, f_{RF}$. The resulting IF signal is subsequently downconverted to DC by a quadrature mixer using an LO frequency at $1/3\, f_{RF}$. Owing to the front-end inductive loading and matching at 2.4 GHz, there is an image rejection of approximately 40 dB at the IF of 800 MHz. Consequently, there is no need for an image rejection filter. At the input of the receiver there is an LNA followed by an RF variable gain amplifier. They both have programmable gain steps. The baseband chain has two stages of combined DC offset DAC and Butterworth

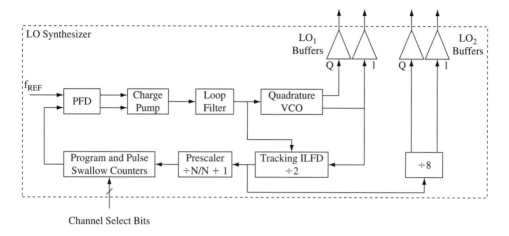

Figure 3.31   Local oscillator synthesizer for the 5 GHz WLAN receiver

filters with 41 dB programmable gain. The overall receiver is implemented in 0.18 μm CMOS technology and has a sensitivity of −73 dBm at 54 Mb/s and a noise figure of 5.5 dB.

Reynolds *et al.* [Reynolds06] presents a superheterodyne architecture which employs a frequency plan such that the variable IF frequency is 1/7 of the RF frequency. In addition, the VCO output frequency is equal to 2/7 of the corresponding RF signal. The image is located at the 5/7 of the RF frequency and this eliminates image-related problems.

3.5.1.4 Hartley and Weaver Image Reject Implementations

In this section we will present briefly the main characteristics of several recently implemented receivers, which follow the Hartley or Weaver image reject architectures. Emphasis is put here on the implementation technology and the achieved figures for the main factors such as image rejection, noise figure and power consumption.

Katzin [Katzin98], presents an image reject receiver operating in the 902–928 MHz ISM band. An image reject I/Q mixer translates the RF signal down to 10.7 MHz. Hartley architecture is implemented in this receiver, where the 90° phase shift following the mixing operation is achieved with a two-section RC polyphase network. After the I/Q phase shifter/combiner the signal goes through 10.7 MHz IF filters and amplifiers to the baseband receiver. The technology of implementation is Si bipolar and the achieved image rejection is 33 dB, whereas the noise figure of the receiver is 4.2 dB.

A similar structure is reported in [Imbornone00] (Figure 3.32) where a dual-band (900 MHz/1.9 GHz) image reject receiver is implemented in SiGe BiCMOS. The implementation includes two parallel front-end chains. The two parallel paths stop at the outputs of the quadrature mixers. These, in turn, converge to a single IF path consisting of a quadrature combiner which implements the 90° shifter in the form of a two-section polyphase filter. The IF frequency is equal to 400 MHz and the reported image rejection is 40 dB, whereas the noise figure varies between 2.8 and 4.1 dB within the edges of the frequency band of operation.

Another approach for image rejection receiver design is given in [Maligeorgos00] in which an image reject mixer is implemented operating in the 5–6 GHz band. The basic block diagram

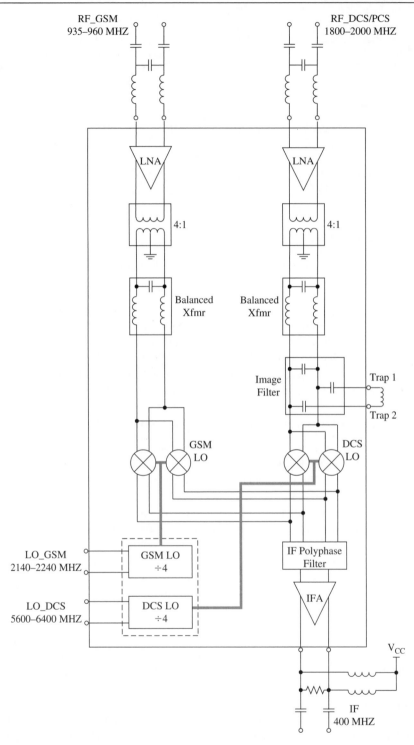

Figure 3.32 Dual-mode image reject mixer receiver in SiGe BiCMOS (reprinted from J. Imbornone, J.-M. Mourant, T. Tewksbury, 'Fully differential dual-band image reject receiver in SiGe BiCMOS', IEEE RFIC Conference, © 2000 IEEE)

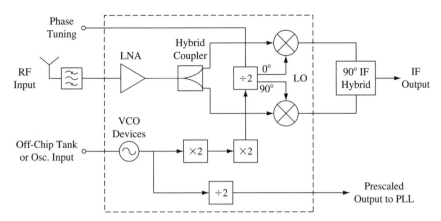

Figure 3.33 A 5.1–5.8 GHz image-reject receiver (reprinted from J. Maligeorgos, J. Long, 'A low-voltage 5–5.8 GHz image-reject receiver with wide dynamic range', IEEE J. Solid-St. Circuits, December 2000, pp. 1917–1926, © 2000 IEEE)

of the receiver is illustrated in Figure 3.33 and it follows the Hartley architecture. A frequency divider from the synthesizer block is used to generate the quadrature LO signals fed to the mixers. Specifically, the LO input at $f_{LO}/2$ is doubled twice and subsequently there is a final two-stage frequency divider which produces I and Q outputs in quadrature. In addition, there is a phase tuning input which regulates the phase relationship between I and Q outputs of the frequency divider that are applied to the LO input. By doing so, very high image rejection can be achieved. A 90° combiner is externally placed at the IF output of the mixers, which is equal to 75 MHz. The integrated receiver has been manufactured in 0.5 μm Si bipolar technology and gives image rejection better than 45 dB across the 5–6 GHz band. The measured noise figure was 5.1 dB and the power dissipation equal to 50 mW.

A low-power image rejection receiver is reported in [Ugajin02] operating at 2 GHz with an IF output frequency in the range of 10 MHz. Quadrature generation of the I and Q LO signals driving the mixers is implemented using a three-stage polyphase filter. Similarly, a three-stage polyphase filter is used at the mixer outputs to produce quadrature I and Q components at the IF signal path. The receiver was implemented in a 0.2 μm fully depleted CMOS/SIMOX technology and gave 49 dB of image rejection with a noise figure of 10 dB.

In another implementation of the Hartley architecture [Pache95], simple RC–CR networks are used in both the LO and IF paths to create the quadrature phase relationship. However, a differential RC–CR network distributed between the two IF paths (I and Q) considerably improves magnitude and phase imbalances. Figure 3.34 shows a block diagram of this implementation proposed by [Pache95]. The RC–CR at the local oscillator and the IF RC–CR blocks have the form illustrated in Figure 3.5(b). Regarding the phase tuning, the system structure allows the needed phase shift between the two paths to be obtained independently of the LO frequency variations and the selected IF frequency. On the other hand, magnitude matching in the two (I and Q) branches is achieved in two ways. First, the connection of the half RC–CR networks in each branch at the IF permits compensation of the attenuation difference in each path. The second element introduced is a voltage controlled gain (VCG) which, connected

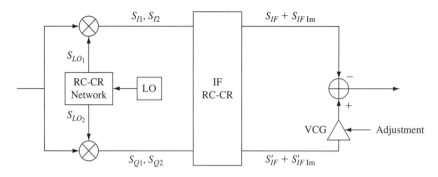

Figure 3.34   An image-reject mixer with gain control and phase shifter distributed in the *I* and *Q* paths [Pache95]

to the LO generator and the IF, permits gain compensation due to f_{LO} and f_{IF}. The following equation gives quantitatively the effects described above in terms of gain imbalance $A_{IF1} - A_{IF2}$ between the two branches:

$$A_{IF1} - A_{IF2} = A_{IF} \cdot \left[\frac{\Delta R}{R} - \frac{\Delta R'}{R'} + \frac{\Delta C}{C} - \frac{\Delta C'}{C'} + \frac{\Delta f_{LO}}{f_{LO}} - \frac{\Delta f_{IF}}{f_{IF}} \right] \qquad (3.27)$$

The above implementation gives an image rejection of 35 dB at an IF frequency of 200 MHz.

Finally, a Weaver-type image rejection receiver is designed and implemented at 2 GHz by Der and Razavi [Der03]. The main parts of the receiver are two pairs of mixers constituting the RF and IF downconverted stages. The first pair downconverts the WCDMA band of 2.11–2.17 GHz at the IF frequency. Gain control is achieved by splitting the mixers of the second downconversion stage into fixed-gain and variable-gain mixers. Phase control is realized by variable delay cells. An important feature of this receiver is that it allows for independent gain and phase calibration at the IF downconverter. The overall receiver is implemented in 0.25 μm CMOS technology and exhibits an NF of 5.2 dB and an image rejection ratio of 57 dB.

3.5.2 The Direct Conversion Architecture

In this section we will present existing development in direct conversion receiver technology. The main features in the design and implementation of representative receivers will be examined in some detail.

A DC receiver for WCDMA applications is developed and tested in [Parssinen99]. Figure 3.35 shows the block diagram of this receiver, which was designed to operate at the 2 GHz band. After the LNA, the RF signal is split to *I* and *Q* branches where it is downconverted to DC using two RF mixers. The two quadrature LO signals (0 and 90°) are produced using a second-order RC polyphase filter. The baseband section performs amplification using gain control and channel selection. It consists of a chain of amplifiers achieving 78 dB of gain control range and a fifth-order Butterworth channel selection filter. This filter can be tuned to three different bandwidths ranging from 5 to 20 MHz. In this way it can allow for chip rates of 4.096, 8.182 and 16.384 Mchips/s. In addition, the baseband section contains a feedback loop which implements the highpass filter that cuts out the DC offset component. The corner frequency is at 2 kHz. Simulations have shown [Parssinen99] that for a processing gain of 6 dB there is almost no deterioration of BER performance. The main reason for this is that, for a

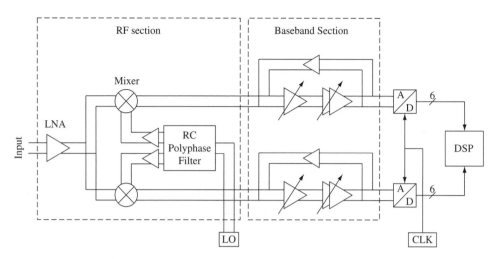

Figure 3.35 A direct conversion receiver for WCDMA (reprinted from A. Parssinen *et al.*, 'A 2-GHz wide-band direct conversion receiver for WCDMA Applications', IEEE J. Solid-St. Circuits, December 1999, pp. 1893–1903, © 1999 IEEE)

wideband system like WCDMA, the 2 kHz high-pass filter cuts off a negligible amount of the signal spectrum. Analog/digital converters with 6-bit resolution follow the baseband section in the receiver chain. The direct conversion receiver was fabricated using 0.35 μm BiCMOS technology. Its noise figure was equal to 5.1 dB whereas the overall voltage gain was equal to 94 dB and the receiver sensitivity was equal to −114 dBm (at 4 Mchips/s and BER = 10^{-3}). The *I/Q* amplitude and phase imbalances were equal to 0.6 dB and 1°, respectively. This creates a reasonably small amount of distortion in baseband signal constellation, which is satisfactory for the WCDMA modulation format. Finally, the measured input IP3 and input IP2 of −9.5 dBm and +38 dBm, respectively, demonstrate a high linearity and low values of envelope distortion.

Another direct conversion receiver front-end implementation is presented in [Liu00]. Figure 3.36 depicts a block diagram of the system which consists of an LNA, *I* and *Q* down-conversion mixers and a quadrature VCO which produces *I* and *Q* outputs with 90° phase difference. Buffer amplifiers are used at the outputs of the VCO in order to improve mixer–VCO isolation and enhance mixer linearity by increasing the power that drives the LO mixer ports. The receiver was designed to operate at the 5 GHz band and was fabricated in 0.25 μm CMOS technology. Measurements of this front-end receiver gave a noise figure of 3 dB, a conversion gain of 9 dB (50 Ω), and input IP3 and IP2 equal to −11 and +16 dBm, respectively. Finally, it must be noted that the LO–RF leakage was −60 dBm, whereas the VCO phase noise was −86 dBc/Hz at 100 kHz away from the carrier.

Direct conversion receivers have been extensively used in wideband radio systems like UWB ([Ismail05], [Roovers05], [Razavi05], [Lo06], [Sandner06]) and IEEE 802.11 WLAN ([Perraud04], [Khorram05], [Cho04]).

The UWB receiver by Roovers *et al.* uses a single-ended topology for the LNA and mixer. Because of the presence of multiple large interferers in UWB, like WLAN systems, the wideband LNA is designed for a moderate gain of 14 dB while most of the gain is realized at the baseband stage (46 dB). The LNA consists of a cascaded input stage connected to an emmiter

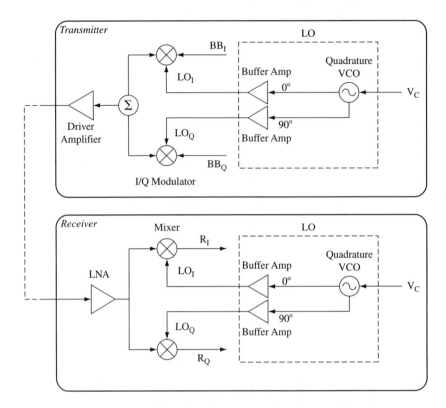

Figure 3.36 A 5 GHz CMOS radio transceiver front-end [Liu00]

follower. To achieve adequate gain a transformer is used for voltage feedback. The mixer is a Gilbert cell with a switching core of eight transistors. The overall receiver was fabricated in 0.25 μm BiCMOS technology and gave a noise figure of 4.5 dB, a maximum power gain of 59 dB and an input IP3 of −6 dBm.

A WLAN direct conversion receiver was reported by Perraud *et al.* [Perraud04]. It is a dual-band receiver (2.4/5 GHz) with two separate LNA paths corresponding to the two frequency bands. The fully differential cascaded LNAs have 1.5 dB noise figure and a gain of 20 dB. A single IQ passive mixer exhibiting high linearity is used for both bands. The LNA–mixer system has high reverse isolation (61 dB) from the mixer to the antenna. The mixers are followed by high-speed current conveyors with good linearity, whereas they have programmable gain (2–14 dB) in 6 dB steps. The overall receiver has a noise figure of 4.5 dB and an overall gain of 60 dB (at 5 GHz). The measured sensitivity is −73/−93 dBm at 54/6 Mbps.

3.6 Transmitter Architectures

The transmitter is responsible for the modulation of the information, the frequency upconversion of the modulated signal to the RF frequency and the amplification of the resulting signal such that the radio signal can meet the requirements for the maximum useful range imposed by the application.

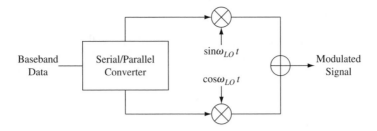

Figure 3.37 Quadrature upconversion for digital modulation

From the above, it is evident that transmitter requirements are different from receiver design tasks. In the design of the transmitter the most important issues are signal preparation (or conditioning) before modulation, modulator design to provide modulation accuracy, transmitted power and spectral content of the transmitted radio signal. Characteristics such as sensitivity, selectivity and noise are of minor importance.

3.6.1 Information Modulation and Baseband Signal Conditioning

We will very briefly address the issue of modulation and signal conditioning to demonstrate the different aspects of the generation of modulated signals. We will treat this subject systematically later in this book.

We can group modulation schemes for wireless transmission in two major categories: nonlinear (constant-envelope) and linear modulation techniques. In nonlinear techniques, information is contained only in the phase of the signal whereas linear modulation maps information in both phase and amplitude. Nonlinear modulation is used in 2G mobile communications (GSM) and Digital Enhanced Cordless Telecommunications. The modulation is called Gaussian Minimum Shift Keying (GMSK). Recent standards like 3G cellular standards (CDMA) and wireless LANs use linear modulations such as QPSK and QAM.

In all digital modulation schemes the information bit stream is fed to the next stages in quadrature (I and Q channels), as shown in Figure 3.37. This scheme is often called vector modulation and its accuracy is determined by the gain and phase imbalances (ΔA and θ) in the I and Q paths, which will produce cross-talk between the two data sequences. Quantitatively, this is given by the ratio of the power at sideband $\omega_{LO} + \omega_{in}$ to the power at $\omega_{LO} - \omega_{in}$ [Razavi98]:

$$\frac{P(\omega_{LO} + \omega_{in})}{P(\omega_{LO} - \omega_{in})} = \frac{1 - (1 + \Delta A) \cdot \cos\theta + \Delta A}{1 + (1 + \Delta A) \cdot \cos\theta + \Delta A} \tag{3.28}$$

Another way to quantitatively describe the modulation accuracy is error vector magnitude (EVM), which represents the error vector between the ideal constellation point vector and the measured vector.

It can be proved that the modulation error is:

$$\text{Modulation error } (\%) = 100 \cdot \frac{\sqrt{\left(|A| \cdot \varphi \cdot \frac{\pi}{180}\right)^2 + (\Delta A)^2}}{|A|} \tag{3.29}$$

where $|A|$ is the amplitude of the vector, and φ, ΔA represent the phase and amplitude errors, respectively, between the ideal vector and the measured one.

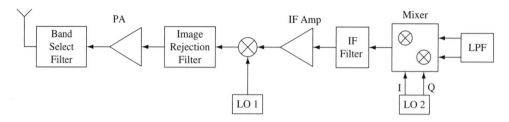

Figure 3.38 Two-stage upconversion transmitter

EVM can be calculated if the amplitude imbalance, phase error and DC offset in the vector modulator are given.

3.6.2 Two-stage Up-conversion Transmitters

Figure 3.38 illustrates the basic topology of a transmitter using two mixing stages [Rudell99]. The first local oscillator LO_2 translates the baseband signal to an IF equal to ω_{LO2} which passes through an IF filter to eliminate its harmonics. Following the IF filter, a second upconversion takes place which will translate the signal to a final RF frequency of $\omega_{LO1} + \omega_{LO2}$. A bandpass RF filter follows the mixer and its main purpose is to eliminate frequency $\omega_{LO1} - \omega_{LO2}$, which is the unwanted sideband of the transmitted carrier. After this filter, there is a power amplifier and finally an RF band-reject filter. The latter is necessary to reject any frequency outside the frequency band of the standard.

From the above discussion it is evident that the role (and consequently the characteristics) of the filters in the transmit path is quite different than that in the receiver path. The IF filter does not need to satisfy the stringent requirements on adjacent channel rejection of the respective IF filters in the receiver. Similarly, the RF filter just after the mixer must only attenuate the other sideband located $2\omega_{LO2}$ away from the RF signal (which is located at $\omega_{LO1} + \omega_{LO2}$). This is an easy task to perform since ω_{LO1} can be chosen to be relatively high. Finally, the RF band-reject filter has to eliminate internally generated out-of-band energy, which is usually considerably smaller than that at the receiving antenna.

On the other hand, upconversion mixers must exhibit high linearity, to confine the number of spurious responses at the output and low noise, to preserve a high signal-to-noise ratio at the transmitting antenna. For example, third-order distortion can result in violation of the transmit spectral mask [Razavi99]. The solution to this problem is to ensure sufficient linearity at the baseband port of the mixer.

Ahola *et al.* [Ahola04] use a two-stage upconversion transmitter. *I–Q* modulation is realized at the IF band permitting for accurate quadrature signal generation. This is achieved by using second-order polyphase filters to produce the LO quadrature signals. The measured spectral mask of the transmitted signal exhibits a margin of approximately 20 dB (outside the useful band) compared with the required specification.

Cho *et al.* [Cho04] also report a two-stage upconversion transmitter at 2.4 GHz for a dual-mode transceiver for Bluetooth and 802.11b. The first pair of mixers upconverts the signal at 800 MHz whereas a second mixing stage uses an LO of 1.6 GHz to bring the transmitted signal up to 2.4 GHz. The large separation between the RF and LO frequencies guarantees no injection pulling of the VCO signal to the power amplifier.

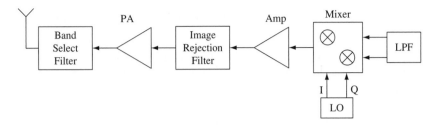

Figure 3.39   Direct upconversion transmitter

3.6.3 Direct Upconversion Transmitters

If the IF filter and the second mixer are removed, we have the direct conversion transmitter architecture, as shown in Figure 3.39. In this case the quadrature (IQ) mixer must operate at radio frequencies. This will alter EVM characteristics resulting in performance deterioration due to higher imbalances in amplitude and phase along the two paths.

Apart from this apparent differentiation in performance characteristics, there are another two aspects, which may lead to further deterioration if they are not carefully addressed. The first is associated with disturbance of the local oscillator caused by the power amplifier, which, owing to its high power and noise, will corrupt the signal of the local oscillator as they operate at the same frequency [Razavi99]. This effect is called 'injection pulling' or simply 'pulling of the VCO'. The way to address this problem is to use two VCOs to derive a transmitting carrier frequency $\omega_1 + \omega_2$ that are much different from ω_1 and ω_2 [Stetzler95].

The other issue stems from the role of the RF filter placed just before the PA which is quite different from that in the case of the two-step upconversion transmitter. Now there is no sideband to eliminate and its role is to reduce the noise power entering the PA and the harmonics from the LO so that the resulting signal is as clean as possible for transmission. If for integration purposes this filter is removed, these tasks have to be undertaken by the RF filter before the antenna. In this case, the filter will have increased insertion loss and there will be more power consumption in the whole system. In addition, the transmitted noise will be higher compared with the case when there is an RF filter before the PA as well.

A dual-band direct upconversion transmitter was implemented by Xu *et al.* [Xu05] for 802.11a/b/g WLAN. To avoid LO leakage from the power amplifiers, an LO calibration loop is used employing a leakage detector and a calibration control unit. This transmitter produced 2.5/1 dBm output power with −31/−32 dB EVM at 2.4/5 GHz.

Finally, [Darabi03] presents a dual-mode 802.11b/Bluetooth transceiver in 0.35 μm technology employing a direct conversion transmitter. To solve the problem of LO pulling by the power amplifier, the frequency planning is designed so that the frequency of the two VCOs is several hundred MHz away from the 2.4 GHz frequency at the output of the PA. The transmitter has adjustable power of −20 to 0 dBm in steps of 2 dB and an output IP3 of 16 dBm.

References

[Abidi95]: A. Abidi, 'Direct-conversion radio transceivers for digital communications', IEEE J. Solid-St. Circuits, December 1995, pp. 1399–1410.

[Ahola04]: R. Ahola, A. Aktas, J. Wilson, K. Rao, F. Jonsson, I. Hyyrylainen, A. Brolin, T. Hakala, A. Friman, T. Makiniemi, J. Hanze, M. Sanden, D. Wallner, Y. Guo, T. Lagerstam, L. Noguer, T. Knuuttila, P. Olofsson,

M. Ismail, 'A single-chip CMOS transceiver for 802.11a/b/g wireless LANs', IEEE J. Solid-St. Circuits, December 2004, pp. 2250–2259.

[Baumberger94]: W. Baumberger, 'A single-chip image rejecting receiver for the 2.44 GHz band using commercial GaAs-MESFET technology', IEEE J. Solid-St. Circuits, October 1994, pp. 1244–1249.

[Behbahani01]: F. Behbahani, Y. Kishigami, J. Leete, A. Abidi, 'CMOS mixers and polyphase filters for large image rejection', IEEE J. Solid-St. Circuits, pp. 873–887, June 2001.

[Boccuzzi98]: V. Boccuzzi, J. Glas, 'Testing the double low-IF receiver architecture', IEEE Personal, Indoor and Mobile Radio Communications (PIMRC) Conference, 1998.

[Cho04]: T.B. Cho, D. Kang, C.-H. Heng, B.S. Song, 'A 2.4 GHz dual-mode 0.18 μm CMOS transceiver for Bluetooth and 802.11b', IEEE J. Solid-St. Circuits, November 2004, pp. 1916–1926.

[Crols95]: J. Crols, M. Steyaert, 'A single-chip 900 MHz CMOS receiver front-end with a high performance low-IF topology', IEEE J. Solid-St. Circuits, December 1995, pp. 1483–1492.

[Crols98]: J. Crols, M. Steyaert, 'Low-IF topologies for high-performance analog front ends of fully integrated receivers', IEEE Trans. Circuits and Systems – II, vol. 45, March 1998, pp. 269–282.

[Darabi00]: H. Darabi, A. Abidi, 'Noise in RF-CMOS mixers: a simple physical model', IEEE J. Solid-St. Circuits, January 2000, pp. 15–25.

[Darabi03]: H. Darabi, J. Chiu, S. Khorram, H. Kim, Z. Zhou, E. Lin, S. Jiang, K. Evans, E. Chien, B. Ibrahim, E. Geronaga, L. Tran, R. Rofougaran, 'A dual mode 802.11b/Bluetooth Radioin 0.35μm CMOS', IEEE International Solid-State Circuits Conference (ISSCC'03), paper 5.1, 2003.

[Der03]: L. Der, B Razavi, 'A 2-GHz CMOS image-reject receiver with LMS calibration', IEEE J. Solid-St. Circuits, February 2003, pp. 167–175.

[Hartley28]: R. Hartley, 'Modulation System', US Patent 1,666,206, April 1928.

[Imbornone00]: J. Imbornone, J.-M. Mourant, T. Tewksbury, 'Fully differential dual-band image reject receiver in SiGe BiCMOS', IEEE RFIC Conference, 2000.

[Ismail05]: A. Ismail, A. Abidi, 'A 3.1- to 8.2 GHz Zero-IF receiver and direct frequency synthesizer in 0.18 μm SiGe BiCMOS for Mode-2 MB-OFDM UWB communication', IEEE J. Solid-St. Circuits, December 2005, pp. 2573–2582.

[Katzin98]: P. Katzin, A. Brokaw, G. Dawe, B. Gilbert, L. Lynn, J.-M. Mourant, 'A 900 MHz image-reject transceiver Si bipolar IC', IEEE RFIC Conference, 1998.

[Khorram05]: S. Khorram, H. Darabi, Z. Zhou, Q. Li, B. Marholev, J. Chiu, J. Castaneda, H.-M. Chien, S.B. Anand, S. Wu, M.-A. Pan, R. Roofougaran, H.J. Kim, P. Lettieri, B. Ibrahim, J. Rael, L. Tran, E. Geronaga, H. Yeh, T. Frost, J. Trachewsky, A. Rofougaran, 'A fully integrated SOC for 802.11b in 0.18 μm CMOS', IEEE J. Solid-St. Circuits, December 2005, pp. 2492–2501.

[Kluge06]: W. Kluge, F. Poegel, H. Roller, M. Lange, T. Ferchland, L. Dathe, D. Eggert, 'A fully integrated 2.4-GHz IEEE 802.15.4-compliant transceiver for ZigBee™ applications', IEEE J. Solid-St. Circuits, December 2006, pp. 2767–2775.

[Lee02]: T. Lee, H. Samavati, H. Rategh, '5-GHz CMOS wireless LANs', IEEE Trans. Microwave Theory Techniques, January 2002, pp. 268–280.

[Lindquist93]: B. Lindquist, M. Isberg, P. Dent 'A new approach to eliminate the DC offset in a TDMA direct conversion receiver', IEEE Vehicular Technology Conference, 1993 (VTC'93), pp. 754–757.

[Liu00]: T.-P. Liu, E. Westerwick, '5-GHz CMOS radio transceiver front-end chipset', IEEE J. Solid-St. Circuits, December 2000, pp. 1927–1933.

[Lo06]: S. Lo, I. Sever, S.-P. Ma, P. Jang, A. Zou, C. Arnott, K. Ghatak, A. Schwartz, L. Huynh, V.-T. Phan, T. Nguyen, 'A dual-antenna phased-array UWB transceiver in 0.18 μm CMOS', IEEE J. Solid-St. Circuits, December 2006, pp. 2776–2786.

[Maligeorgos00]: J. Maligeorgos, J. Long, 'A low-voltage 5–5.8 GHz image-reject receiver with wide dynamic range', IEEE J. Solid-St. Circuits, December 2000, pp. 1917–1926.

[Manstretta01]: D. Manstretta, R. Castello, F. Svelto, 'Low 1/f CMOS active mixers for direct conversion', IEEE Trans. Circuits Systems II, September 2001, pp. 846–850.

[Mashhour01]: A. Mashhour, W. Domino, N. Beamish, 'On the direct conversion receiver – a tutorial', Microwave J., June 2001, pp. 114–128.

[Mehta05]: S. Mehta, D. Weber, M. Terrovitis, K. Onodera, M. Mack, B. Kaczynski, H. Samavati, S. Jen, W. Si, M. Lee, K. Singh, S. Mendis, P. Husted, N. Zhang, B. McFarland, D. Su, T. Meng, B. Wooley, 'An 801.11g, WLAN SoC', IEEE J. Solid-St. Circuits, vol. 40, December 2005, pp. 2483–2491.

[Mirabbasi00]: S. Mirabbasi, K. Martin, 'Classical and modern receiver architectures', IEEE Commun. Mag., November 2000, pp. 132–139.

[Pache95]: D. Pache, J.M. Fournier, G. Billiot, P. Senn, 'An improved 3 V 2 GHz BiCMOS image reject mixer IC', IEEE Custom Integrated Circuit Conference, 1995, pp. 95–98.

[Parssinen99]: A. Parssinen, J. Jussila, J. Ryynanen, L. Sumanen, K. Halonen, 'A 2-GHz wide-band direct conversion receiver for WCDMA Applications', IEEE J. Solid-St. Circuits, December 1999, pp. 1893–1903.

[Perraud04]: L. Perraud, M. Recouly, C. Pinatel, N. Sornin, J.-L. Bonnot, F. Benoist, M. Massei, O. Gibrat, 'A direct-conversion CMOS transceiver for the 802.11a/b/g WLAN standard utilizing a cartesian feedback transmitter', IEEE J. Solid-St. Circuits, December 2004, pp. 2226–2238.

[Razavi97]: B. Razavi, 'Design considerations for direct- conversion receivers', IEEE Trans. Circuits Systems II, June 1997, pp. 428–435.

[Razavi98]: B. Razavi, 'RF Microelectronics', Prentice Hall, Englewood Cliffs, NJ, 1998.

[Razavi99]: B. Razavi, 'RF transmitter architectures and circuits', IEEE Custom Integrated Circuits Conference, 1999, pp. 197–204.

[Razavi01]: B. Razavi, 'A 5.2-GHz CMOS receiver with 62–dB image rejection', IEEE J. Solid-St. Circuits, May 2001, pp. 810–815.

[Razavi05]: B. Razavi, T. Aytur, C. Lam, F.-R. Yang, K.-Y. Li, R.-H. Yan, H.-C. Kang, C.-C. Hsu, C.-C. Lee, 'A UWB CMOS transceiver', IEEE J. Solid-St. Circuits, December 2005, pp. 2555–2562.

[Roovers05]: R. Roovers, D. Leenaerts, J. Bergervoet, K. Harish, R. van de Beek, G. van der Wiede, H. Waite, Y. Zhang, S. Aggarwal, C. Razzell, 'An interference-robust receiver for ultra-wide-band radio in SiGe BiCMOS technology', IEEE J. Solid-St. Circuits, December 2005, pp. 2563–2572.

[Rudell97]: J. Rudell, J.-J. Ou, T. Cho, G. Chien, F. Brianti, J. Weldon, P. Gray, 'A 1.9-GHz wide-band IF double conversion CMOS receiver for cordless telephone applications', IEEE J. Solid-St. Circuits, December 1997, pp. 2071–2088.

[Rudell99]: J. Rudell, J.-J. Ou, R. Narayanaswami, G. Chien, J. Weldon, L. Lin, K-C Tsai, L. Tee, K. Khoo, D. Au, T. Robinson, D. Gerna, M. Otsuka, P. Gray, 'Recent developments in high integration multi-standard CMOS transceivers for personal communication systems', IEEE Custom Integrated Circuit Conference, 1999.

[Samavati00]: H. Samavati, H. Rategh, T. Lee, 'A 5-GHz CMOS wireless LAN receiver front end', IEEE J. Solid-St. Circuits, May 2000, pp. 765–772.

[Sampei92]: S. Sampei, K. Feher, 'Adaptive DC-Offset compensation algorithm for burst mode operated direct conversion receivers', IEEE Vehicular Technology Conference, 1992, pp. 93–96.

[Sandner06]: C. Sandner, S. Derksen, D. Draxelmayr, S. Ek, V. Filimon, G. Leach, S. Marsili, D. Matveev, K. Mertens, F. Michl, H. Paule, M. Punzenberger, C. Reindl, R. Salerno, M. Tiebout, A. Wiesbauer, I. Winter, Z. Zhang, 'A WiMedia/MBOA-compliant CMOS RF transceiver for UWB', IEEE J. Solid-St. Circuits, December 2006, pp. 2787–2794.

[Stetzler95]: T. Stetzler *et al.*, 'A 2.7–4.5 V single chip GSM transceiver RF integrated circuit', IEEE J. Solid-St. Circuits, December 1995, pp. 1421–1429.

[Tasic01]: A. Tasic, W. Serdijn, 'An all-encompassing spectral analysis method for RF front-ends', IEEE Personal Indoor Mobile Radio Communications Conference, PIMRC, 2001.

[Ugajin02]: M. Ugajin, J. Kodate, T. Tsukahara, 'A 1-V 2-GHz RF receiver with 49 dB of image rejection in CMOS/SIMOX', IEICE Trans. Fundamentals, February 2002, pp. 293–299.

[Weaver56]: D.K. Weaver, 'A third method of generation and detection of single-sideband signals', Proc. IRE, December 1956, pp. 1703–1705.

[Xu00]: Z. Xu, S. Jiang, Y. Wu, H-Y Jian, G. Chu, K. Ku, P. Wang, N. Tran, Q. Gu, M-Z. Lai, C. Chien, M. Chang, P. Chow, 'A compact dual-band direct-conversion CMOS transceiver for 802.11 a/b/g WLAN', IEEE International Solid-State Circuits Conference (ISSCC'03), 2005, pp. 98–99.

4

Radio Frequency Circuits
and Subsystems

Figure 4.1 shows a transceiver block diagram incorporating the most important circuits and
subsystems. The main blocks of the transmitter are the modulator and the high-frequency
front-end. The modulator is used to modulate binary data in a form suitable for transmission.
The transmitter front-end is used to upconvert and amplify the modulated signal such that it
can be transmitted through the radio channel. The mixer is used to translate the baseband-
modulated signal into a high frequency signal through multiplication by a high-frequency
sinusoidal or square-wave signal. The succeeding amplifier is used for power amplification
such that the transmitted radio signal has enough power to guarantee satisfactory demodulation
at the receiver. On the other hand, the main function of the receiver front-end is to deliver at the

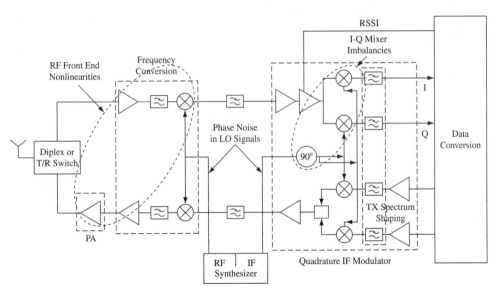

Figure 4.1 Typical radio transceiver block diagram

Digital Radio System Design   Grigorios Kalivas
© 2009 John Wiley & Sons, Ltd

digital demodulator input, a signal with sufficient SNR for successful detection. To do that, the received signal is first amplified by a low-noise amplifier in order to decrease the impact of noise in the best possible way. In turn, the mixer downconverts the radio signal into a lower frequency such that it can be processed by the digital demodulator.

The goal of this chapter is to present the basic properties and characteristics of RF circuits and subsystems such as to give the designer the basic knowledge of the behaviour of the structural elements that can be used to develop a radio front-end. More specifically we focus on issues related to noise, nonlinearities and matching that impact RF circuits and their interfaces.

4.1 Role of RF Circuits

The performance of the overall radio mostly depends on the modulation and demodulation method as well as on the behaviour of the radio channel and the properties of the coding scheme. In addition to these elements, the radio system parameters like noise figure, nonlinearities and local oscillator phase noise, play an important role in determining the system performance within the specified environment and application. More specifically, the application and the wireless network system design define the radio specifications with which the transceiver must comply.

The above two major items, RF design and digital modem parameters, must be considered jointly because RF circuit and system impairments of the radio front-end affect the overall radio performance by deteriorating the equivalent SNR at the detector input.

The major sources of RF impairments are nonlinearities, phase noise and quadrature mixer imbalances. Secondary sources of SNR deterioration are noise and signal leakage from the transmitter to the receiver for full-duplex systems, as well as limited isolation between mixer ports, which can create serious problems to certain receiver architectures like direct conversion receivers. Such problems are mainly DC offset and I–Q mismatches and are presented in detail in Chapter 3. DC offset is created by the self-mixing of the local oscillator signal and the RF signal having the same frequency. I–Q mismatches appear because of mismatches in the in-phase and quadrature branches of the receiver, as will be discussed further below.

Phase noise impacts on the receiver performance in different ways. Figure 4.2 shows a useful signal and a strong inteferer located ΔF away from it. It is illustrated how the integrated phase noise of the adjacent interferer within the useful receiver bandwidth deteriorates receiver

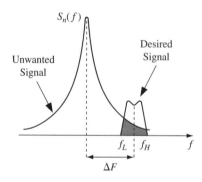

Figure 4.2 Deterioration of SNR due to phase noise of neighbouring interferer

performance. More specifically it will produce an equivalent irreducible SNR, as shown below:

$$\text{SNR} = \frac{P_{\text{sig}}}{P_{n,T}} = \frac{P_{\text{sig}}}{\int_{f_L}^{f_H} S_n(f)df} \tag{4.1}$$

where P_{sig} represents the power of the useful signal and $P_{n,T}$ is the integrated (overall) noise power within the bandwidth $f_H - f_L$ of the useful signal assuming that the phase noise spectral density $S_n(f)$ is approximately constant in this bandwidth. For this reason, the system specification imposes a limit on the phase noise by specifying the maximum level of an interferer that the system should be able to cope with.

The impact of phase noise is also realized by associating the phase noise power spectral density generated by the carrier recovery system (for example a PLL) to the corresponding jitter introduced at the digital demodulator. If the PSD of the phase noise at baseband is $S_{\text{Din}}(f)$ then the jitter is given by:

$$\varphi_{\text{rms}} = \sqrt{\int_0^{B/2} S_{\text{Din}}(f)df} \tag{4.2}$$

This jitter can deteriorate significantly the SNR depending on the modulation type as seen in Table 4.1 [Noguchi86].

Another source of performance deterioration, often identified as I–Q mismatch, can be produced by the amplitude and phase imbalances created at the quadrature mixer at the receiver (see Figure 4.1). As shown in Figure 4.3, these imbalances maybe modelled as amplitude and phase differences feeding the two mixers from the LO input [Valkama01]:

$$x_{\text{LO}}(t) = \cos(\omega_{\text{LO}}t) - ja\sin(\omega_{\text{LO}}t + \varphi) = K_1 e^{-j\omega_{\text{LO}}t} + K_2 e^{j\omega_{\text{LO}}t} \tag{4.3}$$

Table 4.1 Implementation loss due to carrier phase jitter at BER $= 10^{-6}$ [adapted from Noguchi86]

Modulation	8-PSK		16-QAM		64-QAM	
Jittter (degrees rms)	1°	1.5°	1°	1.5°	1°	1.5°
Implementation loss	0.1 dB	0.25 dB	0.2 dB	0.4 dB	1.3 dB	4 dB

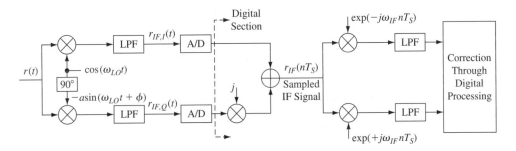

Figure 4.3 Modelling of gain and phase imbalances in quadrature mixers at the lower branch of the receiver

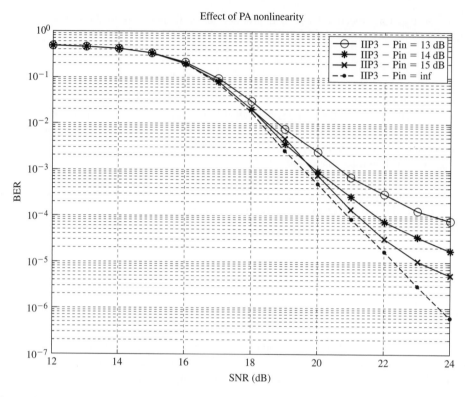

Figure 4.4 SNR deterioration due to PA nonlinearity (reprinted from B. Come *et al.*, 'Impact of front-end non-idealities on bit error rate performances of WLAN-OFDM transceivers', IEEE Radio and Wireless Conference 2000, RAWCON 2000, © 2000 IEEE)

The terms a and φ in the sinusoidal expression for $x_{LO}(t)$ represent the amplitude and phase imbalances, or in terms of K_1 and K_2:

$$K_1 = \frac{1 + ae^{-\varphi}}{2}, \quad K_2 = \frac{1 - ae^{\varphi}}{2} \tag{4.4}$$

It can be shown [Valkama01] that the received RF signal $r(t)$ and the downconverted IF signal $r_{IF}(t)$ are given in terms of the baseband signal $z(t)$ as follows:

$$r(t) = 2\mathrm{Re}\{z(t)e^{j\omega_{LO}t}\}, \quad r_{IF}(t) = K_1 z(t) + K_2 z^*(t) \tag{4.5}$$

The second term represents the unwanted 'image' which creates interference.

Consequently, the signal-to-interference (SIR) ratio is given as:

$$\mathrm{SIR} = \frac{E\{|K_1 z(t)|^2\}}{E\{|K_2 z^*(t)|^2\}} = \frac{|K_1|^2}{|K_2|^2} \tag{4.6}$$

SIR also corresponds to image rejection ratio. Practical values of amplitude and phase imbalances produce SIRs in the order of 20–30 dB [Valkama05], [Razavi98]. These levels can create serious problems in high-constellation QAM modulated signals (16-QAM and higher). For example, in 16-QAM and SIR $= 20$ dB, we get a 4 dB SNR deterioration for symbol error rate in the order of 10^{-2} [Valkama05].

Nonlinearities can also create serious performance degradation especially in linear modulations like QAM. These problems are due to third- and fifth-order intermodulation products, which as discussed in previous chapters can fall within the useful band. These IM products are considered as interference and are quantified in decibels below carrier (dBc).

As we have seen in Chapter 1, error vector magnitude is a parameter that can quantitatively express the impact of nonlinearities on the constellation diagram of a multilevel modulation like QAM. Consequently, EVM can be related to system performance in terms of bit error rate. On the other hand, EVM is associated with the allowed implementation margin.

The simplest (but not always sufficient) way to reduce the impact of nonlinearities in the transmitting power amplifier is to back it off by a few decibels. The input back-off is defined as the value that represents how many decibels lower the power P_{IN} applied at the input of the power amplifier is with respect to the $1 - $ dB compression point CP_{1dB}.

Figure 4.4 shows the SNR deterioration due to power amplifier (PA) nonlinearity when back-off is expressed as the difference of the $1 - $ dB compression point $A_{1dB} = CP_{1dB}$ and input power P_{IN} [Come00]. In this diagram the back-off values $CP_{1dB} - P_{IN}$ are 3.4, 4.4 and 5.4 dB. For back-off equal to 4.4 dB the diagram shows an SNR loss approximately equal to 2 dB.

4.2 Low-noise Amplifiers

4.2.1 Main Design Parameters of Low-noise Amplifiers

As presented in Section 4.1, the position of the LNA in the receiver just after the antenna and the RF filters makes it a critical component with respect to the noise figure of the overall receiver. In addition, it greatly affects the linearity of the receiver. In order to deliver the maximum possible power at the succeeding (or subsequent) stages of the receiver, the LNA input impedance must be conjugate matched to the impedance of the preceding stage which will, most probably, be an RF band selection filter. Moreover, to suppress noise due to subsequent stages, to minimize the noise impact of the RF filter after the antenna and to maintain a high IP3, the LNA must be designed to have a considerable power gain.

In the remaining part of this section we will examine how noise figure, matching and gain are interrelated parameters when considering the design of an LNA. Because matching and noise figure are the most important factors in LNA performance which are mostly affected by the first transistor of the LNA, we concentrate on one-transistor circuit configurations. In addition, it must be noted that the following design principles and considerations concern mostly narrowband amplifiers with operating bandwidths ranging between a few kilohertz and a few tens of megahertz.

4.2.1.1 Noise Figure Considerations and Optimization

Using a transistor noise model, expressions for noise factors as functions of noise currents and voltages can be derived (see Chapter 1). These expressions can be further manipulated to obtain equations of noise factor as a function of the minimum noise factor. Such a general expression, for any kind of transistors, is as follows [Gonzalez96]:

$$F = F_{min} + \frac{R_n}{G_S}\left[(G_S - G_{opt})^2 + (B_S - B_{opt})^2\right] \qquad (4.7)$$

where the minimum noise figure F_{min} is given by:

$$F_{min} = 1 + 2R_n\left(\sqrt{\frac{G_u}{R_n} + G_C^2} + G_C\right) \qquad (4.8)$$

The source admittance is $Y_S = G_S + jB_S$, whereas G_{opt} and B_{opt} represent the optimum values to achieve minimum noise figure.

These expressions are functions of the transistor's two-port noise parameters R_n, G_u and the correlation admittance Y_C:

$$Y_C = G_C + jB_C = \frac{i_{cn}}{U_{in}}, \quad i_n = i_{cn} + i_{un} \qquad (4.9)$$

where i_{cn} and i_{un} represent respectively the correlated and uncorrelated, to noise voltage U_{in}, components of the noise current i_n. R_n is the equivalent noise resistance associated with the spectral density $\overline{U_{in}^2}$ and G_u is the conductance due to the uncorrelated component i_{un} of i_n:

$$\frac{\overline{U_{in}^2}}{\Delta f} = 4kTR_n, \quad \frac{\overline{i_{un}^2}}{\Delta f} = 4kTG_u \qquad (4.10)$$

Considering carefully the noise models of BJT and MOSFET transistors, after computations, the relationship between two-port noise parameters and the device noise parameters is given in Table 4.2.

From all the above, Equation (4.7) leads to the conclusion that, since the noise parameters Y_C, R_n and G_u depend on the device parameters, for minimization of NF the values of G_S

Table 4.2 Two-port noise parameters for bipolar and FET transistors*

FET	Bipolar
$F_{min} = 1 + 2R_n[G_{opt} + G_C] \approx 1 + \dfrac{2}{\sqrt{5}}\dfrac{f}{f_T}\sqrt{\gamma\delta(1 - |c|^2)}$	$F_{min} = 1 + \dfrac{n}{\beta_0} + \dfrac{f}{f_T}\sqrt{K_N},$
	$K_N = \dfrac{2I_C}{V_T}(r_e + r_b)\left(1 + \dfrac{f_T^2}{f^2}\dfrac{1}{\beta_0}\right) + \dfrac{f_T^2}{f^2}\dfrac{n^2}{\beta_0}$
$G_u = \dfrac{\delta\omega^2 C_{gs}^2\left(1 - |c|^2\right)}{5g_{d0}}$	
$G_C \approx 0$	
$B_C = \omega C_{gs}\left(1 + \alpha|c|\sqrt{\dfrac{\delta}{5\gamma}}\right)$	$R_n = n^2\dfrac{V_T}{2I_C} + (r_e + r_b)$
$R_n = \dfrac{\gamma}{\alpha}\cdot\dfrac{1}{g_m}$	$R_{sopt} = R_n\dfrac{f_T}{f}\cdot\dfrac{\sqrt{K_N}}{2K_N + \frac{n^2}{2}}$
	$Y_{Sopt} \cong \dfrac{1}{R_n}\dfrac{f}{f_T}\cdot\left(\dfrac{1}{2}\sqrt{K_N} - j\dfrac{n}{2}\right)$

* For device parameters refer to section 1.5.2

and B_S must be fixed and equal to the values G_{opt} and B_{opt}, respectively. However, it is not usually possible to impose specific values on the input admittance. Consequently, the minimum noise figure does not usually entail maximum power transfer. On the other hand, assuming that matching to the optimum impedance is achieved ($G_S = G_{opt}$, $B_S = B_{opt}$) the NF is reduced when the ratio f_T/f is high. Another experimental observation is that, for a given f_T/f, the NF is minimized for a specific value of collector current [Ryynanen04].

Furthermore, for specific LNA configuration the noise figure will inevitably be a function of input and output impedances. For example, for a common emitter LNA with resistive feedback R_F between base and collector, the noise figure is given by [Ismail04], [Ryynanen04]:

$$NF \approx 10 \log \left\{ 1 + \frac{1}{(g_m R_F - 1)^2} \left[\frac{R_F^2}{R_S} \left(\frac{g_m}{2} + \frac{1}{R_F} + \frac{1}{R_L} \right) + R_S \left(\frac{g_m}{2} + \frac{1}{R_L} + \frac{(g_m R_F - 1)^2}{R_F} \right) \right] \right\}$$

(4.11)

where g_m is the small signal transconductance of the transistor and R_L is the load resistance.

In addition, dependence of noise figure on geometry and bias current can lead to noise figure minimization with respect to reduced power consumption.

4.2.1.2 Input Matching

The concept of matching is associated with achieving maximum power transfer in linear two-port (see Chapter 1). In a typical superheterodyne receiver, the input of the LNA is connected at the antenna through a duplexer or a band-select filter and its output is connected to an image-reject filter followed by a mixer as shown in Figure 4.1. The response of IR filters is very sensitive to the load they are connected to, or to the input impedance. Because front-end filters exhibit 50 Ω resistive impedance, the LNA input and output must be matched to 50 Ω. Using S-parameters, the input reflection coefficient S_{11} and corresponding return loss RL are expressed as:

$$S_{11} = \frac{Z_{in} - Z_0}{Z_{in} + Z_0}, \quad RL = 20 \log |S_{11}|$$

(4.12)

In other receiver architectures (like direct conversion) the output of the LNA may be interfaced directly to the mixer and for this reason, there is no need for output matching.

The above discussion, along with noise figure considerations in Section 4.2.1.1, verify again that input matching does not necessarily result in NF minimization. However, since input matching of the LNA is of great importance for the receiver, the procedure for analysis and design of the LNA of a certain circuit configuration involves determination of the input impedance and calculation of the noise figure under impedance matching conditions.

4.2.1.3 Linearity and Stability

Linearity is an important feature for the LNA because it helps to determine the behaviour of the amplifier when strong signals are present at interfering frequencies. The third-order intercept point, as presented in Chapter 1, is a useful parameter to characterize the nonlinearity of an amplifier. The input IP3 (IIP3) is usually a function of the device IP3 and the circuit configuration. Regarding circuit design, series feedback (resistive or inductive) in common emitter (or common source) configurations can improve linearity. In the following subsections we will briefly consider IIP3 for circuit configurations involving BJT and MOSFET transistors.

Stability is determined from the K stability factor derived from two-port network theory:

$$K = \frac{1 + |\Delta|^2 - |S_{11}|^2 - |S_{22}|^2}{2|S_{21}||S_{12}|}, \quad \Delta = S_{11}S_{22} - S_{12}S_{21} \tag{4.13}$$

For the circuit to be unconditionally stable the condition is $K > 1$ and $|\Delta| < 1$.

From the above formulas one can notice that we can have an unconditionally stable amplifier when we have good input and output matching (low S_{11}, S_{22}) and good reverse isolation (high value for S_{12}). In general, when we do not design for the maximum gain that the device can give and when the frequency of operation is not at the limits of the device abilities, we can have unconditionally stable amplifiers for reasonable matching.

From all the above, we conclude that a reasonable design approach would be to guarantee input matching as a first step. This will result in a circuit configuration for which the *NF* can be determined. Along with that, it can be examined how close the achieved *NF* is to the minimum *NF*, which can demonstrate how suitable the chosen LNA topology is for the requested performance.

4.2.2 LNA Configurations and Design Trade-offs

4.2.2.1 BJT LNAs

The four circuit configurations that can be used as the front-end transistor of the LNA are illustrated in Figure 4.5. Figure 4.5(a) and (b) uses resistance matching, and (c) is connected in common base arrangement. The most promising one, Figure 4.5(d) uses inductive degeneration at the transistor.

For the resistive matching circuit, the noise figure is given by Equation (4.11) above whereas the input resistance is:

$$R_i = \frac{R_F + R_L}{1 + g_m R_L} \tag{4.14}$$

Observing these equations, it is easy to realize that both R_i and *NF* are functions of R_F and R_L. Consequently, they are not independent of each other. To make the noise figure

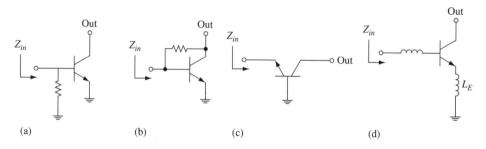

Figure 4.5   Front-end architectures of the LNA with bipolar transistor

independent of Z_{in}, the topology with inductive emitter degeneration has been investigated. For this configuration, the input impedance is:

$$Z_{\text{in}} = r_b + g_m \frac{L_E}{C_\pi} + j\omega L_E + \frac{1}{j\omega C_\pi} \tag{4.15}$$

The noise figure was given in Equation (1.153) in Chapter 1 and is repeated here for convenience:

$$F = 1 + \frac{r_b}{R_S} + \frac{1}{2g_m R_S} + \frac{g_m R_S}{2\beta(0)} + \frac{g_m R_S}{2|\beta(j\omega)|^2} \tag{4.16}$$

In this case, and discarding the last term in the above equation (being very small in comparison to the others), the minimum noise figure is calculated:

$$F_{\text{min}} = 1 + \sqrt{\frac{1 + 2g_m r_b}{\beta(0)}} \tag{4.17}$$

The corresponding optimum source resistance is given as follows:

$$R_{\text{Sopt}} = \sqrt{\beta(0) \frac{1 + 2g_m r_b}{g_m}} \tag{4.18}$$

From the above, the best approach is to design for maximum power transfer (conjugate matching). By tuning out the imaginary part in Z_{in}, we achieve matching by equating R_S to the real part of Z_{in}:

$$R_S = r_b + \frac{g_m}{C_\pi} L_E = r_b + \omega_T L_E \tag{4.19}$$

If the target of the design is to minimize the noise figure ('noise matching'), R_S should be equal to R_{Sopt} above, or a matching circuit should be inserted between the antenna and the LNA input in order to transform the antenna impedance to R_{Sopt}. If R_{Sopt} happens to be close to +50 Ω, no matching circuit is necessary. For example, in the BiCMOS amplifier designed in [Meyer94], $R_{\text{Sopt}} = 45\ \Omega$ at 1 GHz.

To achieve lower noise figure, a capacitor C_x can be connected between the base and the emitter. This approach has been investigated in [Girlando99] in the design of a cascode amplifier (as shown in Figure 4.6) for which the resulting noise figure, when matching conditions hold at the resonant frequency f_0, is:

$$F_{\text{matching}} = 1 + \frac{r_b + r_e}{R_S} \left\{ K^2 + \left[g_m R_S \frac{f_0}{f_T} \frac{1 - K}{K} \right]^2 \right\} + \frac{g_m R_S}{2} \left[\left(\frac{f_0}{f_T} \frac{1}{K} \right)^2 + \frac{1}{\beta_0} \right]$$

$$+ \frac{1}{2\beta_0 g_m R_S} \left(\frac{f_T}{f_0} K \right)^2 + \frac{4 R_S}{R_C} \left(\frac{f_0}{f_T} \frac{1}{K} \right)^2 \tag{4.20}$$

where ω_0 and K are given by:

$$\omega_0 \approx \sqrt{\frac{1}{(L_b + L_e)(C_\pi + C_\mu)}}, \quad K = \frac{C_\pi + C_\mu}{C_\pi + C_\mu + C_x} \tag{4.21}$$

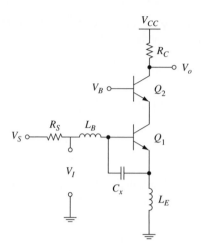

Figure 4.6 BJT cascode amplifier with capacitor connected between base and emitter

Capacitor C_x serves to implement a parallel-series matching network that helps to minimize the overall noise figure as implied by Equation (4.20). More specifically, due to the fact that K is lower than unity, the impact of r_b, r_e and I_B of the transistor on the overall noise figure is minimized [Girlando99]. Although the noise figure contributed by R_C and I_C increases, their contribution to the overall NF in Equation (4.20) is lower than that from r_b, r_e and I_B terms.

IP3 can be determined by calculating coefficients a_1 and a_2 in Equation (1.116).

$$y(t) = \sum_{n=0}^{3} a_n x^n(t) \tag{4.22}$$

In a common emitter configuration the collector current can be expressed as:

$$I_C = I_S \exp \frac{V_{BE}}{V_T} \left[1 + \frac{U_i}{V_T} + \frac{1}{2} \left(\frac{U_i}{V_T} \right)^2 + \frac{1}{6} \left(\frac{U_i}{V_T} \right)^3 + \cdots \right] \tag{4.23}$$

where V_T is the threshold voltage of the transistor.

The input IIP_3 is:

$$IIP_3 = \sqrt{\frac{4}{3} \left| \frac{c_1}{c_3} \right|} = \sqrt{8} V_T \tag{4.24}$$

4.2.2.2 MOSFET LNAs

For the MOSFET LNAs we have similar circuit configurations as for bipolar transistors. Figure 4.7(a) shows these arrangements. The shunt input resistance and the feedback resistance amplifiers (a, b) suffer from noise figure considerably higher than the minimum NF, as discussed in Section 4.2.1.1. An interesting configuration is the common-gate amplifier [Figure 4.7(c)], which exhibits input impedance that is resistive in nature ($1/g_m$ looking at the

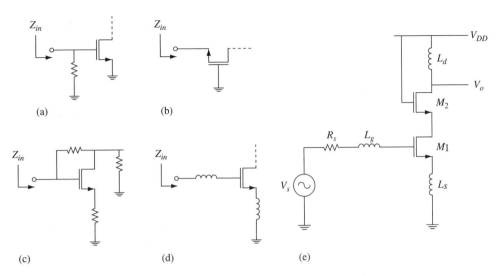

Figure 4.7 (a), (b), (c), (d) Front-end architectures of the LNA with field effect transistor; (e) LNA with inductive source degeneration

source of the device). Furthermore, the noise factor of this arrangement has a lower limit as follows [Lee98]:

$$F \geq 1 + \frac{\gamma}{a}, \quad a = \frac{g_m}{g_{do}} \tag{4.25}$$

The value of a is close to 1 for long channel devices, whereas it is reduced to 0.5 for short channel devices.

We examine below in some detail the inductive source degeneration LNA circuit [Figure 4.7(d)]. If in addition to emitter degeneration an inductor L_g is included at the gate [Figure 4.7(e)], calculation of noise factor under power and impedance matching conditions gives [Janssens02]:

$$F \approx 1 + \frac{\gamma}{a} \left(\frac{f_0}{f_T}\right)^2 \left(\frac{2}{5} + g_m R_S\right) + \frac{a\delta\left(1 - |c|^2\right)}{5g_m R_S} \tag{4.26a}$$

where δ is a coefficient related to gate noise and c is the correlation coefficient of the noise gate current and noise drain current (see Chapter 1):

$$|c| = \left|\frac{\overline{i_g i_d^*}}{\sqrt{\overline{i_g^2} \cdot \overline{i_d^2}}}\right| \tag{4.26b}$$

The two factors of the above equation represent the drain noise and the correlated gate noise, respectively. It is interesting to note that $g_m R_S$ appears as product in both expressions. Consequently, reduction of g_m is compensated for by equal increase of R_S and vice versa. Furthermore, increasing the product $g_m R_S$ increases the drain noise but reduces the correlated gate noise factor.

Figure 4.8 shows a circuit including the intrinsic gate-source capacitance C_{gs} along with an external capacitance C_d. This is introduced by the designer to alter the input quality factor Q of the circuit [Andreani01]. The input impedance of the above circuit is:

$$Z_{\text{in}} = g_m \frac{L_S}{C_T} + j \left(\omega L_T - \frac{1}{\omega C_T} \right), \quad L_T = L_g + L_S, \quad C_T = C_d + C_{gs} \tag{4.27}$$

Furthermore, considering the noise currents at the output contributed by the source resistance, the drain-noise current, the gate-noise current and their correlation, we get an improved overall noise figure [Andreani01]:

$$F = 1 + \frac{\beta_1 (Q_{\text{in}}^2 + 0.25) P^2 \left(\frac{\gamma}{a} \right)^{-1} + \frac{\gamma_1}{4} \left(\frac{\gamma}{a} \right) + \sqrt{\frac{4}{45} c P}}{R_S Q_{\text{in}}^2 g_m} \tag{4.28}$$

where Q_{in} is the quality factor of the input circuit and P is the ratio of C_{gs} to the total capacitance between the gate and source terminals. These are given as follows:

$$Q_{\text{in}} = \frac{1}{2 R_S \omega_0 C_T}, \quad P \equiv \frac{C_{gs}}{C_T}, \quad C_T = C_d + C_{gs} \tag{4.29}$$

Furthermore, the values for γ_1 and β_1 for long channel devices are $\gamma_1 = 1, \beta_1 = 8/45$ [Andreani01].

Simulations and careful consideration of Equation (4.28) show that, to minimize the noise figure, we design for the highest possible Q_{in}. For this highest practically feasible Q_{in}, an optimum transistor width W_{opt} can be obtained to achieve the lowest possible noise figure [Andreani01].

The IIP_3 for the common gate MOSFET is expressed as [Janssens02] [Lu03]:

$$IIP_3 = [IIP_3]_{\text{dev}} \times (\omega_0 C_{gs} R_S)^2 = 2 \sqrt{\left| \frac{1}{3 K'_{3gm}} \right|} \times (\omega_0 C_{gs} R_S)^2 \tag{4.30}$$

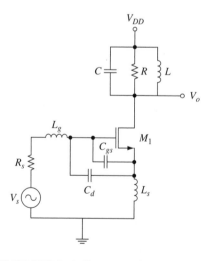

Figure 4.8   CMOS LNA including external gate-source capacitance C_d

where K'_{3gm} is the normalized third-order nonlinearity coefficient of the device and is a function of the mobility reduction, the velocity saturation and the term $(V_{GS} - V_T)$.

4.3 RF Receiver Mixers

4.3.1 Design Considerations for RF Receiver Mixers

The basic function of a mixer circuit is to translate the carrier frequency of an RF input signal to a lower or higher frequency through multiplication by a local oscillator, which is the second input of the mixer. The LO input is usually a sinusoidal signal.

However, it is not possible to have analogue ideal multiplication. Hence, the general principle of mixer operation is based on multiplication of the information bearing RF signal with a squarewave periodic signal. This is equivalent to a switching operation, as indicated in Figure 4.9, where the switch is controlled by a squarewave $S_{LO}(t)$. During the positive semi-period, the switch is connected to position 1 whereas during the negative semi-period it is connected to position 2.

The frequency of the switching function is equal to the LO frequency. If the input of this system is fed with an RF signal $V_{RF}(t)$, its output V_{IF} is expressed as a summation of terms, the first one of which contains the useful downconverted signal:

$$
\begin{aligned}
V_{IF}(t) &= V_{RF} \cos \omega_{RF} t \left[\frac{4}{\pi} \left(\cos \omega_{LO} t - \frac{1}{3} \cos (3\omega_{LO} t) + \frac{1}{5} \cos (5\omega_{LO} t) + \cdots \right) \right] \\
&= \frac{2}{\pi} V_{RF} [\cos (\omega_{RF} - \omega_{LO}) t + \cos (\omega_{RF} + \omega_{LO}) t] \\
&\quad - \frac{2}{3\pi} V_{RF} [\cos (\omega_{RF} - 3\omega_{LO}) t + \cos (\omega_{RF} + 3\omega_{LO}) t] + \cdots
\end{aligned}
\tag{4.31}
$$

The receiver mixer is used to downconvert the RF signal delivered to it by the LNA. Consequently, it must have low noise figure, good linearity and sufficient gain. Finally, it is necessary to have good LO-to-RF port isolation in order to eliminate problems created from self-mixing of the LO frequency when used in direct conversion receivers.

4.3.1.1 SSB and DSB Noise Figure

In downconversion both the useful signal band and the image frequency band are translated down to the same IF band. Even if there is no active signal at the image frequency, its thermal

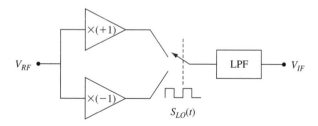

Figure 4.9 System model of mixer operation

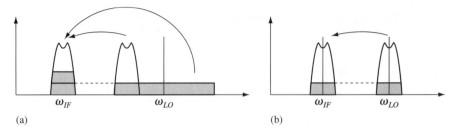

Figure 4.10 SSB and DSB noise figure in downconversion operation

noise is downconverted to the IF band. Hence, the IF band at the output of the mixer contains the useful signal, the white noise from the useful signal band and the white noise from the image frequency band, as shown in Figure 4.10(a). Hence, for a hypothetical noiseless mixer, the resulting SNR at the output of the mixer is half that of the input. This is defined as the SSB noise of the mixer. On the other hand, when only the noise within the band of the useful signal is translated down to IF, there is no noise increase at the output of the mixer. This happens when we use a direct conversion receiver where the LO frequency is the same as the RF carrier frequency. This situation is illustrated in Figure 4.10(b) and is defined as the DSB noise of the mixer. Usually the noise figure of a mixer is associated with the DSB noise figure.

Regarding gain, the voltage conversion gain G_V is defined as the ratio of the useful output voltage (at IF frequency) to the input RF voltage. Power conversion gain G_P can be defined in a similar manner. However, in general and depending on the source and load impedances, the square of the voltage conversion gain is not equal to the power gain ($G_V^2 \neq G_P$).

Nonlinearity in frequency mixers is defined as the departure of the transfer function from the ideal linear relationship [$V_{O(IF)} = KV_{IN(RF)}$ or $I_{O(IF)} = KV_{IN(RF)}$]. For example, as we see below, an active mixer operates as a transducer to transform $V_{IN(RF)}$ into an output current. In this case, it is the output current which determines the nonlinearity of the mixer. IIP3 is a useful parameter to express quantitatively mixer nonlinearity since it can be used to calculate the overall nonlinearity of the receiver.

Good isolation between the three ports is a very important aspect due to the fact that the mixer is used to translate an input frequency to another frequency (higher or lower). For example, when part of the LO signal is fed through to the RF input of the mixer (due to poor LO-to-RF isolation), it can create serious problems because by going through the LNA it can reach the antenna. Serious problems of different nature can also be created by poor LO-to-IF and RF-to-IF isolation.

4.3.2 Types of Mixers

There are two types of mixers, active mixers and passive mixers. The basic building block of active mixers is the Gilbert cell consisting of three differential transistor pairs. Passive mixers can be implemented using transistors operating as switches. They can also be based on diode circuits.

4.3.2.1 Active Mixers

Multiplier-type mixers using one or more differentially connected transistor pairs are the most commonly used type of active mixers. Figure 4.11(a) shows a single-balanced multiplier-type

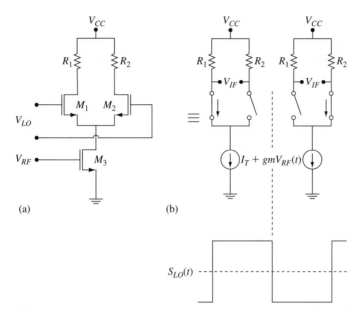

Figure 4.11 Single-balanced MOSFET mixer and its equivalent switching operation

mixer using MOSFETs. The LO input is differential whereas the RF signal is injected at the bottom transistor. The basic principle of operation is to transform RF input voltage $V_{RF}(t)$ into a time-variant current $I_{RF}(t)$:

$$I_{RF}(t) = I_{BS} + i_{RF}(t) = I_{BS} + g_m v_{RF}(t) \qquad (4.32)$$

On the other hand, the LO-driven differential pair operates in a switching manner. During the positive semi-period of V_{LO} all the drain current goes through M1 while M2 is off whereas, while V_{LO} is negative, M1 is off and all the current goes through M2. Figure 4.11(b) shows this operation in a pictorial manner substituting the transistors by switches. It is easy to see now that the circuit operates as illustrated in Figure 4.9 in the previous section and performs frequency translation by multiplying current $I_{RF}(t)$ by the switching LO current function:

$$I_{out}(t) = I_{RF}(t)S_{LO}(t) = [I_{BS} + g_m V_{RF} \cos{(\omega_{RF} t)}] \left\{ \frac{4}{\pi} \sum_{n=0} \frac{(-1)^n \cos{[(2n+1)\omega_{LO} t]}}{2n+1} \right\}$$

$$(4.33)$$

Discarding all other higher frequency harmonics except the ones at $\omega_{RF} \pm \omega_{LO}$ and ω_{LO} we have:

$$I_{out}(t) = \frac{4}{\pi} I_{BS} \cos{(\omega_{LO} t)} + \frac{2}{\pi} g_m V_{RF} \cos{(\omega_{RF} - \omega_{LO})t} + \frac{2}{\pi} g_m V_{RF} \cos{(\omega_{RF} + \omega_{LO})t} \quad (4.34)$$

The first term of the above equation represents the LO feed-through to the IF output port, the second term is the useful downconversion term whereas the third one is filtered out by IF filtering.

Since multiplication takes place in the current domain while the performance parameters are calculated in voltage terms, it is convenient to use the transconductor concept to express the time-variant current $i_{RF}(t)$. Hence, depending on the circuit topology the time-variant current is:

$$i_{RF}(t) = G_m V_{RF}(t) \tag{4.35}$$

For example, in a common-gate configuration with V_{RF} connected at the source of the MOSFET through a resistor R_S, we have:

$$G_m = \frac{1}{R_S + \frac{1}{g_m}} \tag{4.36}$$

In that case, g_m in the above expression of $I_{out}(t)$, is substituted by G_m.

The conversion gain for the single-balanced mixer is:

$$G_V = \frac{V_{out|IF}}{V_{RF}} = \frac{I_{out|IF} R_L}{V_{RF}} = \frac{2}{\pi} g_m R_L \tag{4.37}$$

Figure 4.12 shows a double balanced MOSFET mixer. Without elaborating, it is easy to show that:

$$i_{D1}(t) = I_S + \frac{1}{2} g_m V_{RF}(t), \quad i_{D2}(t) = I_S - \frac{1}{2} g_m V_{RF}(t) \tag{4.38}$$

$$i_{RF}(t) = i_{D1}(t) - i_{D2}(t) = g_m V_{RF}(t) \tag{4.39}$$

From the above we see that the incremental current $i_{RF}(t)$ is the same for both single-balanced and double-balanced mixers. It is also easy to show that the voltage conversion gain is exactly the same as in single-balanced topology above. The output voltage and conversion gains are the same for both bipolar and MOSFET mixers as calculated above.

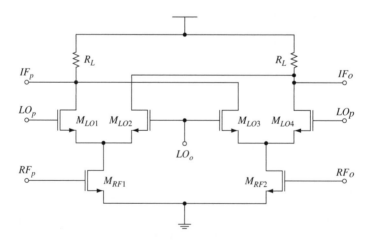

Figure 4.12   Double-balanced MOSFET mixer

4.3.2.2 Passive Mixers

Diode Mixers
Using the nonlinear transfer function of RF diodes, switching mixers can be implemented. Figure 4.13 shows a two-diode switching mixer with transfer function given by:

$$V_O(t) = V_i(t)S(t) + V_{LO}(t) = V_i(t)\left[\frac{4}{\pi}\sum_{n=0}^{\infty}\frac{\sin(2n+1)\omega_{LO}t}{2n+1}\right] + V_{LO}(t) \qquad (4.40)$$

where the switching function is:

$$S(t) = \begin{cases} 1, & V_{LO}(t) > 0 \\ -1, & V_{LO}(t) < 0 \end{cases} \qquad (4.41)$$

$V_{LO}(t)$ can be eliminated by using a double-balanced four-diode mixer as illustrated in Figure 4.14. In this case it can be shown that the output signal, when the diode resistance (at the ON state) is zero, is [Smith00]:

$$V_O(t) = V_i(t)S(t) = V_i(t)\left[\frac{4}{\pi}\sum_{n=0}^{\infty}\frac{\sin(2n+1)\omega_{LO}t}{2n+1}\right] \qquad (4.42)$$

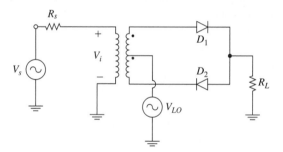

Figure 4.13   Two-diode switching mixer

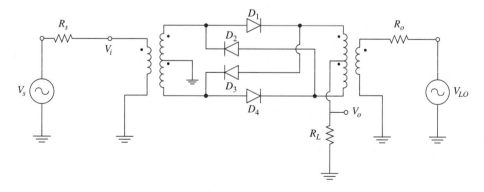

Figure 4.14   Double-balanced mixer with diodes

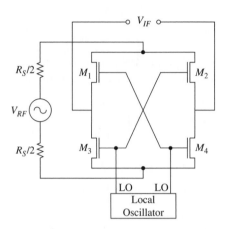

Figure 4.15 Passive mixer using MOS transistors

Furthermore, assuming ideal switching and transformer coupling, at the output we have the useful downconverted signal, the upper sideband, and spurious responses located above and below the odd harmonics of the local oscillator.

Mixers with Transistors
A similar performance with the four-diode mixer can be achieved if four MOS transistors are connected as switches as shown in Figure 4.15. The final equation is almost the same as Equation (4.42). The main difference is concerned with the nature of the switching function which is associated with the time variable nature of equivalent transconductance g_T seen from the IF terminal [Lee98].

4.3.3 Noise Figure

We use a single-balanced mixer to present the noise contributors and determine the noise figure of the circuit. The main sources of noise in this circuit are depicted in Figure 4.16. The noise term $\overline{I_{M3}^2}$ is the noise produced by the transconductor M_3 when the effect of switching in the LO differential pair is taken into account. This noise has two components, one due to the gate noise current $\overline{i_g^2}$ and one to the drain noise current $\overline{i_d^2}$ of the MOSFET. The other important noise component $\overline{I_S^2}$ is due to the switching pair M_1–M_2 noise contribution from the drain noise currents of the two transistors.

Below we present in more detail the noise contribution due to the transconductor and to the switching pair.

4.3.3.1 Noise due to the Transconductor $\overline{I_{M3}^2}$

Because of the switching operation we get a time-varying PSD $S_{nM3}(\omega, t)$. The time-average of this will produce the required PSD due to the transconductor noise [Heydari04]:

$$\overline{S_{nM3}(\omega, t)} = \sum_{k=-\infty}^{\infty} |p(k)|^2 S_{nM3}(\omega - k\omega_{LO}) \tag{4.43}$$

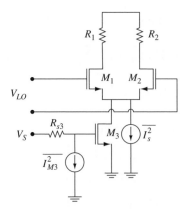

Figure 4.16   Single-balanced $\overline{I_{M3}^2}$ mixer including noise sources

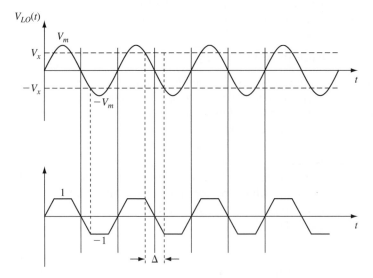

Figure 4.17   Trapezoidal waveform representing the switching process due to the local oscillator input $V_{LO}(t)$

where $p(t)$ is the periodic trapezoidal waveform representing the switching process of the M_1–M_2 pair, as shown in Figure 4.17. The fact that this waveform is not a squarewave reflects the situation where the transition from the ON to the OFF state of transistors M_1 and M_2 is not instantaneous. Consequently, there is a period of time Δ during which both transistors are ON. When the LO voltage $V_{LO}(t)$ is within $+V_x$ and $-V_x$, the two transistors are both contributing current to the overall tail current.

Using the transistor noise model and after tedious mathematical manipulations the term inside the summation $S_{nM3}(\omega)$ can be calculated for the gate noise current $\overline{i_{ng}^2}$ and is given as [Heydari04]:

$$S_{nM3}(\omega) = 4KTg_{m3}^2 R_{SG3} \tag{4.44}$$

Calculation of the average PSD in Equation (4.43) produces:

$$\overline{S_{nM3}(\omega,t)} = 4KTg_{m3}^2 R_{SG3} \sum_{k=-\infty}^{\infty} |p(k)|^2 \cong 4KTg_{m3}^2 R_{SG3} \tag{4.45}$$

The last is true because it can be shown that $\sum_{k=-\infty}^{\infty} |p(k)|^2$ is approximately equal to 1 [Heydari04].

Similarly, the term $S'_{nM3}(\omega)$ due to the drain noise current $\overline{i_{nd}^2}$ is determined to be:

$$\overline{S'_{nM3}(\omega)} = 4KT\gamma g_{do} \tag{4.46}$$

The overall noise voltage PSD at IF output from M_3 and load resistor R_L is:

$$\overline{S_{nM3T}(\omega,t)} = (4KTg_{m3}^2 R_{SG3} + 4KT\gamma g_{do})R_L^2 + 4KT \times (2R_L) \tag{4.47}$$

We must note here that we assume the correlation between noise sources $\overline{i_{ng}^2}$ and $\overline{i_{nd}^2}$ has not been taken into account. However, in most cases there is a correlation factor between the two noise sources that makes the analysis more complicated.

4.3.3.2 Noise due to the Switching Pair $\overline{I_{Sn}^2}$

The switching pair contributes noise only for the time period during which transistors M_1 or/and M_2 are ON. Ideally, only one of the two transistors is ON at a certain period of time. However in practical situations where both transistors M_1 and M_2 are ON for a period of time Δ, they both contribute to the IF output noise from the switching pair during that time period. In this case, noise from the switching pair can be higher than the noise from the transconductor. Therefore two effective ways to reduce this noise term are: (a) to use high LO amplitude; and (b) to reduce the drain currents I_{D1} and I_{D2}. Practically, an amplitude of 1V of $V_{LO}(t)$ is sufficient to guarantee instantaneous switching (reduce Δ to zero) in MOS transistor pairs [Razavi98].

Through mathematical formulation we can calculate the summation of the individual drain noise currents of M_1 and M_2:

$$\overline{I_{Sn}^2} = S_{Sn}(\omega,t) = \overline{I_{D1}^2} + \overline{I_{D2}^2} \tag{4.48}$$

However this is a time-varying function and the average (in time) PSD is calculated [Heydari04]:

$$\overline{S_{Sn}(\omega,t)} = 4KT\left(\frac{2I_{SS}}{\pi V_0}\right)\left\{\gamma\left(\frac{\omega_p}{\omega_z}\right)^2 + \left[\left(\frac{1.16I_{SS}}{\pi V_x}\right)R_{SGM1M2} + \gamma\left(1 - \left(\frac{\omega_p}{\omega_z}\right)^2\right)\right] \right.$$
$$\left. \times \left(\frac{\omega_p}{\omega_{LO}}\right)\tan^{-1}\left[\frac{\omega_{LO}}{\omega_p}\right]\right\} \tag{4.49}$$

where ω_p and ω_z are the pole and zero of the PSD and R_{SGM1M2} represents the series resistance of the high frequency model of the two transistors.

From all the above, the resulting SSB noise figure of the mixer can be determined:

$$NF_{SSB} = \frac{S_{nIF}(\omega)}{GS_{nRs}(\omega)} = \frac{\overline{S_{nM3T}(\omega,t)} + \overline{S_{Sn}(\omega,t)} + 4KT/R_L}{\left(\frac{2}{\pi}g_{m3}R\right)^2(4KTR_S)} \tag{4.50}$$

where the power gain G is given as:

$$G = \left(\frac{2}{\pi}g_{m3}R\right)^2$$

In general, and in agreement to the above considerations, to reduce the noise term due to the switching pair, for both bipolar and MOS single-balanced mixers mostly two rules must be taken into account: reduce the collector (or drain) currents and modify the size of the devices to accommodate higher LO swing.

4.3.4 Linearity and Isolation

In single-balanced and double-balanced active mixers the RF input transistors and their IP_3 characteristics determine the overall circuit nonlinearity. In addition, like in low noise amplifiers, the existence of feedback can improve linearity. Furthermore, resistive degeneration at the bottom differential pair of the Gilbert cell can satisfy the requirement for high IIP3 at the mixer stage of the receiver.

Comparing passive and active mixers in terms of linearity behaviour, the former are superior, especially for strong LO input. In contrast, the required LO driving power for the Gilbert cell is considerably lower compared with passive mixers, thus resulting in low power consumption.

Regarding LO-to-RF isolation double balanced mixers have inherently better performance. However, isolation also depends on the type of the interface of the mixer to the LO input and to the IF output as well. Considering the interface to the IF output, the main problem comes from the fact that in many cases the following IF filter is single ended. In that case, the solution can be a balanced-to-unbalanced transformer circuit interconnecting the mixer output to the filter input. On the other hand, LO feedthrough to the RF input is produced by capacitive coupling between the LO and RF inputs. One way to address this issue is to insert a buffer between the local oscillator output and the differential LO input to the mixer [Toumazou02].

4.4 Oscillators

4.4.1 Basic Theory

As presented in Section 4.1, the impact of phase noise of the local oscillator on transceiver performance results in considerable deterioration in the BER performance of the composite system. Frequency generation at the LO is based on RF oscillators, which constitute a major part of the overall LO system. To reduce phase noise, the oscillator is locked through a feedback system such as a phase-locked loop. In addition, it must provide the capability for varying the LO frequency in a programmable fashion. Furthermore, RF oscillators must exhibit low phase noise and must have the ability to tune over a considerable frequency range.

In general, oscillators are electronic circuits transforming DC power into signal power in the form of a periodic waveform. Harmonic oscillators produce a sinusoidal waveform and they consist of a resonant tank which is the basic element of oscillations and a mechanism used to counteract the losses which would lead to cessation of oscillation. Consequently, along with the resonant tank, oscillators include a positive feedback mechanism or a negative resistance element to support oscillations.

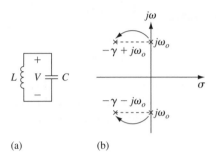

(a)   (b)

Figure 4.18   (a) Ideal LC oscillator; (b) location of the roots of lossless and lossy oscillator

However, ideal oscillators consist only of a lossless resonant tank. More specifically, an ideal LC tank, as shown in Figure 4.18(a), is governed by the following differential equation:

$$\frac{d^2V}{dt^2} + \frac{1}{LC}V = 0 \tag{4.51}$$

where $1/LC$ represents the square of the resonant frequency:

$$\omega_0 = 1/\sqrt{LC}$$

In the Laplace transform domain the corresponding equation is:

$$s^2 + \omega_0^2 = 0 \tag{4.52}$$

which gives roots s_1, s_2 located at the imaginary axis in the cartesian coordinate system as shown in Figure 4.18(b):

$$s_{1,2} = \pm j\omega_0 \tag{4.53}$$

Regarding nonideal oscillators, losses in circuit elements modify the characteristic equation and the corresponding roots as follows:

$$s^2 + 2\gamma s + \omega_0^2 = 0 \tag{4.54a}$$

$$s_{1,2} = -\gamma \pm j\sqrt{\omega_0^2 - \gamma^2} \approx -\gamma \pm j\omega_0 \tag{4.54b}$$

The roots now move to the left-half plane (LHP) of the imaginary axis at the complex plain as illustrated in Figure 4.18(b). The positive feedback or the negative resistance mechanisms will help the system bring the roots back on the imaginary axis.

4.4.1.1 Feedback Approach to Oscillation

Figure 4.19 shows a positive feedback system including an amplifying element A_U in the forward path and a network A_B in the feedback path. Its transfer function (system gain) is:

$$A_T = \frac{V_{out}}{V_i} = \frac{A_U}{1 - A_U A_B} \tag{4.55}$$

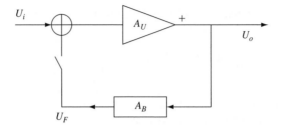

Figure 4.19 Amplifier with positive feedback input

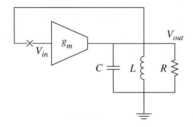

Figure 4.20 Positive feedback oscillator with transconductor

The above equation gives the Barkhausen criterion that must be fulfilled for oscillations to be sustained:

- The phase shift around the loop must be exactly 180°: $\angle\{A_U A_B\} = 180°$.
- The overall gain around the loop must be exactly equal to 1: $|A_U A_B| = 1$.

A representative example of a positive feedback oscillator is illustrated in Figure 4.20, where a transconductor is connected to an RLC resonator consisting of three elements (resistor, R, inductance, L, capacitance, C) connected in parallel. If the feedback connection is disrupted at point V_{in} we have for the transfer function:

$$T(s) = \frac{V_{out}}{V_{in}} = g_m Z_t(s) \tag{4.56}$$

where impedance Z_T is given as follows:

$$Z_T(s) = \frac{LRs}{LCRs^2 + Ls + R} \tag{4.57}$$

The Barkhausen criterion is:

$$|T(j\omega)| + 1 = 0$$

$$\angle\{T(j\omega)\} = 180° \tag{4.58}$$

which results in:

$$1 + g_m Z_T(s) = 0 \Rightarrow \frac{LCRs^2 + L\left(1 - g_m R\right)s + R}{LCRs^2 + sL + R} \tag{4.59}$$

The roots s_1, s_2 of the denominator will reveal the properties of the system:

$$s_{1,2} = \frac{g_m R - 1}{2RC} \pm \sqrt{\frac{(1 - g_m R)^2}{4(RC)^2} - \frac{1}{LC}}$$

(4.60)

From the above expressions for s_1, s_2 we observe that for $g_m = 1/R$ the two poles are purely imaginary, satisfying the condition for oscillation. The value of the transconductance g_m depends on the voltage across the tank and for this reason in the beginning the circuit elements are selected such that $g_m > 1/R$. In this case, the poles are located at the right-half plane (RHP), resulting in exponential increase of the amplitude of oscillations. This results in decrease of the value of g_m. Subsequently, the poles are shifted to the left-half plane (LHP), resulting in continuous decrement of the oscillations. This, in turn, will bring increase of the value of g_m bringing again the poles at the RHP as the phenomenon repeats. This behaviour of the circuit results in sustained oscillations.

4.4.1.2 Negative Resistance Approach to Oscillation

The circuit of Figure 4.21(a) shows a circuit on which addition of an active nonlinear element (like a tunnel diode) of negative resistance across the resonant tank results in elimination of the losses of the inductor and capacitor. This is translated to bringing the poles from the LHP to the RHP, ensuring in this way the start of oscillations. The mechanism that sustains oscillations is very close to the one described above for the feedback oscillators.

Figure 4.21(b) shows an I–V characteristic of such a diode where one can notice the region of negative slope producing the negative resistance of the element. If the I–V characteristic is $I = f_1(V)$ and the voltage and current variables translated to the centre of the I–V diagram are $U' = V - V_{DD}$ and $i' = I - f_1(V_{DD}) = f_2(U')$, the Kirchoff current law gives the following differential equation and corresponding characteristic equation [Pederson91]:

$$LC\frac{d^2 U'}{dt^2} + L\frac{d}{dt}\left[\frac{U'}{R} + f_2(U')\right] + U' = 0$$

(4.61)

$$(LC)s^2 + L\left(\frac{1}{R} - a\right)s + 1 = 0$$

(4.62)

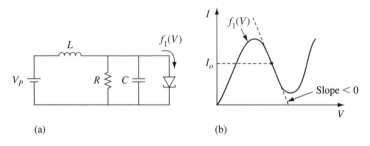

(a)   (b)

Figure 4.21   Oscillator employing nonlinear diode for negative resistance generation

where a is the slope of the I–V characteristic defined as:

$$a = -\frac{df_2(U')}{dU'}\bigg|_{U'=0} \tag{4.63}$$

Consequently, the roots s_1, s_2 of the characteristic equation are [Pederson91]:

$$s_{1,2} = -\left(\frac{G-a}{2C}\right) \pm j\sqrt{\frac{1}{LC} - \left(\frac{G-a}{2C}\right)^2} \tag{4.64}$$

From the above equation it is clear that the relation between the slope a of the I–V characteristic and the losses (represented by R) make the operation of the circuit as oscillator feasible. Specifically, for $a > 1/R$ the roots lie in the right-half plane and, as discussed above, this produces exponentially growing oscillations.

4.4.1.3 Quality Factor

Another concept directly related to a resonant circuit is that of the quality factor Q. As noted above, a type of oscillator consists of an active device connected to a resonant circuit in the form of a parallel RLC. The active device provides the energy necessary to overcome the losses of the resonant circuit and sustain oscillation.

The Q is associated with the losses of the resonant circuit. Formally, Q represents the ratio of the average energy stored in the electric and magnetic fields to the energy loss per time period. Mathematical manipulation shows that Q is also equal to the ratio of the resonant frequency ω_0 to the 3-dB bandwidth of the frequency response:

$$Q = \frac{\omega_0}{\Delta\omega_{3dB}} \tag{4.65}$$

Qualitatively, this corresponds to how sharp is the magnitude of the circuit frequency response. Calculations of the stored and dissipated energy result in the following formula for a parallel RLC resonant circuit [Pozar05]:

$$Q = \frac{R}{\omega_0 L} = \omega_0 RC \tag{4.66}$$

4.4.2 High-frequency Oscillators

4.4.2.1 Topologies of LC Oscillators

The negative transconductance of an LC-tank oscillator is inversely proportional to the quality factor Q_T of the LC tank. This, along with the bandpass nature of the LC resonant circuit, makes LC oscillators having considerably lower phase noise compared with other approaches like the ring oscillators, which involve switching mechanisms [Craninckx95]. In Section 4.4.3 the phase noise and signal quality of oscillators are considered in more detail.

Figure 4.22(a) illustrates one of the basic oscillator topologies called the Colpitts oscillator. It comprises an LC tank connected at the collector of a transistor. To implement a positive feedback loop the LC tank is also connected to the emitter of the transistor. In this connection, to satisfy the Barkhausen criterion requiring an open loop gain equal to 1, an impedance

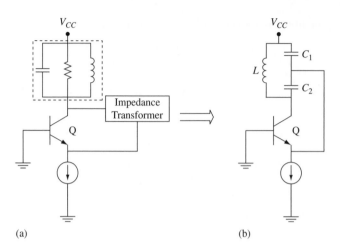

Figure 4.22 LC oscillator topology and equivalent Colpitts configuration [Razavi98]

transformation must take place to increase the impedance seen by the emitter [Razavi98]. This can be done by using a capacitive divider as shown in Figure 4.22(b). In this way the tank is transformed into an equivalent parallel LC having equivalent resistance $R_T \approx (1 + C_1/C_2)^2/g_m$ and capacitance $C_T \approx C_1 C_2/(C_1 + C_2)$ [Razavi98].

The moderate increase of the value of the inductance will increase the amplitude of the output voltage. When an inductive divider is used to connect the feedback path to the emitter of the transistor, we have the Hartley oscillator, which is obtained by interchanging capacitors and inductors in a Colpitts oscillator. Basic transistor noise considerations dictate that, for reduced noise, the size of the transistor must be increased while at the same time the bias current must be kept low [Gray01].

4.4.2.2 Voltage-controlled Oscillators

To turn an LC oscillator into a VCO, a varactor diode is inserted in the resonance tank. Varactors are reverse-biased RF diodes in which the capacitance changes as a function of an applied DC voltage. Figure 4.23(a) shows a tank implementation with the varactor in parallel and the varactor equivalent circuit. The junction resistance R_j takes high values. Figure 4.23(b) illustrates the junction capacitance as a function of the applied voltage. Tuning the diode in the almost-linear part of the curve will give an approximately linear VCO characteristic.

As presented in Chapter 1, VCOs are an essential part of a PLL. In addition, the VCO frequency changes linearly as a function of the tuning voltage V_t:

$$\omega_0 = \omega_{FR} + K_{VCO} V_t \tag{4.67}$$

where K_{VCO} is the tuning sensitivity of the VCO and ω_{FR} is the free running VCO output frequency.

Furthermore, the output signal of a VCO is expressed as:

$$y(t) = A \cos \left[\omega_{FR} t + K_{VCO} \int_{-\infty}^{t} V_t dt \right] \tag{4.68}$$

indicating that the VCO can be used for implementing direct frequency modulation.

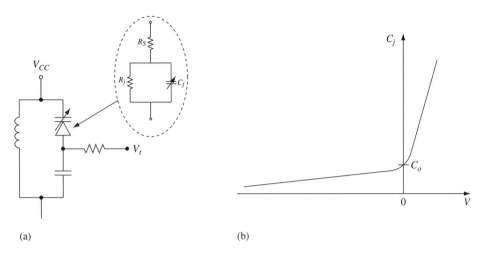

Figure 4.23 (a) VCO tank with varactor equivalent circuit; (b) varactor characteristic as function of input voltage

4.4.2.3 Design of LC Oscillators

Because of superior phase noise properties, harmonic oscillators using LC resonators are more popular than other configurations. A preferred topology for implementation of LC oscillators is the cross-coupled differential transistor pair illustrated in Figure 4.24, also known as the negative g_m oscillator. The reason for this name is that this layout gives a negative input resistance equal to $R_{\text{in}} = -2/g_m$, looking into the collectors of the two transistors [Razavi98]. According to the above, if the resonant circuit is designed such that its equivalent resistance satisfies $R_{\text{eq}} \geq |R_{\text{in}}|$ then the resulting circuit becomes an oscillator. Cross-coupled pairs have frequently been used over the last decade to design Colpitts oscillators [Li05], which exhibit lower phase noise compared with other approaches.

4.4.3 Signal Quality in Oscillators

The signal quality in oscillators is directly related to the phase noise and the total quality factor of the circuit, as explained below. Figure 4.25(a) shows that in a feedback oscillator there can be two sources of noise: the first one is introduced in the RF signal path and the other appears at the control port. We employ this simple model to demonstrate the effect of a noise term $n(t)$ appearing at the input signal of the system. Noise at the control input induces mostly FM noise [Razavi98] and for reasons of brevity will not be considered here.

If $G(s)$ denotes the open-loop transfer function, the closed loop transfer function is:

$$\frac{Y(s)}{N(s)} = \frac{G(s)}{1 - G(s)} \tag{4.69}$$

and $G(\omega) = |G(\omega)| \exp{(j\varphi)}$.

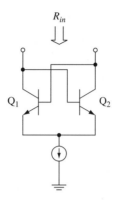

Figure 4.24 Cross-coupled transistor pair (negative $-g_m$) oscillator

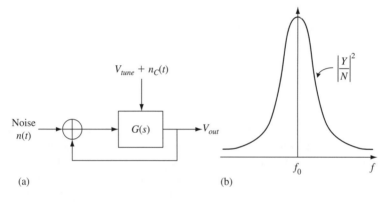

(a)   (b)

Figure 4.25 (a) Noise mechanisms in a feedback oscillator; (b) corresponding phase noise spectrum

At a frequency $f = f_0 + \Delta f$ located Δf away from the resonant frequency (frequency of oscillation), after some manipulation, it can be shown that Equation (4.69) gives a phase noise spectrum as shown in Figure 4.25(b) [Razavi98]:

$$L_T(\Delta\omega) = \left| \frac{Y(j\omega)}{N(j\omega)} \right|^2 = \frac{\omega_0^2}{4(\Delta\omega)^2} \left[\frac{\omega_0}{2} \left(\frac{d\varphi}{d\omega} \right)^2 \right]^{-1} = \frac{1}{4Q^2} \left(\frac{\omega_0}{\Delta\omega} \right)^2 \qquad (4.70)$$

In a similar fashion, the transconductance can be expressed as [Kinget99]:

$$g_m = \frac{1}{R_{\text{Peq}}} = \frac{1}{Q_T \sqrt{\frac{L}{C}}} \qquad (4.71)$$

Figure 4.26 illustrates the equivalent circuit of an LC resonator. R_{Peq} is the equivalent parallel resistance of the resonator to account for losses of reactive elements [Kinget99]:

$$R_{\text{Peq}} = R_\text{P} || R_{\text{CP}} || R_{\text{LP}} \qquad (4.72a)$$

$$R_{\text{LP}} = \omega L Q_\text{L}, \quad R_{\text{CP}} = Q_\text{C}/\omega C \qquad (4.72b)$$

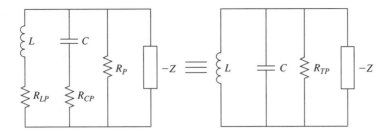

Figure 4.26 Equivalent circuits of LC resonator with negative impedance

where Q_L, Q_C represent quality factors of the branches including L and C, respectively. The total quality factor Q_T of the resonator is given by:

$$Q_T = \frac{R_{\mathrm{Peq}}}{Z_0} \text{ or } \frac{1}{Q_T} = \frac{Z_0}{R_P} + \frac{1}{Q_L} + \frac{1}{Q_C} \tag{4.73}$$

where Z_0 is defined as an equivalent characteristic impedance of the resonator:

$$Z_0 = \omega L = \sqrt{\frac{L}{C}} \tag{4.74}$$

For the overall noise calculation, the noise current of the negative transcoductance and the noise current due to losses within the resonator are taken into account. The resulting expression is [Kinget99]:

$$L_T(\Delta\omega) = kT \left(\frac{\sqrt{L/C}}{Q_T} \right) \left(\frac{\omega_0}{\Delta\omega} \right)^2 \frac{1+\alpha}{V_{\mathrm{RMS}}^2} \tag{4.75}$$

where α is an excess factor associated with noise contributed by the negative resistance. It is interesting to note that this expression is similar to Equation (4.70) above.

From all the above it is concluded that the phase noise of an oscillator is directly associated with the quality factor Q_T and the amplitude of oscillation V_{RMS}.

4.5 Frequency Synthesizers

4.5.1 Introduction

As show in Figure 4.27, the synthesizer is used to produce high frequency sinusoidal signals of high quality and accuracy, to be used for downconversion of RF received signals to baseband. Similarly, the synthesizer is used to perform upconversion of the baseband modulated signals at the transmitter. The quality and accuracy of the generated signals are directly related to the impact of phase noise and frequency stability on the communication system performance as presented in Section 4.1. Another important feature of the synthesizer is the minimum step between generated frequencies, also called frequency resolution. This is dictated by the application and is related to the number of channels required by the radio system.

In most cases, synthesizers, in order to fulfill the above requirements, constitute feedback systems which generate the output frequencies by locking the output of a VCO to a frequency

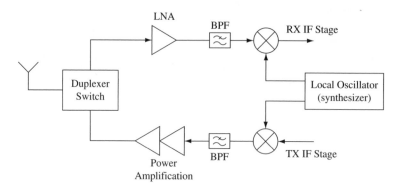

Figure 4.27   Front-end transceiver system

reference input f_R. This is done by using a phase-locked loop as the basic system for frequency locking.

The target application usually dictates the requirements for a synthesizer, which are frequency resolution, frequency stability, frequency range, speed of frequency change (settling time), phase noise and spurious responses.

Frequency resolution and output frequency range are the principle factors directly dictated by the application. For example, in a GSM application the channel bandwidth (and hence the resolution) is 200 kHz and the system operates in the 890–915 MHz band. On the other hand, wireless LANs (WLAN) require a frequency resolution of 20 MHz operating at the 2.4 or 5.5 GHz band. The settling time of the synthesizer plays an important role in the system performance and is defined as the time needed for the system to settle to a new frequency within a specified accuracy (for example 2% of the final frequency). This is determined by the order of the PLL and, as we shall see below, is inversely proportional to the two parameters of the loop; the natural frequency ω_n and the damping factor ζ.

As explained in Section 4.1, phase noise deteriorates the receiver performance significantly because the phase noise spectrum of the LO will appear in the useful band through downconversion of a close and strong interferer (see Figure 4.2). The shape of phase noise is determined by the configuration and bandwidth of the synthesizer, as we will see later in this section. Finally, a sideband spur of the LO can lead to received SNR deterioration. This occurs because, in a similar arrangement of the useful signal and an interferer to that shown in Figure 4.28, the sideband of the LO can be downconverted on the same ω_{IF}, leading to performance deterioration.

Because of the great importance of the frequency synthesizers in conjunction with PLLs in transceiver systems, there are various textbooks devoting lengthy parts to frequency synthesizers. In this section we present the most important aspects, design principles and implementation approaches of frequency synthesizers in reasonable depth. In presenting the synthesizer architectures in Section 4.5.3 we follow [Razavi98], which considers in a concise way the most important approaches in synthesizer design.

4.5.2 Main Design Aspects of Frequency Synthesizers

To present the basic principles associated with frequency resolution and settling speed we use the simpler and most popular form of the indirect synthesizer, the integer-N synthesizer. It is

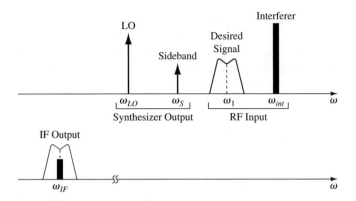

Figure 4.28   Effect of synthesizer sideband on IF signal at the receiver (adapted from [Razavi97])

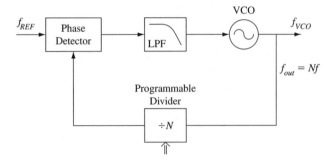

Figure 4.29   Typical integer-N synthesizer

depicted in Figure 4.29 and comprises a PLL having a programmable divider inserted in its feedback loop. The output frequencies synthesized by this system are given by:

$$f_{\text{out}} = Nf_{\text{REF}} = N_{\text{Low}}f_{\text{REF}} + nf_{\text{REF}}, \quad n = 0, 1, 2, \ldots, n_F \qquad (4.76)$$

The above equation indicates that the lower output frequency is $N_{\text{Low}}f_{\text{REF}}$, whereas the highest is given by $(N_{\text{Low}} + n_F)f_{\text{REF}}$. To impose the low and the high output frequencies, a pulse swallow frequency divider can be used, as we shall see further below in this chapter. However, the main issue is that the resolution of the synthesizer is equal to f_{REF}, which is also the channel spacing for the application in which the system is used.

On the other hand, to guarantee stability, the PLL bandwidth must be considered jointly with f_{REF}. For example, in a type-II PLL, this bandwidth should be approximately $f_{\text{REF}}/10$ [Razavi98]. Furthermore, as explained in Chapter 1, the bandwidth of the loop is inversely proportional to the settling time of the system. Hence, the requirement of a fast settling loop is contradictory to narrow channel spacing.

From the above it is deduced that a frequency synthesizer is mainly characterized by its output frequency range, frequency resolution and speed of locking to a new frequency. To briefly examine the relation of frequency resolution and output range we consider the most frequently used integer-N architecture for which f_{out} is given by: $f_{out} = Nf_{REF}$. From this, one can easily see that f_{out} is limited by the maximum and minimum values of N and by the value of the input reference frequency f_{REF}. To enhance the range of programmable ratios, a digital circuit can be used that can implement a multiplication factor N consisting of two or more terms. In addition, to enhance the operation frequency of the synthesizer, apart from a high-frequency VCO, a prescaler must be used. The prescaler is a frequency divider operating at high frequency ranges (RF and microwave). Figure 4.30 depicts such a synthesizer. The programmable ratios are realized using the 'programmable division hardware' (PDH) block which has two inputs and three outputs. The first input from f_{REF} represents the output from a crystal oscillator needed to set the frequency resolution of the synthesizer. f_{REF} may be further divided to produce the required resolution when integer-N architecture is presented, as we will see below. The second input to the programmable division block comes from the prescaler, which is necessary in order to divide the high frequency of the VCO by a factor of P such that it can be processed by the lower frequency division mechanism. On the other hand, two of the outputs of the PDH block are used as inputs to the phase detector. These inputs will be locked in frequency and phase during the operation of the synthesizer. The third output is an input control signal to the prescaler to possibly change its division ratio. From the above discussion it can be concluded that the PDH block is used to perform frequency translation of the signals of the crystal oscillator and prescaler/VCO such that the resulting signals can be used by the PD to implement the required operation range and resolution of the output frequency.

To investigate settling time in relation to frequency step and PLL bandwidth, we first note that a change from N to $N + N_m$ on the division ratio is equivalent to a frequency step from f_{REF} to $f_{REF}(1 + N_m/N)$ [Razavi98]. Using the transient response of the PLL (see Chapter 1), it is easy to show that, in order to settle to within $100a\%$ of the final value of frequency the settling time is [Razavi98]:

$$t_s \approx \frac{1}{\zeta\omega_n} \ln\left(\frac{N_m}{N|a|\sqrt{1-\zeta^2}}\right) \tag{4.77}$$

In broadband systems (HIPERLAN/2,802.11a/b/g) the requirements for fast settling in frequency steps with regard to wide system bandwidth (1–20 MHz) allow for the design of synthesizers with wide loop bandwidth. This is because settling time is inversely proportional to $\zeta\omega_n$ [$t_s \approx K/(\zeta\omega_n)$] and $\zeta\omega_n$ is directly related to the loop bandwidth. Consequently, high loop bandwidth guarantees fast settling (low t_s).

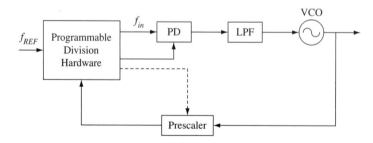

Figure 4.30   Integer-N synthesizer with prescaler

However, for narrowband systems (GSM, CDMA, etc.), the channels are located a few tens of kilohertz apart. In this case, loop bandwidths (which are also related to channel spacing) are also low and it is hard to satisfy the settling speed requirements by employing single-loop integer-N synthesizers. In addition, low loop bandwidth can result in high overall phase noise because, as mentioned above, the phase noise of the VCO is not filtered beyond the loop bandwidth and results in deterioration of the overall receiver performance.

From all the above we can conclude that basic synthesizer parameters such as frequency resolution, loop bandwidth and settling time must be looked at in conjunction to each other because they are interrelated and only in this way a satisfactory system can be designed.

4.5.3 Synthesizer Architectures

4.5.3.1 Integer-N Frequency Synthesizer

The very basic form of a realistic high-frequency integer-N synthesizer is the one depicted in Figure 4.31. In addition to the necessary $\div N$ divider, it includes a prescaler used to divide the high-frequency output of the VCO by P (as mentioned above) and a reference divider. Prescalers usually give division ratios higher or equal to 2. The reference divider is used to produce fine frequency resolution (in the order of 1–100 kHz) from the basic crystal oscillator, which usually produces relatively higher frequencies (in the order of 1–100 MHz). The following equation gives the output frequency:

$$f_{out} = NP\frac{f_{in}}{R} = NP f_{REF}, \quad P \geq 2 \tag{4.78}$$

From this equation we observe that the minimum frequency increment at the output is not f_{REF} but $P f_{REF}$.

To obtain frequency resolution equal to f_R a pulse swallow divider has to be used as illustrated in Figure 4.32. Both the programmable counter (counting up to C) and the swallow counter (counting up to S, $S < C$) start counting together. At this stage (while both counters count) the prescaler is set to $P + 1$. When S completes a period it goes to zero and commands the prescaler to divide by P. Up to that point, $(P + 1)S$ VCO periods have been counted. From that point until the programmable counter reaches zero there are $P(C - S)$ VCO periods. In conclusion, the whole pulse swallow counter produces a pulse (feeding the PD) for every $(N + 1)S + N(C - S)$ VCO periods. Consequently f_{out} is:

$$f_{out} = (PC + S)f_{REF} \tag{4.79}$$

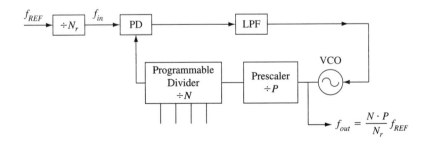

Figure 4.31 Basic high-frequency integer-N synthesizer

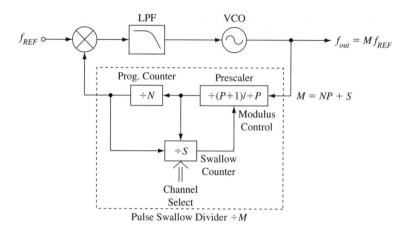

Figure 4.32 Architecture of pulse shallow divider in integer-N synthesizer [Razavi97]

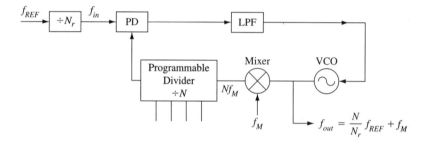

Figure 4.33 Translation of the output frequency of the synthesizer with a mixer

Since P and S are not fixed but programmable, the minimum frequency increment at the output is now f_{REF}.

Another approach to translate the output frequency of the VCO to a lower frequency is by using a mixer, as shown in Figure 4.33. The output of the VCO is mixed down to produce a frequency Nf_{in} which, divided by N, gives the signal that is applied to the PD input connected to the feedback signal. Hence the output frequency is given as:

$$f_{\text{out}} = Nf_{in} + f_{\text{mix}} \tag{4.80}$$

The mixing produces spurious signals located at frequencies $f_{\text{MP}} = \pm n_1 f_{\text{out}} \pm n_2 f_{\text{mix}}$. These by-products can be easily removed by the filtering action of the PLL.

4.5.3.2 Fractional-N and $\Sigma\Delta$ Fractional-N FS

If the division ratio is made variable such that for K1 periods of the VCO frequency the divider divides by N and for K2 periods it divides by $N + 1$, then we have a fractional-N synthesizer. This means that the output frequency can be increased or decreased by fractions of f_{REF}:

$$f_{\text{out}} = (N + k_F)f_{\text{REF}}, \quad 0 < k_F < 1 \tag{4.81}$$

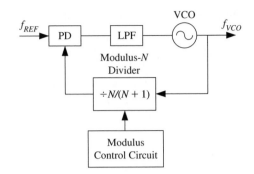

Figure 4.34   Fractional-N synthesizer

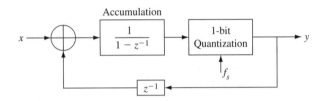

Figure 4.35   A first-order Sigma–Delta modulator

Figure 4.34 illustrates the basic form of a fractional-N synthesizer. The pattern according to which the division ratio is changed from N to $N+1$ is dictated by the modulus control circuit. Owing to the variable division ratio, even when the system is in lock, the pulses at the output of the phase detector become wider for a period of time [Razavi98]. This means that at the output of the LPF a sawtooth waveform is produced with period $1/(k_F f_{REF})$ [Razavi98]. Such a waveform at the input of the VCO will modulate it producing sidebands at frequencies $mk_F f_{REF}, m = 1, 2, 3, \ldots$

To eliminate such spurs at the synthesizer output, the controlling mechanism of the modulus division must be randomized. This can be done by using a $\Sigma\Delta$ modulator as a modulus control circuit. A $\Sigma\Delta$ modulator of first order is depicted in Figure 4.35 with output $y(z)$ given by:

$$y(z) = x(z) + H_{\text{noise}}(z) \cdot q(z) \tag{4.82}$$

where $q(z)$ is the quantization noise, $x(z)$ is the input signal and $H_{\text{noise}}(z)$ is the noise transfer function. $y(t)$ represents the output of the Modulus Control Circuit in Figure 4.34.

Because of the z-transfer function [Equation (4.82) above] quantization noise $q_N(t)$ is of highpass nature. Consequently, the instantaneous output frequency $f_0(t)$ applied to the PD in Figure 4.34 and the corresponding quantization noise $n_F(t)$ at the same point are expressed as follows:

$$f_0(t) = \frac{f_{VCO}}{N + y(t)} = \frac{f_{VCO}}{N + X + q_N(t)} \tag{4.83}$$

$$n_F(t) = f_0(t) - \frac{f_{VCO}}{N + X} = -\frac{f_{VCO}}{N + X} \frac{q_N(t)}{N + X + q_N(t)} \tag{4.84}$$

It is easy to show that the PSD of the noise $n_F(t)$ of $f_{in}(t)$ of the synthesizer is [Razavi98]:

$$S_{n_F}(f) = \left(\frac{f_{VCO}}{N+X} \cdot \frac{1}{N} \right)^2 |Q_N(f)|^2 \tag{4.85}$$

As mentioned above, the PSD $Q_N(f)$ of $q_N(t)$ is located at high frequencies and for this reason the spurs created by the $\Sigma\Delta$ modulator on the frequency spectrum are high-frequency spurs, eliminated by the lowpass filter of the synthesizer.

4.5.3.3 Dual Loop Synthesizers

An effective way to modify the inflexible relationship between reference frequency, channel spacing and bandwidth of the integer-N architecture is to employ dual-loop synthesizers in which mixers are also necessary to produce the output frequency. One such example is illustrated in Figure 4.36. This is a serial-connected dual-loop synthesizer and the frequency resolution is set by the lower loop, which also constitutes the low frequency loop. The output of VCO1 is mixed with the output of VCO2. In this configuration, the mixer is part of the high frequency loop and its output is divided before it feeds the PD of the second PLL. The most important feature of this configuration is reduced phase noise due to the ability to optimize the upper PLL bandwidth without having to sacrifice frequency resolution, which is set independently by the lower loop. The relation between output frequency and reference frequencies of the two loops is:

$$f_{out} = M \cdot f_{REF1} + N \cdot \left(\frac{M}{R \cdot X} \right) \cdot f_{REF2} \tag{4.86}$$

Nevertheless, this approach cannot be used to reduce settling time because the lower PLL has to settle first before the system is operational. The mixer will produce harmonics but they are located away from the useful frequencies and are filtered by the PLL.

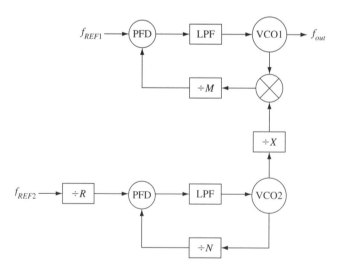

Figure 4.36 Series-connected dual-loop frequency synthesizer

Another configuration is the parallel-connected dual-loop as in Figure 4.37. The output frequency of this system is:

$$f_{\text{out}} = M f_{\text{REF1}} - N f_{\text{REF2}} \tag{4.87}$$

The mail advantage of this arrangement is that f_{REF1} and f_{REF2} can be kept quite high, helping to maintain high loop bandwidths and high settling times. On the other hand their difference ($f_{\text{REF1}} - f_{\text{REF2}}$) can be made small. In this way, both requirements for fine resolution and high settling speeds can be satisfied.

The weak point in dual loop architectures is that care must be taken during the design such that spurious frequencies produced by cross-mixing or harmonic generation are eliminated. In the series configuration, these spurs are usually located out of the band of operation. In the parallel architecture careful frequency planning can bring the spurious products in a band such that they can be easily removed by filtering.

4.5.3.4 Direct Synthesizers

Direct digital synthesizers (DDS) use digital techniques to generate a digital signal which can be transformed to sinusoid by means of digital to analogue (D/A) conversion and filtering. A counter is used to count from 0 to M. The successive counter outputs are applied in a ROM, as shown in Figure 4.38, which has stored values corresponding to the values of a sinusoidal function. In this way, a quantized sinusoid is produced at the output of the DAC. The quantized waveform is smoothed out using an LPF. To be able to produce variable frequency outputs while using the same input clock, an accumulator structure can be employed as shown in Figure 4.39. The basic principle behind its operation is that, by producing fewer samples per period, a higher output frequency can be obtained. This is shown pictorially in Figure 4.40.

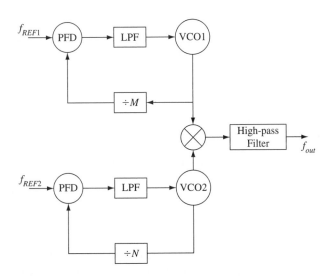

Figure 4.37 Parallel configuration of dual-loop synthesizer

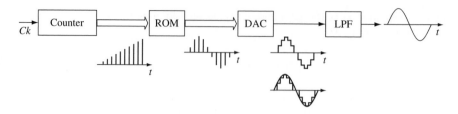

Figure 4.38 Architecture of direct digital synthesizer

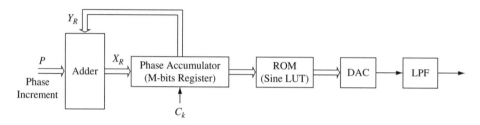

Figure 4.39 Accumulator structure of direct digital synthesizer

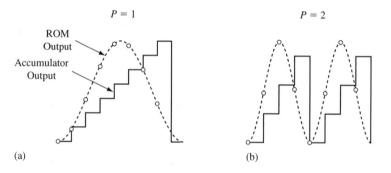

Figure 4.40 Basic principle for variable frequency generation in direct digital synthesis (reprinted from B. Razavi, 'RF Microelectronics', Prentice Hall, Englewood Cliffs, NJ, 1998, copyright © 1998 Prentice Hall Inc.)

The adder in Figure 4.39 is incremented by P in every clock cycle. The value applied to the accumulator each time is:

$$X_R(k) = Y_R(k\text{-}1) + P, \text{ for } X_R(k) \le 2^M \tag{4.88}$$

With these in mind it can be shown that the generated sinusoid has a frequency [Razavi98]:

$$f_{out} = P\frac{f_{Ck}}{2^M} \tag{4.89}$$

The maximum value of f_{out} is equal to $f_{Ck}/2$ whereas f_{out} takes its minimum value for $P = 1$.

Although in principle, direct digital synthesis is a straightforward clean way to produce sinusoids of various frequencies, it has serious limitations when it comes to implementation.

These are related to spur generation, the size of ROM and the most basic issue concerning the limitations on the highest possible generated frequency.

The number of bits of the ROM results in limited accuracy on the amplitude of the samples and consequrently in quantization error, which is shown to be a periodic function of time [Razavi97]. Also, this will produce spurs in the frequency spectrum of the synthesizer. It has been shown that, when the system generates the maximum possible output frequency, these spurs are S_{DS} below signal power [Tierney71]:

$$S_{DS} = \frac{2^{2-2k}}{3} \tag{4.90}$$

Another problem is associated with the mapping of a wide accumulator word to a smaller width of the ROM words. The accumulator word must be wide in order to produce frequency steps as small as possible. The width of ROM words should match that of the accumulator. If this is not done, there is an error due to the misalignment between the input phase and the amplitude produced at the output of the ROM which in turn results in the corruption of the generated sinusoid. This will produce spurs due to the periodic nature of this error. To address this problem, compression techniques are used for storing the data needed to reproduce the sinusoid in the ROM. In this way, the width of the ROM word can match that of the accumulator.

The main advantages of DDS are low phase noise, equal to that of the crystal used as the system clock, very good frequency resolution and very fast switching in frequency. The latter is due to the fact that there is no feedback loop in the system, which would limit the frequency settling speed. The major weakness of DDS is related to the difficulty of implementation in integrated form when the desired frequency is high. The clock speed should be twice the required output frequency (from Nyquist theorem). To relax filtering requirements, the clock speed should be increased to three or even four times the output frequency. In addition, the DAC speed, accuracy and spurious-free dynamic range are major issues for the DAC design. Finally, the common issue of all the above is power dissipation, which can be prohibitively high.

Another form of synthesizer that has recently attracted much interest involves direct analogue synthesis. It is mostly associated with a system based on one PLL synthesizer, the output of which is used as input to a system of single sideband mixers and dividers to produce the required frequency bands. This technique is mostly used in recent applications concerning wideband frequency generation for Ultra-Wideband technology standards.

4.5.4 *Critical Synthesizer Components and their Impact on the System Performance*

4.5.4.1 PLLs and Synthesizers using Charge-pumps

To convert the logic levels at the output of the PFD into an analogue DC signal to enter the LPF, charge-pumps are used. They produce pulses of current i_P flowing in or out of a capacitor connected in parallel at the output of the charge pump. Figure 4.41 depicts the simplified electronic circuit along with the equivalent model of a typical charge pump. When the UP input is ON, switch S_1 is closed, delivering a positive current pulse at the capacitor. When the DN is ON, we have a negative current. When both are OFF, there is no current flowing through the capacitor. This is a tri-state charge pump and is used in conjunction with a tri-state PFD the same as the one described in Chapter 1.

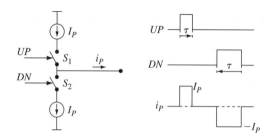

Figure 4.41   Charge pump controlled by U and D pulses

Figure 4.42 shows such a system and depicts its behaviour when two pulses of a certain phase difference are applied at its input. As explained in Chapter 1, a positive transition at input A places output UP at high level and a positive transition at input B produces a logic high at the DN output. Because A is leading in time with respect to B in the Figure 4.42, and according to the operation of the PFD (see chapter 1), when B assumes a logic high, both UP and DN are forced by the AND gate to go to zero. Assuming that the gate delay is almost zero, the positive level at DN has an infinitesimally small duration (it is an impulse). Hence the output voltage V_{out} increases every time UP is at a high level and remains unchanged in time periods where both UP and DN are zero, as shown in the figure. This means that with a static positive phase error (as in the figure) V_{out} can increase indefinitely, suggesting that the composite PFD-charge-pump has an infinite gain at DC, or in other words, its transfer function is $K_{\text{P-CP}}/s$ where $K_{\text{P-CP}}$ is the sensitivity of the composite PFD-charge-pump circuit. Consequently, the closed loop transfer function is [Razavi98]:

$$H(s) = \frac{K_{\text{P-CP}}K_{\text{VCO}}}{s^2 + K_{\text{P-CP}}K_{\text{VCO}}} \tag{4.91}$$

This means that the charge-pump PLL with a capacitor C_P is unstable. The problem can be solved by inserting a resistor in series with C_P converting it to a typical LPF.

Although a charge-pump PLL is a sampled time system and assuming that only small changes take place in the PLL state, we can consider the average behaviour and use continuous systems transfer functions, as presented in Chapter 1, to evaluate its performance [Gardner80]. In addition, the following stability condition must be satisfied [Gardner79], [Paemel94]:

$$\omega_n^2 < \frac{4}{\left(RC_P + \dfrac{2\pi}{\omega_{\text{in}}}\right)^2} \tag{4.92}$$

When the duration of the pulse at the output of the phase detector is τ, the phase error and the average current of the charge-pump are given as [Paemel94]:

$$\theta_e = \frac{2\pi\tau}{T}, \quad \overline{i_P} = \frac{I_P\theta_e}{2\pi} \tag{4.93}$$

Using now classical continuous-system PLL analysis, it is easy to obtain the following expressions for the transfer function and ω_n and ζ:

$$H(s) = \frac{K_V I_P R_2 s + \dfrac{K_V I_P}{C}}{s^2 + K_V I_P R_2 s + \dfrac{K_V I_P}{C}} = \frac{2\zeta\omega_n s + \omega_n^2}{s^2 + 2\zeta\omega_n s + \omega_n^2} \tag{4.94}$$

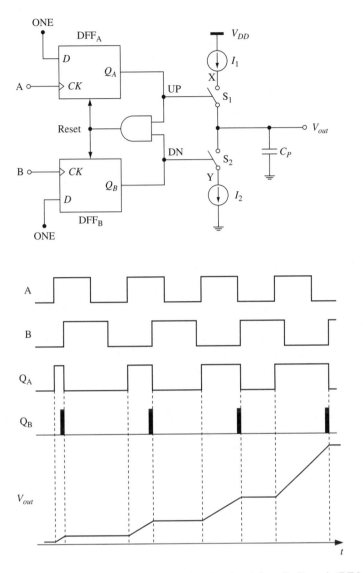

Figure 4.42 Phase frequency detector and its operation (reprinted from B. Razavi, 'RF Microelectronics', Prentice Hall, Englewood Cliffs, NJ, 1998, copyright © 1998 Prentice Hall Inc.)

$$\omega_n = \sqrt{\frac{K_V I_P}{C}} \qquad (4.95)$$

$$\zeta = \frac{R_2}{2}\sqrt{K_V I_P C} \qquad (4.96)$$

Modern charge pumps employ two transistors to act as the *UP* and *DN* current sources as described above [Zargari04], [Lam00]. Differential topologies using four transistors driven by the *UP*, *DN*, $\overline{UP}$ and $\overline{DN}$ outputs of the PFD are also very popular [Rategh00], [Verma05].

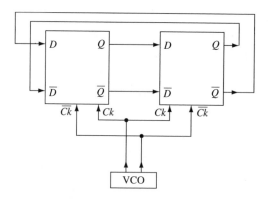

Figure 4.43 Digital frequency divider

4.5.4.2 High Frequency Dividers (Prescalers)

The most critical part of the frequency division subsystem is the prescaler, the first circuit following the VCO. As mentioned above, it is used to translate the high output frequency of the VCO to values that can be handled by digital programmable dividers. The factor that can greatly affect the system performance is balanced design, to guarantee proper operation and low phase noise. The approaches to design prescalers are mostly digital and use latches with feedback loop and dynamic division techniques. However, recently, regenerative dividers based on analogue techniques have gained increasing interest.

In digital dividers, the main building elements are two latches connected in a feedback loop, as shown in Figure 4.43. Their clock inputs are driven by the VCO. It is important for the Ck and $\overline{Ck}$ inputs to be perfectly complementary in time transitions. Differential outputs from the VCO and perfect matching of the two latches are the critical points for proper operation.

Two current-mode flip-flops connected in a feedback topology are frequently used to design high-frequency prescalers [Zargari04]. A common concern in most designs is to keep the number of transistors as low as possible to reduce contribution of the prescaler to phase noise [Lee04]. Another similar approach is to use injection locking prescalers [Rategh00].

4.5.5 Phase Noise

The main noise components contributing to the overall phase noise at the output of the PLL are associated with input noise at the phase detector and generated phase noise at the VCO. Figure 4.44 shows a PLL with these two noise sources appearing at the PD input and at VCO output. The noise disturbance at the input may be due either to AWGN at the input signal or to phase noise generated by the reference oscillator. These disturbances can be modelled as $n(t)$ or $\varphi_{nIN}(t)$ as given in the following equations:

$$y_{in}(t) = A \sin \left[\omega_r t + \varphi_I(t) \right] + n(t) \tag{4.97a}$$

$$y_{in}(t) = A \sin \left[\omega_r t + \varphi_I(t) + \varphi_{ni}(t) \right] \tag{4.97b}$$

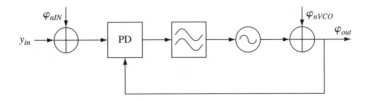

Figure 4.44 Models of noise sources in phase-locked loops

4.5.5.1 Impact of Noise at the Input of the PLL

Gardner showed that when the amount of noise is small, both noise disturbances at the input [additive Gaussian noise $n(t)$ or phase noise $\varphi_n(t)$] can be modelled as equivalent phase jitter $\overline{\varphi_{n1}^2}$ of value:

$$\overline{\varphi_{ni}^2} = \sigma_{n'}^2 = \frac{P_n}{2P_S} \tag{4.98}$$

where P_n and P_S represent the noise and signal power, respectively, at the input of the PLL. In terms of input SNR it can be expressed as:

$$\overline{\varphi_{ni}^2} = \frac{1}{2SNR_i} \text{ rad}^2 \tag{4.99}$$

Consequently, the output phase noise spectrum $\Phi_{no}(f)$ is:

$$\Phi_{no}(f) = \Phi_{n'}(f)|H(j\omega)|^2 \tag{4.100}$$

where $\Phi_{n'}(f)$ is the spectral density of the noise at the output of the PD and $H(j\omega)$ is the closed-loop transfer function of the PLL which for second-order type-II PLL is

$$H(s) = \frac{s(\omega_n^2/\omega_z) + \omega_n^2}{s^2 + 2\zeta\omega_n s + \omega_n^2} \tag{4.101}$$

where ω_z is the zero of the filter transfer function.

Having in mind that this is a lowpass transfer function, one can conclude that noise at the input of the PLL passes unaltered at low frequencies whereas it is attenuated when the noise component is of high frequency. Hence, the PLL passes the phase noise of the reference source unaltered, up to a bandwidth corresponding to the transfer function corner frequency.

4.5.5.2 Impact of the Phase Noise of the VCO

Taking into account only the φ_{nVCO} noise term at the noise model of Figure 4.44 it is clear that the transfer function is:

$$\frac{\Phi_{out}(s)}{\Phi_{nVCO}(s)} = \frac{s^2}{s^2 + 2\zeta\omega_n s + \omega_n^2} \tag{4.102}$$

This corresponds to the phase error transfer function $H_e(s)$ and is of highpass nature. Consequently the phase noise of the VCO is filtered by the effect of the PLL and only the noise beyond the bandwidth of the PLL goes through to the output unaltered.

4.5.5.3 Reference Feed-through

To explain spurs created by reference feed-through we follow [Razavi98]. Each time the PD performs a frequency–phase comparison, regardless of whether the input and output phases are equal, the charge-pump creates short current pulses to correct for mismatches in gate-drain capacitance of the two transistors and charge flow of the two switches. Consequently, a periodic waveform of short pulses $g(t)$ modulates the VCO creating a voltage disturbance at the VCO output equal to:

$$ K_{VCO} \int g(t)\mathrm{d}t = K_{VCO} \int \left[\frac{\Delta V \Delta t}{T_R} + \sum_{n \neq 0} \alpha_n \cos\left(n\omega_{REF}t + \theta_n\right) \right] \tag{4.103} $$

4.6 Downconverter Design in Radio Receivers

In this section we consider briefly the block LNA-mixer-local oscillator. On one hand, the interface between the LNA and the mixer can greatly vary, depending on the chosen receiver architecture. On the other hand, depending on the application, the LO frequency band must be carefully chosen such that no problems arise from intermodulation products and other spurious responses induced by nonlinearities.

4.6.1 Interfaces of the LNA and the Mixer

Figure 4.45 shows the basic arrangements of a heterodyne and a direct conversion receiver, as presented in Chapter 3. From Figure 4.45 it can be seen that for the heterodyne architecture there is an image reject filter (IRF) between the LNA and the mixer necessary to eliminate images appearing at the image frequency band. Because these filters should exhibit specific input and output impedances, matching circuits are needed at the output of the LNA and the input of the mixer to interface to the input and output of the IRF, respectively. These matching circuits along with the image reject filter complicate the design of the front-end while at the same time they increase considerably the power consumption. However, if an IRF is necessary, a single-ended Gilbert mixer can be used, as in Figure 4.46. This can be directly connected to the output of the IRF because, apart from being single-ended, it can give low input impedance in the order

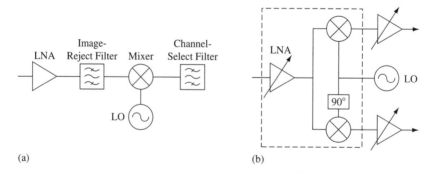

(a)   (b)

Figure 4.45   Basic heterodyne and direct conversion receiver

of 50 Ω, necessary to interface with the image filter. On the other hand, it is preferable, even in heterodyne architectures, to remove the image reject filter and implement filtering at the input of the LNA along with the matching network. Depending on the application, this solution can be very satisfactory. For example, in [Mehta05] a 40 dB image suppression is achieved at the 2.4 GHz band without the need of image filter.

In direct conversion receivers, the LNA and mixer are interfaced directly within the chip. Very frequently, double-balanced Gilbert mixers are used for increased isolation and consequently the LNA must have differential output. A popular way to do that is by connecting to the output transistor a single-to-differential converter, as shown in Figure 4.47.

Another way to connect the LNA is to interface it directly to the mixer such that a common power supply arrangement can be used as well. Figure 4.48 depicts such an LNA–mixer configuration. One disadvantage of this configuration is that it is not feasible when very low supply voltage is to be used.

Considering the output of the downconverter, an important issue is the output matching of the mixer, which is usually connected to an intermediate-frequency or a low-frequency filter. An IF filter usually has a relatively high input impedance (300–1000 Ω). However, in many applications it is necessary to connect the mixer to a 50 Ω impedance (for example in test circuits). In this case, an emitter follower (or source follower in MOS transistors) can be connected at the output of the mixer cell to act as a buffer and to reduce the resistance to 50 Ω. This buffer also permits the mixer to retain a high load impedance and consequently exhibit a high gain.

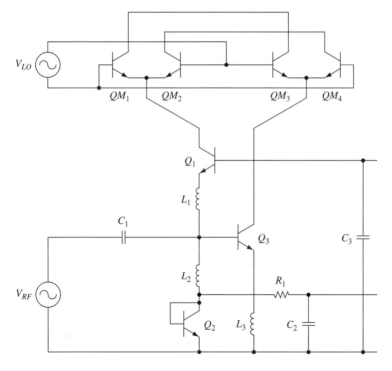

Figure 4.46   Single-ended Gilbert mixer

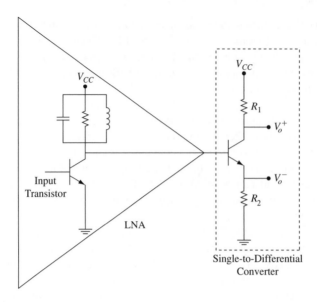

Figure 4.47   Interface of LNA output to single-to-differential converter

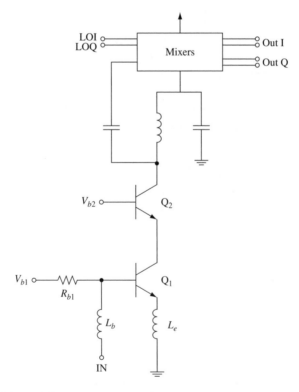

Figure 4.48   LNA–mixer interface with common power supply

4.6.2 Local Oscillator Frequency Band and Impact of Spurious Frequencies

Nonlinearities introduced by various RF circuits in the front-end result in generation of spurious responses at the output of the front-end. Concentrating on the downconverter block, we can express the mixer nonlinearity by expanding the transfer function in a power series:

$$U_{\text{out}} = \sum_{n=1}^{N} a_n U_{\text{in}}^n \tag{4.104}$$

where it was assumed that the highest order of the nonlinearity is N.

To examine the effect of nonlinearity, an input consisting of two tones is applied at the input of the mixer:

$$U_{\text{in}} = A_1 \cos(\omega_1 t) + A_2 \cos(\omega_2 t) \tag{4.105}$$

It can be shown [Manassewitsch05] that the term $\alpha_M U_{\text{in}}^M$ with nonlinearity exponent M will produce terms of the form:

$$\cos[n\omega_1 + m\omega_2], \quad n, m \text{ integers: } |n| + |m| \leq M \tag{4.106}$$

From the above it can be deduced that a specific IF frequency f_{IF} of a given system, apart from the useful downconversion relation $|\omega_{\text{RF}} - \omega_{\text{LO}}|$, can be produced by the spurs created by f_{RF} and f_{LO} as shown below:

$$n f_{\text{RF}} + m f_{\text{LO}} = f_{\text{IF}} \quad \text{or} \quad n \frac{f_{\text{RF}}}{f_{\text{IF}}} + m \frac{f_{\text{LO}}}{f_{\text{IF}}} = 1 \tag{4.107}$$

The above formula is the basis of creating an intermodulation (IM) product chart [Manassewitsch05], [Egan03], which is the representation of the above equation in rectangular coordinates $f_{\text{LO}}/f_{\text{IF}}$ vs $f_{\text{RF}}/f_{\text{IF}}$ for various values of n and m satisfying Equation (4.106). Figure 4.49 shows an IM-product chart for $M = 5$. The lines on the chart represent Equation (4.107). The shaded rectangles represent the locus of possible values for input, output and LO frequencies. These correspond to the range of values of f_{RF} and f_{LO} dictated by the application. The lines of the IM product chart crossing the shaded rectangles correspond to generated spurious responses falling inside the useful output frequency band (in-band spurs).

System requirements impose specific maximum levels for these spurious products. After identification from the IM-product chart, the in-band spurs must be reduced below specified levels by careful system design.

4.6.3 Matching at the Receiver Front-end

From the previous sections in LNA and downconverter design it is understood that matching is an important issue in the receiver chain. As deduced from the above sections we can investigate matching in three different levels. We can have matching for minimum noise, matching for maximum power transfer and matching for optimum impedance.

Noise matching entails finding the optimum impedance to achieve the minimum noise figure. To do that, it is sufficient to derive closed-form expressions of the noise factor as function of noise voltage V_n, noise current I_n and impedance elements $(R_{\text{in}}, X_{\text{in}})$ of the specific circuit and set its derivative to zero:

$$\frac{\partial F}{\partial Z} = \frac{\partial f(\overline{V_n^2}, \overline{I_n^2}, R_{\text{in}}, X_{\text{in}})}{\partial Z} = 0 \tag{4.108}$$

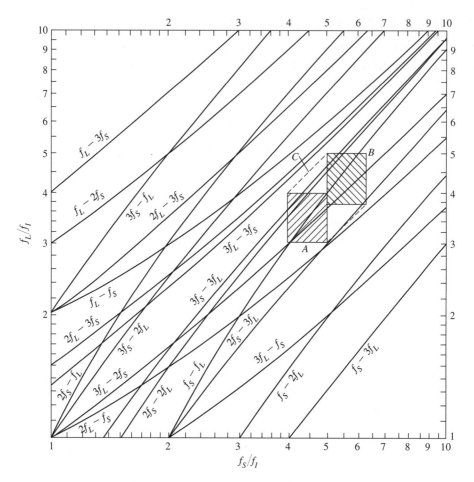

Figure 4.49 Intermodulation product chart (reprinted from W. Egan, 'Frequency Synthesis by Phase-Lock', copyright © 1981 John Wiley & Sons Inc.)

This procedure will determine optimum (in terms of noise) values for R_{in} and X_{in} and in this way we can obtain the minimum possible noise figure for the specific circuit:

$$F_{min} = f(\overline{V_n^2}, \overline{I_n^2}, R_{iopt}, X_{iopt}) \qquad (4.109)$$

However, in the design of functional blocks like the downconverter, noise figure minimization or reduction is accounted for during the circuit design stage (as shown in previous sections). Hence, when the downconverter is considered as a block, it is important to look at impedance matching.

In the previous section and in the context of LNA–mixer interface, we considered mainly the matching issues concerning it. Here, we will briefly consider matching to external elements at the input of the LNA, which represents the input of the downconverter block.

The various LNA architectures correspond to different input impedances Z_{in}. However, practically, the configuration of the first transistor of the LNA determines the value of Z_{in}. For

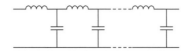

Figure 4.50 L-section network for matching

example, when we have a common-base arrangement of the input transistor (see Figure 4.5) then the input impedance is:

$$Z_{\text{in}} = \frac{1}{1/Z_{iT} + g_m} \tag{4.110}$$

where Z_{iT} and g_m are the transistor input impedance and transconductance respectively.

We focus on inductive degeneration which is the most frequently used approach. This is mainly because it can achieve low noise figure without affecting impedance matching, which can be designed independently [Ismail04]. In this case, input impedance is given by:

$$Z_{\text{in}} = s(L_s + L_g) + \frac{1}{sC_{gs}} + \left(\frac{g_{m1}}{C_{gs}}\right)L_s \tag{4.111}$$

At the resonance frequency Z_{in} becomes:

$$Z_{\text{in}} \approx \omega_T L_s, \quad \omega_0 = 1 \Big/ \sqrt{C_{gs}(L_s + L_g)} \tag{4.112}$$

This represents narrowband matching. Matching can be improved by connecting one or more L-sections, as shown in Figure 4.50.

4.7 RF Power Amplifiers

4.7.1 General Concepts and System Aspects

One of the most critical components in the transmitter is the power amplifier (PA). Located at the final stage before the transmitting antenna, it is used to deliver the required signal power for transmission. It is important to have enough power at the transmitting antenna in order to satisfy the link budget requirement, which will result in the specified system performance. In addition, the PA linearity should comply with the requirements of the specific standard being implemented. Apart from that, its output spectrum should be below the transmitted spectral mask imposed by the standard. Finally, since the PA must deliver a high output power, it should also have high efficiency in order to reduce the dissipated power, which is by far the highest of the overall system. To outline the requirements for the design of the PA, these are: high output power, high efficiency, controlled linearity and output spurs.

The general configuration of a PA circuit is as shown in Figure 4.51. The final stage consists of one transistor capable of producing high power. Its drain is connected to an RF choke in order to be able to have the maximum possible drain voltage swing. Hence we can assume that the peak-to-peak voltage swing is close to $2V_{DD}$. The load R_L corresponds to the transmitting antenna and is close to 50 Ω. The power delivered to the load is:

$$P_{\text{RF}} = \frac{V_{DD}^2}{2R_{\text{in}}} \tag{4.113}$$

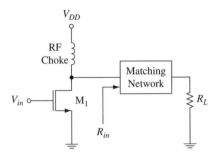

Figure 4.51 Interconnection of matching network to power amplifier output

Consequently, to deliver the maximum possible RF power to the antenna, impedance R_{in} looking at the drain towards the load will be very small [in the order of a few ohms, see Equation (4.113)]. This is why a matching network (impedance transformer) needs to be connected between the drain of the transistor and the load. A consequence of the above is high output current.

The impedance transformation can be done by either an RF transformer or an LC-matching network, as shown in Figure 4.50. The former involves substantial losses. An LC-matching circuit has very low losses and, for this reason, it is the most suitable solution.

An important parameter to quantify the efficiency of the amplifier is the power-aided efficiency (PAE) defined as follows:

$$\eta_{PAE} = \frac{P_{out} - P_{in}}{P_{DC}} \tag{4.114}$$

where P_{in} and P_{out} are the power of the input RF signal and that at the output of the power amplifier respectively. P_{DC} represents the dissipated power.

Regarding linearity and spectral spurs issues, we must note that the two-tone test is not sufficient to predict interference form adjacent channels (ACI). Hence, it is more suitable to measure spectral regrowth when a modulated signal is applied to the input of the PA [Razavi98].

4.7.2 Power Amplifier Configurations

4.7.2.1 Class A

Assuming an ideal I–V curve of the power transistor like the one in Figure 4.52, which has very small knee voltage, we locate the load line as illustrated. The maximum current and voltage swings are I_{max} and $V_{max} = 2V_{DD} - V_{kn}$ where V_{kn} is the knee-voltage of the I–V curve. Consequently the resistance for optimum power match is the ratio of the voltage-to-current amplitude:

$$R_m = \frac{(2V_{DD} - V_{kn})/2}{I_{max}/2} \approx \frac{(2V_{DD})/2}{I_{max}/2} \tag{4.115}$$

where we assume that V_{kn} is very low. It must be noted here that any other resistance lower ($R_L = R_m/k$) or higher ($R_H = kR_m$) than the optimum resistance R_m by a ratio of k will reduce the RF output power by the same factor [Cripps83].

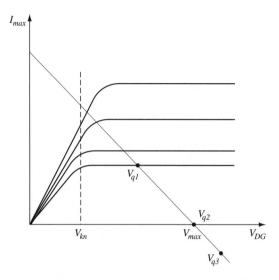

Figure 4.52 I–V characteristic of a power transistor

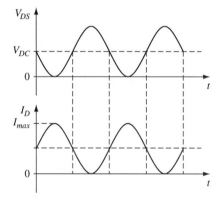

Figure 4.53 Output voltage and current waveforms in class A amplifier

Indeed, the output RF power is:

$$P_{\text{RFout}} = \frac{1}{2}V_{\text{DC}}I_{\text{DC}} = \frac{1}{2}\frac{V_{\text{max}}}{2}\frac{I_{\text{max}}}{2} = \frac{V_{\text{max}}I_{\text{max}}}{8} \qquad (4.116)$$

On the same I–V curve we see the load line for AC operation. When the operating point is set at V_{q1} located at the middle of the distance $V_{\text{max}} - V_{kn}$, the amplifier operates in class A. Ideally we have linear operation and the output voltage and current waveforms ($V_{\text{DS}}, I_{\text{D}}$) are pure sinusoids, as illustrated in Figure 4.53, indicating that the transistor is ON during the whole period (i.e. the conduction angle is equal to $\theta = 2\pi$). However, due to large amplitude signals at the drain the resulting waveforms $U_{\text{D}}(t)$ and $I_{\text{D}}(t)$ contain higher harmonics. Figure 4.54

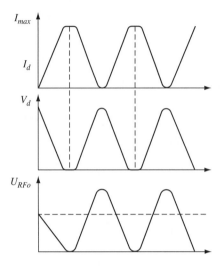

Figure 4.54 Drain current and voltage waveforms and RF voltage at the output load (U_{RFo})

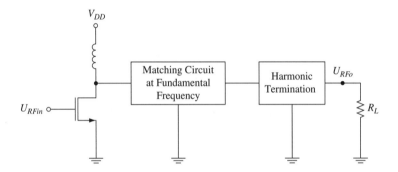

Figure 4.55 General form of the matching network at the output of the amplifier

shows the current and voltage waveforms at the drain and the RF voltage at the load. The current waveform is wider at the top, reflecting nonlinear distortion when the transistor, at that time instant, operates close to the saturation region. On the other hand, the RF output voltage $U_{RFo}(t)$ is a perfect sinusoid. This is due to a harmonic termination circuit used to eliminate (short circuit) the higher harmonics of the drain voltage waveform. Figure 4.55 shows a general form of the matching networks at the output of the PA. There is a matching circuit to match the optimum impedance at the output of the PA to the load R_L and a harmonic termination circuit used to eliminate the remaining harmonics from the output signal. Finally, since the device is biased at the middle of the maximum possible voltage and current swings, the PAE of a class A amplifier is:

$$\eta_{PAE} = \frac{P_{RFo}}{P_{DC}} = \frac{(V_{max}I_{max})/8}{\frac{V_{max}}{2}\frac{I_{max}}{2}} = \frac{1}{2} \tag{4.117}$$

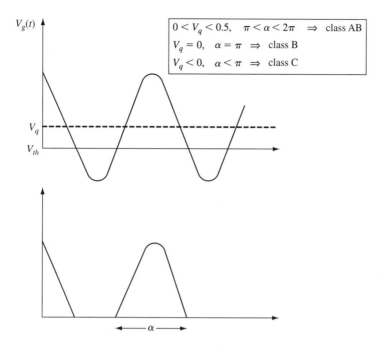

$$
\begin{array}{ll}
0 < V_q < 0.5, \quad \pi < \alpha < 2\pi & \Rightarrow \quad \text{class AB} \\
V_q = 0, \quad \alpha = \pi & \Rightarrow \quad \text{class B} \\
V_q < 0, \quad \alpha < \pi & \Rightarrow \quad \text{class C}
\end{array}
$$

Figure 4.56 Relation of the duration of current pulses and class of operation

4.7.2.2 Class B, Class AB and Class C

By reducing the normalized bias point V_q of the active device below the value 0.5 ($V_q = V_Q/V_{max}$) the amplifier is not ON all the time but only for the part of the input sinusoidal RF signal for which $U_g(t) = V_q + U_{RFin}(t)$ is into the active region. Figure 4.56 shows this operation and the resulting current pulse, the duration of which is α. If we set $V_q = 0$ the device is ON for 50% of the time and the duration of the current pulses is equal to $\pi/2$. In this case, the PA operates under class B.

From the above it is understood that reduction of V_q results in higher efficiency compared with class A operation because the device in not always ON. However, in class B we have current only for half the period and therefore the resulting power of the output RF signal is considerably reduced. One way to increase the output power is to use a push–pull configuration with two transistors employing an RF transformer to combine the current pulses from the two transistors, as shown in Figure 4.57.

Considering Figure 4.56, again the classes of operation are easily deduced. In class A we have $V_q = 0.5$ and the device is always ON, producing ideally sinusoidal drain current. In class B we have $V_q = 0$ for which the device is ON half of the period and produces half sinusoidal current pulses at the drain. In between these two classes we have class AB for which $0 < V_q < 0.5$ and the conduction angle is between 0 and π. Finally, when V_q becomes negative, the transistor is ON for less than half the signal period and the conduction angle is smaller than π. This is class C of operation.

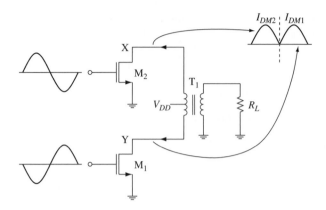

Figure 4.57 Push–pull configuration to combine current pulses [Razavi98]

By expressing the drain current as a function of conduction angle α, which is used as the independent variable, closed formed expressions can be derived for the average drain current and the amplitude of the fundamental frequency:

$$I_d(\theta) = \begin{cases} I_q + I_{pk} \cos\theta, & -\alpha/2 < \theta < \alpha/2 \\ 0, & |\theta| > \alpha/2 \end{cases} \tag{4.118}$$

The conduction angle is associated with I_q and I_{pk} by:

$$\cos\left(\frac{\alpha}{2}\right) = -\frac{I_q}{I_{pk}}, \quad I_{pk} = I_{max} - I_q \tag{4.119}$$

Consequently $I_d(\theta)$ becomes:

$$I_d(\theta) = \frac{I_{max}}{1 - \cos(\alpha/2)}[\cos\theta - \cos(\alpha/2)] \tag{4.120}$$

To determine the DC current and its harmonics I_n, the Fourier integrals are calculated. By doing so, the values of I_{DC} and the current I_1 at the fundamental frequency are [Cripps83]:

$$I_{DC} = \frac{I_{max}}{2\pi}\left[\frac{2\sin(\alpha/2) - \alpha\cos(\alpha/2)}{1 - \cos(\alpha/2)}\right], \quad I_1 = \frac{I_{max}}{2\pi}\left(\frac{\alpha - \sin\alpha}{1 - \cos(\alpha/2)}\right) \tag{4.121}$$

Figure 4.58(a, b) shows output RF power and efficiency as a function of the conduction angle for classes A, AB, B and C. Figure 4.58(a) shows that class A has the lowest efficiency (50%) and class B the highest, 75% between classes A, AB and B, as expected. However, the output power for these three classes is almost constant at 0 dB. Figure 4.58(b) shows that, depending on the conduction angle, class C can have an efficiency close to 1. However, at high efficiencies the RF output power is decreased dramatically.

4.7.2.3 Class E

Operation of Class E amplifiers is based on the principle that the transistor acts as a switching element and not as a current source controlled by the input voltage, as was the case for classes

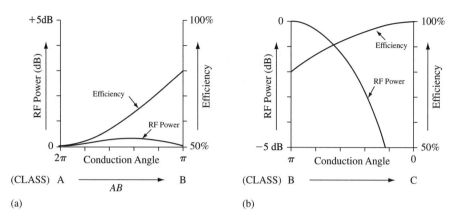

Figure 4.58 Efficiency and output power as a function on conduction angle (reprinted from S. Cripps, 'RF Power Amplifier for Wireless Communications', copyright © 1999 Artech House Inc.)

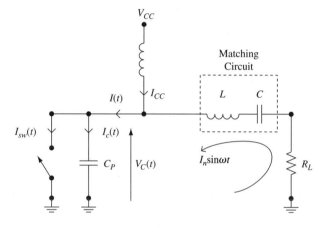

Figure 4.59   Class E amplifier circuit configuration

AB, B and C. Although the notion of switching entails digital operation, the resulting signals are analogue.

The basic configuration of class E is shown in Figure 4.59 and includes a switch corresponding to the transistor, a capacitor in parallel with the switch and an output matching circuit. During the time period T_{sw} when the switch is closed, the current goes through it, whereas when the switch is open (at T_c) the current flows through the capacitor C_P. We assume that the time of switching ON and OFF can be set by the designer in an independent manner. The overall current and output voltage as well as the current through the switch I_{oSw} and the capacitor I_{oC} are depicted in Figure 4.60.

We follow in brief the analysis of Cripps [Cripps83] to concentrate on some qualitative characteristics and quantitative results. The composite current around the output loop flows either through the switch for T_{sw} seconds or through the capacitor for T_c seconds and is given as:

$$I(\theta) = I_{RF} \sin \theta + I_{DC} \tag{4.122}$$

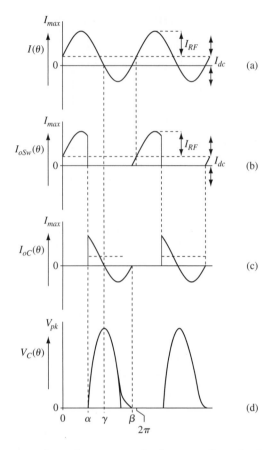

Figure 4.60 Switch current and capacitor current and voltage waveforms for class E amplifier (reprinted from S. Cripps, 'RF Power Amplifier for Wireless Communications', copyright © 1999 Artech House, Inc.)

Furthermore, we assume that the switch closes at time instant β when $I(\theta)$ is zero just before it starts taking positive values. It remains closed until a time instant α determined by the designer. Determination of the relationships between I_{RF}, I_{DC}, α and β will give the design steps for a class E amplifier:

$$\frac{I_{DC}}{I_{RF}} = \frac{\cos\alpha - \cos\beta}{\alpha - \beta} \quad \text{and} \quad \sin\beta = \frac{\cos\beta - \cos\alpha}{\alpha - \beta} \tag{4.123}$$

$$I_{DC} = \frac{I_{pk}}{1 + 1/\varepsilon_I}, \quad I_{RF} = \frac{I_{pk}}{1 + \varepsilon_I} \tag{4.124}$$

where

$$\varepsilon_I = \frac{I_{DC}}{I_{RF}} = -\sin\beta, \quad I_{pk} = I_{RF} + I_{DC} \tag{4.125}$$

The voltage across the capacitor is [Cripps99]:

$$V_C(\theta) = \frac{1}{\omega C_P}[I_{DC}(\theta - \alpha) + I_{RF}(\cos\alpha - \cos\theta)], \quad \alpha < \theta < \beta$$

$$V_C(\theta) = 0, \quad \beta - 2\pi < \theta < \alpha \tag{4.126}$$

In addition to the above, the supply voltage V_{DC} is set equal to the mean value of the voltage across the capacitor C_P [Cripps99]:

$$V_{DC} = \frac{I_{RF}}{2\pi\omega C_P}[0.5(\beta - \alpha)(\cos\beta + \cos\alpha) - \sin\beta + \sin\alpha] \tag{4.127}$$

The above expression can be used to calculate C_P when V_{DC} is given and after choosing α (which determines a value for β as well).

The value of the load resistor R_L is associated with the in-phase component V_{Ci} of $V_C(\theta)$:

$$R_L = \frac{V_{Ci}}{I_{RF}}, \quad V_{Ci} = \frac{1}{2}\frac{I_{RF}}{\pi\omega C_P}\sin\beta[(\beta - \alpha)(\cos\beta + \cos\alpha) + \sin\alpha - \sin\beta] = 2V_{DC}\sin\beta \tag{4.128}$$

Finally the RF power is expressed as:

$$P_{RF} = \frac{I_{RF}V_{Ci}}{2} = I_{RF}V_{DC}\sin\beta \tag{4.129}$$

From the above design procedure of a class E amplifier, we see that the decisive step is to choose the parameter α. By choosing the value for α, the above equations can be used to design the amplifier and determine the values of all components in the circuit.

Concluding on class E amplifiers, we note that their main advantage is that efficiency can be equal to 1. This can be deduced from Figure 4.60, where it is seen that there is no time period during which both the current through the switch (I_{oSw}) and the output voltage are different from zero. Furthermore, the highly nonlinear nature of the output voltage is a major shortcoming of this circuit. However, by applying linearization techniques, as presented below, improved performance can be achieved.

4.7.3 Impedance Matching Techniques for Power Amplifiers

To consider impedance matching for PAs we must distinguish two cases depending on the kind of transistor output impedance. The first involves linear output impedance whereas the second addresses a more realistic situation according to which the output impedance of the transistor is nonlinear. The latter is related to the fact that the transistor must operate under large signal conditions. In the case where the transistor is characterized by linear impedance with zero reactive part, as was explained in Section 4.7.1, the transistor output impedance R_o must be transformed into a very small load impedance R_L to anticipate for high output power.

When the transistor impedance is linear but complex it is rather straightforward to insert between the complex Z_{out} and R_L a T-network or a P-network of reactive elements to achieve matching to a resistive load. Figure 4.61 depicts one example where a capacitive C_o is tuned out by L_1 whereas C_1 and L_2 are used to match R_L to R_o.

Considering the second case where a nonlinear complex impedance must be connected to the load, a 'load-pull technique' must be used [Cripps83] in order to find the optimum complex load that will produce the maximum output power. This is done by connecting an impedance-tuning circuit at the output of the transistor. The output of the tuner is connected to a power meter.

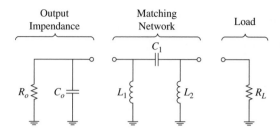

Figure 4.61 Matching of a complex output impedance to a resistive load R_L

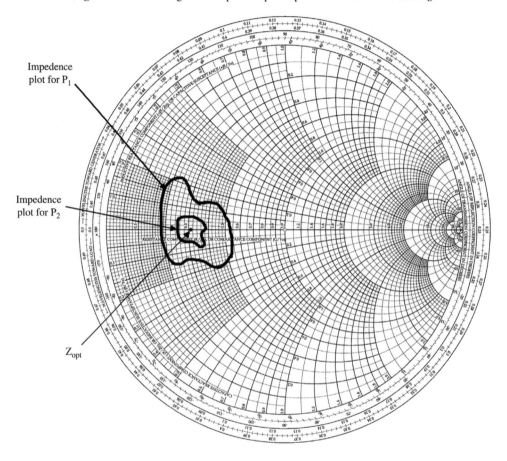

Figure 4.62 Optimum impedance at the center of contours of constant power [Razavi98]

By controlling externally the impedance tuning circuit, various impedance values are realized. When this is done, care must be taken to retain the same output power. In this way, constant power contours can be drawn in a Smith chart. It can be shown [Cripps99] that as power is increased the radius of the contour decreases. The centre of these contours represents a single point corresponding to the optimum impedance value Z_{opt} in the sense of maximum output power. Figure 4.62 shows the contours of constant power and the point corresponding to Z_{opt}.

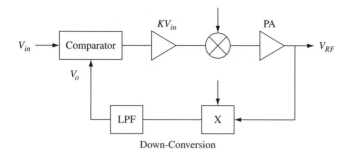

Figure 4.63 Feedback technique for power amplifier linearization

It must be noted that the load-pull extraction procedure can be realized without implementing an actual test procedure but using only simulation techniques.

4.7.4 Power Amplifier Subsystems for Linearization

Many wireless applications employ linear modulation (QPSK and QAM) and consequently require linear RF amplification at the transmitter. A straightforward way to satisfy this requirement is to use class A PAs, which operate only with 30–50% efficiency. To handle this problem, linearization techniques can be used, in which a system is used to transform the nonlinear RF output signal from the PA into a linear signal suitable for transmission. Linearization methods are mostly based on feedback and feedforward techniques. Additional linearization techniques are linear amplification using nonlinear components (LINC) and amplifier modulation using its envelope.

Feedback techniques use feedback to make the downconverted version of the output signal V_{RF} of the PA similar to the voltage input V_{in} of the loop. This technique is illustrated in Figure 4.63 and in practice it implements most of the gain at low frequency. Hence the resulting, amplified low-frequency signal KV_{in} need only be upconverted and amplified further using the PA.

In feedforward techniques, the nonlinear output signal is split into the linear part and the nonlinear portion (called also the 'error signal'). The nonlinear part is amplified separately and added to the linear signal, producing a linearized version of the transmitted signal.

The LINC method is based on the principle of separation of the information signal in two-phase modulated signals with constant envelope. These two signals can be amplified separately and subsequently added to produce the composite amplified information signal.

All mentioned approaches require extra hardware, increasing the cost and the power consumption of the transceiver. Therefore, their application in commercial systems is limited.

References

[Andreani01]: P. Andreani, H. Sjoland, 'Noise optimization of an inductively degenerated CMOS low noise amplifier', IEEE Trans. Circuits Systems II, vol. 48, September 2001, pp. 835–841.

[Come00]: B. Come, R. Ness, S Donnay, L. Van der Perre, W. Eberle, P. Wambacq, M. Engels, I. Bolsens, 'Impact of front-end non-idealities on bit error rate performances of WLAN-OFDM transceivers', IEEE Radio and Wireless Conference 2000, RAWCON 2000.

[Craninckx95]: J. Craninckx, M. Steyaert, 'Low-noise voltage-controlled oscillators using Enhanced LC -tanks', IEEE Trans. Circuits Systems II, vol. 42, December 1995, pp. 794–804.

[Cripps83]: S.C. Cripps, 'A theory of the prediction of GaAs FET load-pull power contours', IEEE MTT-Symposium, 1983, pp. 221–223.

[Cripps99]: S. Cripps, "RF Power Amplifiers for Wireless Communications", Artech House, Norwood, MA, 1999.

[Egan81]: W. Egan, 'Frequency Synthesis by Phase Lock', John Wiley and Sons Inc., New York, 1981.

[Gardner79]: F. Gardner, 'Phaselock Techniques', 2nd edn. John Wiley & Sons Inc., New York, 1979.

[Gardner80]: F. Gardner, 'Charge-pump phase-lock loops', IEEE Trans. Commun., vol. COM-28, November 1980, pp. 1849–1858.

[Girlando99]: G. Girlando, G. Palmisano, 'Noise figure and impedance matching in RF cascode amplifiers', IEEE Trans. Circuits Systems II, vol. 46, November 1999, pp. 1388–1396.

[Gonzalez96]: G. Gonzalez, 'Microwave Transistor Amplifiers, Analysis and Design', 2nd edn, Prentice Hall, Englewood Cliffs, NJ, 1996.

[Gray01]: P. Gray, P. Hurst, S. Lewis, R. Meyer, 'Analysis and Design of Analog Integrated Circuits', 4th edn, John Wiley & Sons Inc., New York, 2001.

[Heydari04]: P. Heydari, 'High-frequency noise in RF active CMOS mixers', Proc. 2004 Conference on Asia South Pacific Design Automation, pp. 57–60.

[Ismail04]: A. Ismail, A. Abidi, 'A 3–10 GHz low-noise amplifier with wideband LC-ladder matching network', IEEE J. Solid-St. Circuits, vol. 39, December 2004, pp. 2269–2277.

[Janssens02]: J. Janssens, M. Steyaert, 'CMOS Cellular Receiver Front-ends from Specification to Realization', Kluwer Academic, Dordrecht, 2002.

[Kinget99]: P. Kinget, 'Integrated GHz voltage controlled oscillators'. In 'Analog Circuit Design: (X)DSL and Other Communication Systems; RF MOST Models; Integrated Filters and Oscillators', W. Sansen, J. Huijsing, R. van der Plassche (eds), Kluwer Academic, Dordrecht, 1999, pp. 353–381.

[Lam00]: C. Lam, B. Razavi, 'A 2.6-GHz/5.2-GHz Frequency Synthesizer in 0.4-μm CMOS Technology', IEEE J. Solid-St. Circuits, vol. 35, May 2000, pp. 788–794.

[Lee98]: T. Lee, 'The Design of CMOS Radio-Frequency Integrated Circuits', Cambridge, Cambridge University Press, 1998.

[Lee04]: S.-T. Lee, S.-J. Fang, D. Allstot, A. Bellaouar, A. Fridi, P. Fontaine, 'A quad-band GSM-GPRS transmitter with digital auto-calibration', IEEE J. Solid-St. Circuits, vol. 39, December 2004, pp. 2200–2214.

[Li05]: X. Li, S. Shekhar, D. Allstot, 'G_m-boosted common-gate LNA and differential colpitts VCO/QVCO in 0.18-μm CMOS', IEEE J. Solid-St. Circuits, vol. 40, December 2005, pp. 2609–2619.

[Lu03]: R. Lu, 'CMOS low noise amplifier design for wireless sensor networks', Master's thesis, University of California, Berkeley, CA, 2003.

[Manassewitsch05]: V. Manassewitsch, 'Frequency Synthesizers, Theory and Applications', 3rd edn, John Wiley & Sons Inc., New York, 2005.

[Mehta05]: S. Mehta, D. Weber, M. Terrovitis, K. Onodera, M. Mack, B. Kaczynski, H. Samavati, S. Jen, W. Si, M. Lee, K. Singh, S. Mendis, P. Husted, N. Zhang, B. McFarland, D. Su, T. Meng, B. Wooley, 'An 802.11g WLAN SoC', IEEE J. Solid-St. Circuits, vol. 40, December 2005, pp. 2483–2491.

[Meyer94]: R. Meyer, W. Mack, 'A 1-GHz BiCMOS RF front-end IC', IEEE J. Solid-St. Circuits, vol. 29, March 1994, pp. 350–355.

[Noguchi86]: T. Noguchi, Y. Daido, J. Nossek, 'Modulation techniques for microwave digital radio', IEEE Commun. Mag., vol. 24, October 1986, pp. 21–30.

[Paemel94]: M. Van Paemel, 'Analysis of a charge-pump PLL: a new model', IEEE Trans. Commun., vol. 42, 1994, pp. 2490–2498.

[Pederson91]: D. Pederson, K. Mayaram, 'Analog Integrated Circuits for Communication', Kluwer Academic, Dordrecht, 1991.

[Pozar05]: D. Pozar, 'Microwave Engineering', 3rd edn, John Wiley & Sons Inc., New York, 2005.

[Razavi97]: B. Razavi, 'Challenges in the design of frequency synthesizers for wireless applications', IEEE 1997 Custom Integrated Circuits Conference (CICC97), 1997, pp. 395–402.

[Rategh00]: H. Rategh, H. Samavati, T. Lee, 'A CMOS frequency synthesizer with injection-locked frequency divider for a 5-GHz wireless LAN receiver', IEEE J. Solid-St. Circuits, vol. 35, May 2000, pp. 780–787.

[Razavi98]: B. Razavi, 'RF Microelectronics', Prentice Hall, Englewood Cliffs, NJ, 1998.

[Ryynanen04]: J. Ryynanen, 'Low-noise amplifiers for integrated multi-mode direct conversion receivers', Ph.D. dissertation, Helsinki University of Technology, June 2004.

[Smith00]: J. Smith, 'Modern Communication Circuits', 2nd edn, McGraw Hill, New York, 2000.

[Tierney71]: J. Tierney, C. Rader, B. Gold, 'A digital frequency synthesizer', IEEE Trans. Audio Electroacoustics, vol. 19, March 1971, pp. 48–57.

[Toumazou02]: C. Toumazou *et al.*, 'Trade-offs in Analog Circuit Design: The Designer's Companion', Kluwer Academic, Dordrecht, 2002.

[Valkama01]: M. Valkama, M. Renfors, V. Koivunen, 'Advanced methods for I/Q imbalance compensation in communication receivers', IEEE Trans. Signal Processing, vol. 49, October 2001, pp. 2335–2344.

[Valkama05]: M. Valkama, M. Renfors, V. Koivunen, 'Blind signal estimation in conjugate signal models with application to I/Q imbalance compensation', IEEE Signal Processing Lett., vol. 12, November 2005, pp. 733–736.

[Verma05]: S. Verma, J. Xu, M. Hamada, T. Lee, 'A 17-mW 0.66-mm^2 direct-conversion receiver for 1-Mb/s cable replacement', IEEE J. Solid-St. Circuits, vol. 40, December 2005, pp. 2547–2554.

[Zargari04]: M. Zargari, M. Terrovitis, S. Jen, B. Kaczynski, M. Lee, M. Mack, S. Mehta, S. Mendis, K. Onodera, H. Samavati, W. Si, K. Singh, A. Tabatabaei, D. Weber, D. Su, B. Wooley, 'A single-chip dual-band tri-mode CMOS transceiver for IEEE 802.11a/b/g wireless LAN', IEEE J. Solid-St. Circuits, vol. 39, December 2004, pp. 2239–2249.

5

Synchronization, Diversity and Advanced Transmission Techniques

This chapter covers a demanding area in receiver design (synchronization and carrier recovery) along with special techniques that enhance system performance. Synchronization and carrier recovery systems are necessary for coherent demodulation and their careful design guarantees improved performance and minimization of the implementation loss associated with it. In addition, transmitter and receiver diversity constitutes one of the dominant approaches to combining high bandwidth with increased performance in modern wireless systems. Furthermore, orthogonal frequency division multiplexing is the chosen transmission technique for many modern broadband wireless systems including WLANs, personal area networks (PANs) and broadcasting. Finally, spread spectrum offers enhanced performance with robustness and multiuser capabilities much needed in 4G and PAN applications.

5.1 TFR Timing and Frequency Synchronization in Digital Receivers

5.1.1 Introduction

Figure 5.1 shows the block diagram of a multilevel receiver (M-PSK or QAM) where the positions of carrier recover (CR) and timing recovery (TR) blocks are also illustrated. The carrier recovery block produces a carrier with the same frequency and phase as the local oscillator at the transmitter. The recovered carrier is used to downconvert the received signal before entering the matched filters (MF). Timing recovery is responsible for estimation of the correct timing instant $(nT + \hat{\tau})$ at which the received downconverted signal at the output of the matched filters will be sampled. If the basic signalling pulse is not a rectangular pulse a signal pulse generator must be inserted to multiply by the locally recovered carrier before downconversion. This generator is driven by the timing synchronization block [Proakis01].

Digital Radio System Design   Grigorios Kalivas
© 2009 John Wiley & Sons, Ltd

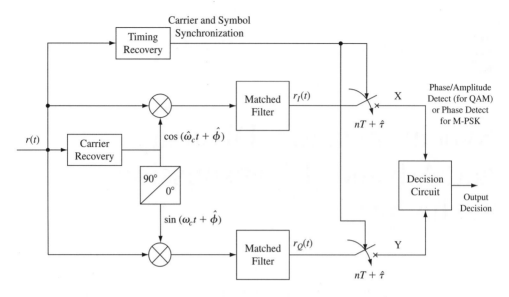

Figure 5.1   Block diagram of general coherent receiver

Let the transmitted QAM or M-PSK modulated signal be:

$$s(t) = \text{Re}\{S(t)\exp[j2\pi f_C t]\}$$
$$S(t) = \sum_n a_n g(t - nT_S) \tag{5.1}$$

where a_n represent the transmitted symbols whereas $g(t)$ is a signalling pulse of unit energy.

Transmission through the radio channel will result in the following passband signal at the input of the receiver in Figure 5.1:

$$r_{\text{pb}}(t) = s_{\text{pb}}(t - \tau) + n_{\text{pb}}(t) \tag{5.2a}$$

Multiplication of $r_{\text{pb}}(t)$ by the quadrature LO signals $\cos(2\pi f_{CR} t + \phi_R)$ and $\sin(2\pi f_{CR} t + \phi_R)$ and subsequent filtering can be shown to give the baseband signals [Proakis01], [Mengali97]:

$$
\begin{aligned}
r(t) &= r_I(t) + jr_Q(t) + n_{\text{bb}}(t) = S(t - \tau)\exp[j(2\pi\Delta ft + \varphi)] + n_{\text{bb}}(t) \\
&= S(t - \tau)[\cos(2\pi\Delta ft + \varphi) + j\sin(2\pi\Delta ft + \varphi)] + n_{\text{bb}}(t)
\end{aligned} \tag{5.2b}
$$

where Δf represents the difference between the transmitter and receiver LO frequencies. In addition φ_R is the cumulative phase offset introduced to the received signal from the channel and the receiver local oscillator and τ is a delay mostly introduced by the channel response at the received signal:

$$\varphi = \varphi_R - 2\pi f_C \tau \tag{5.3}$$

$r_I(t)$ and $r_Q(t)$ are quadrature components of the baseband received signal $r(t)$.

The last block of the receiver detects the transmitted signal by estimating a phase difference for M-PSK modulation. For QAM modulated signals this block estimates euclidean distances

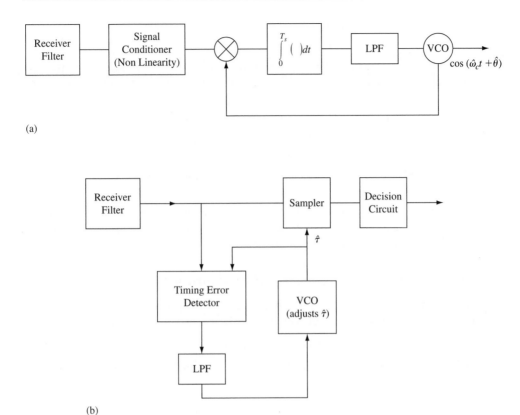

Figure 5.2 (a) Feedback frequency/phase recovery, (b) feedback timing synchronization

between the received signal and the M possible transmitted symbols. This, in turn will be used to make the decision for the transmitted symbol.

The general approaches for achieving frequency/phase and timing synchronization are based on feedback and feed-forward techniques. In the former, there is a feedback mechanism between the error detection and error correction blocks which aids to synchronize the signal in frequency or time. Figure 5.2(a) and (b) illustrates the basic functionality for feedback frequency/phase recovery and for timing synchronization, respectively. The former contains a signal-conditioning block to help remove information from the received signal. The frequency/phase error is detected by a mixer/phase-detector, which feeds the feedback mechanism (a filter and a VCO) to produce the recovered and synchronized carrier.

Figure 5.2(b) shows a timing recovery (TR) system with feedback. The basic differences are in the timing error detector instead of a mixer/phase detector and in the output of the VCO, which has to be a time-synchronized periodic pulsed signal to drive the sampling device properly.

Figure 5.3 shows the arrangement of feed-forward estimators for time, frequency and phase. One can see that the basic difference between these structures and the FB CR and TR systems is that in feedback recovery an error is estimated which is used by a feedback mechanism to achieve synchronization. In feed-forward systems the timing or frequency/phase estimation is used to directly adjust the sample time, frequency or phase of the received signal respectively.

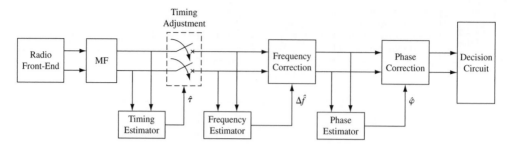

Figure 5.3 Feed-forward timing and frequency/phase recovery

It must be noted that we mainly use the baseband domain throughout Section 5.1. Furthermore, we follow Proakis [Proakis01] for feedback synchronizers and Mengali and D'Andrea [Mengali97] when dealing with feed-forward carrier recovery.

5.1.2 ML Estimation (for Feedback and Feed-forward) Synchronizers

Denoting by φ and τ the carrier phase offset of the receiver LO and the timing delay with which the transmitted signal arrives at the receiver, the received signal can be expressed as:

$$r(t) = s(t; \varphi, \tau) + n(t) \tag{5.4a}$$

where $s(t; \varphi, \tau)$ denotes the complex envelope of the transmitted signal affected by φ and T:

$$s(t; \varphi, \tau) \triangleq e^{j\varphi} \sum a_n g(t - nT_S - \tau) \tag{5.4b}$$

Furthermore, by expanding $r(t)$ into orthonormal functions the vector $\mathbf{r} = [r_1, r_2, \ldots, r_N]$ is obtained, whereas the parameters to be estimated can be represented by vector $\boldsymbol{\psi} = [\varphi, \tau]$ as well. To obtain the maximum likelihood (ML) estimate of $\boldsymbol{\psi}$ the conditional p.d.f. $p(\mathbf{r}|\boldsymbol{\psi})$ must be maximized. It can be shown [Proakis01] that this is equivalent to maximizing the likelihood function $\Lambda(\boldsymbol{\psi})$:

$$\Lambda(\boldsymbol{\psi}) = \exp\left\{-\frac{1}{N_0}\int_{T_0}[r(t) - s(t; \boldsymbol{\psi})]^2 dt\right\} \tag{5.5a}$$

To obtain the ML estimate we take the logarithm of the maximum likelihood function (the log-likelihood function Λ_L) which, after some mathematical manipulation, becomes [Proakis01]:

$$\Lambda_L(\varphi) = \frac{2}{N_0}\int_{T_0} r(t)s(t; \varphi, \tau) dt \tag{5.5b}$$

To get the ML phase estimate we assume that the delay τ is known or has been estimated and we eliminate this parameter from the ML function resulting in Λ_L as follows:

$$\Lambda_L(\varphi) = \frac{2}{N_0}\int_{T_0} r(t)s(t; \varphi) dt \tag{5.5c}$$

If the transmitted signal is (as given in Equation (5.1) $S(t) = \sum_n a_n g(t - nT_S)$, the baseband signal at the receiver is $r(t)$ as given by Equation (5.2b).

The output of the matched filter is:

$$y(t') = \sum r(t) \cdot g(t' - t) \tag{5.6}$$

If in addition to phase and timing offsets (φ, τ) we have frequency offset Δf imposed by the local oscillator at the receiver, then the output of the matched filter is:

$$y(t') = \sum r(t) \cdot g_M(t' - t) \exp[-j2\pi\Delta f(t - t')] \tag{5.7}$$

In light of the above, it can be shown [Mengali97] that Λ_L takes the form:

$$\int_0^{T_0} r(t)s(t)dt = \exp(-j\varphi) \sum_k a_k \int_0^{T_0} r(t)[\exp(-j2\pi\Delta ft)]g(t - kT_S - \tau)dt$$

$$\approx \exp(-j\varphi) \sum_{k=1}^{L_0-1} a_k^* y(k) \tag{5.8}$$

where $y(k)$ are the samples (at time instant $kT_S + \tau$) of $y(t)$ with integral form as given below

$$y(t) \equiv \int_0^\infty r(\lambda)[\exp(-j2\pi\Delta f\lambda)g(\lambda - \tau)d\lambda] \tag{5.9}$$

Consequently, the log-likelihood function can now be written [Mengali97]:

$$\Lambda_L(\varphi) = \frac{2}{N_0} \left[e^{-j\varphi} \sum_{k=1}^{L_0-1} a_k^* y(k) \right] \tag{5.10}$$

where $L_0 = T_0/T_S$ represents the observation interval length normalized to the symbol duration.

Below, using an example from [Proakis01], we show briefly how the ML principle is used to obtain a feedback-type synchronizer. We assume an unmodulated carrier $\cos(\omega_C t)$ is transmitted and the received passband signal $r(t)$ consists of the transmitted sinusoidal with a phase offset φ plus a narrowband noise term $n(t)$:

$$r(t) = A\cos(\omega_C t + \varphi) + n(t) \tag{5.11}$$

Subsequently we obtain:

$$\Lambda_L(\varphi) = \frac{2}{N_0} \int_{T_0} r(t)\cos(\omega_C t + \varphi)dt \tag{5.12}$$

Setting the derivative of $\Lambda_L(\varphi)$ with respect to φ equal to zero we get [Proakis01]:

$$\left. \begin{array}{l} \text{ML condition}: \displaystyle\int_{T_0} r(t)\sin(\omega_C t + \hat{\varphi}_{ML}) = 0 \\[4mm] \hat{\varphi}_{ML} = -\tan^{-1}\left\{ \displaystyle\int_{T_0} r(t)\sin(\omega_C t)dt \Big/ \displaystyle\int_{T_0} r(t)\cos(\omega_C t)dt \right\} \end{array} \right\} \tag{5.13}$$

The first equation (ML condition) suggests a general implementation of ML phase estimation in the form of a PLL. The VCO is commanded by the filtered output of the phase detector to produce $\sin(\omega_C t + \hat{\varphi}_{ML})$ which in turn is applied at one of the inputs of the phase detector.

More specifically, if the input of the PLL consists of a sinusoidal signal plus narrowband noise as in Equation (5.11), it can be shown [Gardner79] that the variance of the phase error $\Delta\varphi$ at the output of the PD is

$$\sigma_\varphi^2 = \frac{N_0 B_L}{A^2} = \frac{1}{2\text{SNR}_L} \tag{5.14}$$

where N_0 is the PSD of noise, B_L is the equivalent noise bandwidth of the PLL (see Chapter 1) and SNR_L is the SNR of the loop. Figure 5.4 shows the variance of the phase error as a function of SNR_L for first and second order loops.

5.1.3 Feedback Frequency/Phase Estimation Algorithms

We now proceed to examine carrier synchronization techniques when a modulated received signal is used for frequency/phase recovery.

5.1.3.1 Decision-directed Techniques

In this case we assume that the sequence of data symbols a_i has been recovered and that only the phase of the carrier is still unknown. If the received signal is as in Equation (5.5), we use a log-likelihood function $\Lambda_L(\varphi)$ as in Equation (5.4). Substituting in Equation (5.4) for $r(t)$ and $s(t)$ as given in Equation (5.5) with $\tau = 0$, after some mathematical manipulations it can be shown that [Proakis01]:

$$\hat{\varphi}_{ML} = -\tan^{-1}\left[\text{Im}\left(\sum_{n=0}^{L_0-1} a_n^* y_n\right) \Big/ \text{Re}\left(\sum_{n=0}^{L_0-1} a_n^* y_n\right)\right] \qquad (5.15)$$

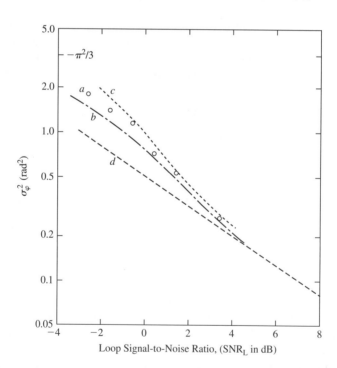

Figure 5.4 The variance of phase error as a function of loop SNR, (a) experimental data for second order loop, (b) exact nonlinear analysis for first order loop [Viterbi66], (c) approximate nonlinear analysis for second order loop [Lindsey73], (d) linear approximation (reprinted from F. Gardner, 'Phaselock Techniques', 2nd edn, copyright ©1979 by John Wiley & Sons, Inc.)

where the output of the matched filter during the nth interval is:

$$y_n = \int_{nT}^{(n+1)T} r(t)g^*(t - nT_S)dt \tag{5.16}$$

where the sampled output of the matched filter during the nth interval is

$$s(t) = S(t) \cdot \cos(\omega_C t + \varphi) \tag{5.17}$$

We must keep in mind that a QAM signal is transformed to PAM after splitting into in-phase and quadrature-phase branches. After some mathematical calculations, it can be shown that the error signal at the input of the loop filter is:

$$e(t) = [S^2(t)/2]\sin \Delta\varphi + [S(t)/2][n_C(t)\sin \Delta\varphi - n_S(t)\cos \Delta\varphi] + \text{terms centred at } 2\omega_C \tag{5.18}$$

The loop filter eliminates the high frequency terms and uses the first component of Equation (5.18), which contains the error $\Delta\varphi$ to drive the VCO and close the PLL recovery loop.

The subsystem used for carrier recovery of an M-PSK modulated signal having the form $A\cos(\omega_C t + \theta_i(t))$ [with $\theta_i = 2\pi(i - 1)/M$] has many similarities to Figure 5.5. The main difference is that both the in-phase and quadrature branches have delay and detection blocks such that at the end the information phase $\hat{\theta}_i$ is estimated. Subsequently, $\cos \hat{\theta}_i$ and $\sin \hat{\theta}_i$ are generated to create an error function which has the following low-frequency components [Proakis01]:

$$e_{LF}(t) = -\frac{A}{2}\sin(\varphi - \hat{\varphi}) + \frac{n_C(t)}{2}\sin(\varphi - \hat{\varphi} - \hat{\theta}_i) + \frac{n_S(t)}{2}\cos(\varphi - \hat{\varphi} - \hat{\theta}_i) \tag{5.19}$$

Because there are no products of noise in both the above error functions $e_{LF}(t)$, there is no power loss when using decision-directed carrier recovery with PLLs.

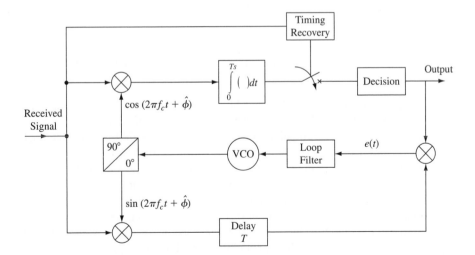

Figure 5.5 Phase/frequency recovery using a PLL system

5.1.3.2 Nondecision-directed Techniques

If the received data are considered random variables, to obtain the ML equations for CR, averaging of the likelihood function must be employed. This is known as the nondecision aided approach to synchronization and probability density functions of the information symbols are required.

In this subsection passband transmitted and received signals are used to show more clearly the resulting PLL structure of the ML estimator. However, the forms of the resulting equations and estimator blocks are the same as those derived when considering baseband signals [Mengali97].

For an amplitude-modulated signal $s(t)$ (like PAM) as in Equation (5.17), the ML estimate is obtained by differentiating the average $\bar{\Lambda}_L(\varphi)$ which, in general, is given by:

$$\bar{\Lambda}_L(\varphi) = \ln\left[\int_{-\infty}^{\infty} \Lambda(\varphi)p(A)\mathrm{d}A\right] \tag{5.20}$$

The p.d.f. can have different forms depending on the type of modulation. Very frequently it is a zero-mean Gaussian random variable with unit variance:

$$p(A) = \frac{1}{\sqrt{2\pi}}e^{-A^2/2} \tag{5.21}$$

In that case, the corresponding log-likelihood function is given by [Proakis01]:

$$\bar{\Lambda}_L(\varphi) = \left[\frac{2}{N_0}\int_0^T r(t)\cos\left(\omega_C t + \varphi\right)\mathrm{d}t\right]^2 \tag{5.22}$$

At this point we note that the Gaussian distribution for the data symbols is realistic when the cross-correlation of $r(t)$ with $s(t; \varphi)$ is relatively small. This results in a quadratic form of the log-likelihood function.

We continue by averaging the likelihood function over an observation interval $T_0 = L_0 T_S$ under the assumption that the transmitted data symbols during this interval are statistically independent and identically distributed. In that case we obtain the equation satisfying the ML criterion:

$$\sum_{n=0}^{L_0-1}\int_{nT}^{(n+1)T} r(t)\cos\left(\omega_C t + \hat{\varphi}\right)\mathrm{d}t\left[\int_{nT}^{(n+1)T} r(t)\sin\left(\omega_C t + \hat{\varphi}\right)\mathrm{d}t\right] = 0 \tag{5.23}$$

The above equation can be translated to a synchronization loop as shown in Figure 5.6. Three important observations are in order: (a) the structure has the form of a dual PLL where the third multiplier eliminates the sign information from the modulated signal; (b) the summation block operates like the loop filter of a single-loop PLL and can be implemented as a digital filter; and (c) the overall structure has the form of a Costas loop about which we will discuss below.

Two popular nondecision directed loops for PAM carrier synchronization are the squarer loop and the Costas loop. Figures 5.7 and 5.8 illustrate the structure of these circuits. For a received signal having the form of Equation (5.17), the squarer has removed the sign of the modulated signal at the output of the bandpass filter tuned at $2\omega_C$, at which its amplitude is proportional to $A^2(t)$. Furthermore, the output of the locked VCO is divided in frequency by 2, in order to obtain the phase-synchronized carrier.

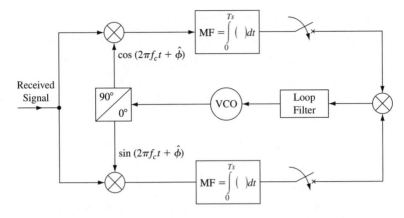

Figure 5.6 Synchronization loop for nondecision-directed technique using PLL

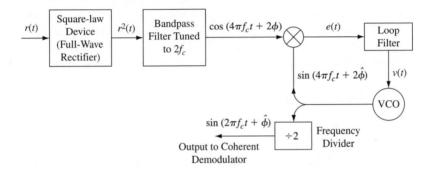

Figure 5.7 Carrier recovery using square-law nonlinearity

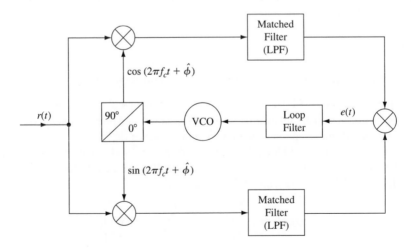

Figure 5.8 Carrier recovery using Costas loop

The following equation gives the error signal at the output of the PD [Lindsey71]:

$$e(t) = K\{[A^2(t) - n_C^2(t) - n_S^2(t) - 2A(t)n_S(t)]\sin[2\Delta\varphi(t)]$$
$$+ n_C(t)[2A(t) - 2n_S(t)]\cos[2\Delta\varphi(t)]\} \tag{5.24}$$

Regarding Costas loop, we observe that the error signal here is formed by the multiplication of the filtered outputs of the upper (I-loop) and lower (Q-loop) loops. After some mathematical manipulation this error signal at the input of the loop filter is given by [Proakis01]:

$$e(t) = \frac{1}{8}[A^2(t) + n_C^2(t) + 2A(t)n_C(t) - n_S^2(t)]\sin[2\Delta\varphi(t)]$$

$$-\frac{1}{4}n_S(t)[A(t) + n_C(t)]\cos[2\Delta\varphi(t)] \tag{5.25}$$

From the above expression, the first term $A^2(t) \cdot \sin 2(\hat{\varphi} - \varphi)$ is the desired term, which will lock the VCO in order to obtain the required phase-recovered carrier.

One important thing to note here concerns the effect of noise. If we assume that at the input of both configurations we have an additive noise term along with the signal $[s(t) + n(t)]$, then it can be easily seen there exist two components associated with noise at the output which have the form (*signal*) × (*noise*) and (*signal*) × (*noise*). Because these components are centred at the same frequency as the useful signal entering the loop ($2\omega_C$), we have the same noise deterioration for both systems. It can be shown that the variance of the phase error is [Proakis01]:

$$\sigma_\varphi^2 = \frac{1}{SNR_L SL}, \quad SL = \frac{1}{\left(1 + \frac{B_{BPF}/2B_{eq}}{SNR_L}\right)} \tag{5.26}$$

where SNR_L is the SNR of the loop, B_{BPF} is the bandwidth of the BPF at the output of the squarer and B_{eq} represents the loop equivalent noise bandwidth. SL represents the loss due to the squarer operation and is associated to the noise terms at the output of the BPF.

For N-phase modulated carrier (for example N-PSK) it can be shown that an N-phase Costas loop can be used to recover the carrier [Lindsey72]. However, it can be seen that such a structure would exhibit high implementation complexity. Figure 5.9 gives the proposed system in [Leclert83], which has a low complexity and a very similar structure to quadriphase lock loops presented above. In addition, it was demonstrated [Leclert83] that it does not have any false lock points.

5.1.4 Feed-forward Frequency/Phase Estimation Algorithms

As briefly presented above, FF frequency/phase estimators operate in the digital domain and use preamble sequences to obtain estimates for the frequency and phase of the incoming signal. Consequently the estimator operates on known data sequence $\{a_n\}$.

Figure 5.10 shows a general block diagram of a frequency and phase estimator which also compensates for the values of f and φ after estimation. Using Equation (5.2) we arrive at the following likelihood function [Mengali97], [Proakis01]:

$$\Lambda_L(\varphi, f) = \frac{1}{N_0}\left[\exp(-j\varphi)\sum_{k=1}^{L_0-1} a_k y_k(\mathbf{\psi})\right] \tag{5.27}$$

where y_k correspond to samples of the received signal at the output of the matched filter and $\mathbf{\psi} = [\varphi, f]$ is the parameter vector. a_k is the data symbol at the kth time instant.

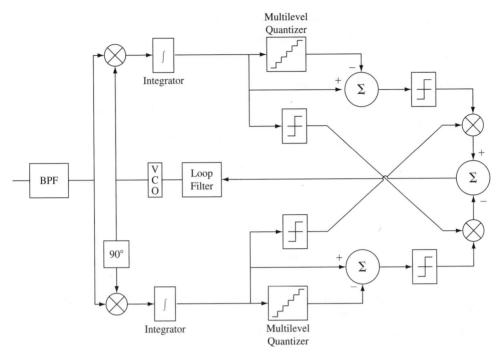

Figure 5.9 Low complexity M-PSK carrier recovery (reprinted from A. Leclert, P. Vandamme, 'Universal carrier recovery loop for QASK and PSK signal sets', vol. 31, pp. 130–136, ©1983 IEEE)

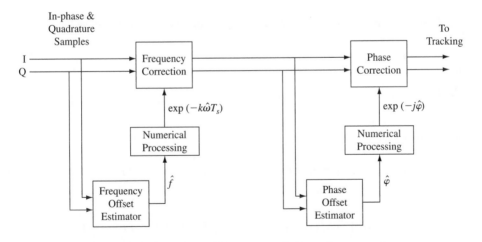

Figure 5.10 Feed-forward frequency and phase-offset correction

In more detail, from Equation (5.27), the joint phase/frequency ML criterion can be shown to be:

$$\{\hat{\varphi}, \Delta\hat{f}\} = \arg\left(\max_{\Delta f,\theta}\left[\sum_{k=0}^{L-1} a_k^* \exp\left(-j\hat{\varphi}\right)\exp\left(-j2\pi kT\,\Delta\hat{f}\right)x_k^*\right]\right) \tag{5.28}$$

This two-dimensional ML estimator can be split iton two (one for frequency and one for phase) by first considering φ as a uniformly distributed random phase.

5.1.4.1 Frequency Offset Estimation

For feed-forward estimation of the frequency offset a group of samples N_P from the preamble can be used to feed the receiver. For frequency estimation after some nontrivial mathematical manipulation the following log-likelihood function is obtained [Mengali97]:

$$\Lambda_{\Gamma}(\tilde{f}) = \left| \sum_{k=0}^{L_0-1} a_k y_k \right| \tag{5.29}$$

where

$$y(t) \equiv \int_{-\infty}^{\infty} r(x) \exp\left(-j2\pi\Delta f x\right) g(x-t) \mathrm{d}x \tag{5.30}$$

where $y(t)$ is actually the output of the matched filter when the input signal is $r(t)\exp\left(-j2\pi\tilde{f}t\right)$.

Maximization of $\Lambda_{\Gamma}(f)$ above is cumbersome and other methods of lower complexity are reported. Following [Luise95] an equivalent likelihood function is:

$$\Lambda_L(\Delta f) \equiv \left| \sum_{k=1}^{N} (a_k y_k) \exp\left(-j2\pi\Delta\tilde{f}T_S k\right) \right|^2 = \sum_{i=1}^{N}\sum_{m=1}^{N} (a_i y_i)(a_m y_m)^* \exp\left[-j2\pi\Delta\tilde{f}T_S(i-m)\right] \tag{5.31}$$

$N(N \leq L_0 - 1)$ is a parameter to be decided from simulations. Furthermore we have:.

$$z_k \overset{\Delta}{=} a_k^* y_k \tag{5.31a}$$

$$y_k \overset{\Delta}{=} \int_{-\infty}^{\infty} r(t)g(t - kT_S - \tau)\mathrm{d}t, \qquad r(t) = \mathrm{e}^{j(2\pi\Delta f + \varphi)} \sum_{n} a_n g(t - nT_S - \tau) + n(t) \tag{5.31b}$$

It can be shown that

$$y_k = a_k \exp j[2\pi\Delta f(kT_S + \tau) + \varphi] + \int_{-\infty}^{\infty} n(t)g(t - kT_S - \tau)\mathrm{d}t \tag{5.31c}$$

$$z_k = \exp j[2\pi\Delta f(kT_S + \tau) + \varphi] + a_k^* \int_{-\infty}^{\infty} n(t)g(t - kT_S - \tau)\mathrm{d}t \tag{5.31d}$$

A good Δf estimate of reduced complexity can be found by taking the derivative of Equation (5.31), which results in the following expression [Luise95]:

$$\mathrm{Im}\left\{ \sum_{k=1}^{M} R(k) \exp\left(-j2\pi k T_S \Delta\tilde{f}\right) \right\} = 0 \tag{5.32}$$

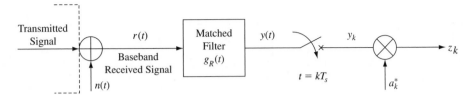

Figure 5.11 Sampled sequence Y_k used as input to the feed-forward frequency estimator

where $M \leq N - 1$ and $R(k)$ represents an estimate for the autocorrelation of the sampled sequence $a_k y_k$. It is defined as:

$$R(k) \equiv \frac{1}{N-k} \sum_{i=k+1}^{N} (a_i y_i)(a_{i-k} y_{i-k})^*, \quad 0 \leq k \leq N - 1 \tag{5.33}$$

Equation (5.32) can lead to the following frequency offset estimate:

$$\hat{\Delta f} \cong \frac{1}{2\pi T} \frac{\sum_{k=1}^{M} \text{Im}[R(k)]}{\sum_{k=1}^{M} k \text{Re}[R(k)]} \tag{5.34}$$

Observing Equations (5.31) and (5.33) we see that the sampled sequence z_k used as the input to this estimator is illustrated in Figure 5.11 and is expressed as:

$$z_k = a_k^* y_k = \exp j(k 2\pi \Delta f T_S) + n'(t) \tag{5.35}$$

Finally, since $R(k)$ has an exponential format [Luise95], further simplifications result in a computationally simpler estimator:

$$\hat{\Delta f} \approx \frac{1}{\pi T(M + 1)} \arg \left[\sum_{k=1}^{M} R(k) \right] \tag{5.36}$$

The frequency offset $\hat{\Delta f}$ limits under which the Luise algorithm operates properly are $\Delta f < 1/T(M + 1)$. Furthermore, the error variance of the estimator is close to the Cramer–Rao lower bound for values of M close to $N/2$. Figure 5.12 shows a possible realization where on top there is a FIR-like structure to calculate the quantity within the arg function. Subsequently, a look-up table can be used which assigns angles to the summation quantities.

5.1.4.2 Phase Offset Estimation

The estimator for the phase offset uses a group of N_P samples from the preamble which have been previously gone through the frequency correction system, as shown in Figure 5.13. These samples $y_{F,k}$ have the form:

$$y_{F,k} = a_k \exp \left[j(k 2\pi \Delta f T_S) + \varphi \right] + n_k \tag{5.37}$$

where $\Delta f = f - \hat{f}$ is the remaining error in frequency after correction in frequency offset has been applied. a_k represents the symbols of the preamble and n_k is additive white Gaussian noise.

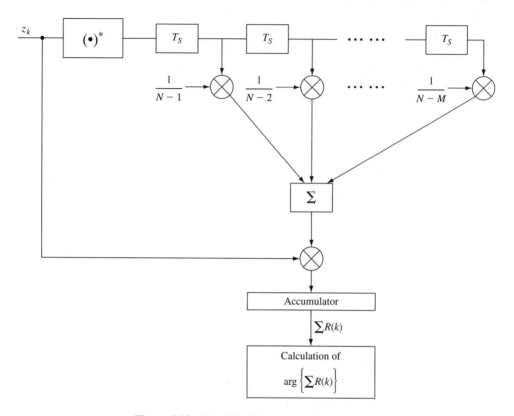

Figure 5.12 Simplified frequency offset estimator

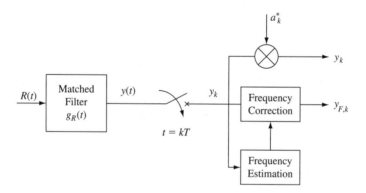

Figure 5.13 Block diagram producing samples $y_{F,k}$ used for the feed-forward phase estimator

It can be shown that the maximum-likelihood function for the estimation of the phase offset is:

$$\Lambda_L(\varphi) = \text{Re}\left[\sum_{k=K}^{K+N_P-1} a_k^* y_{F,k} \exp\left[-j(k2\pi\Delta f\,T_S) + \varphi\right] \right] \qquad (5.38)$$

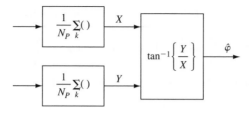

Figure 5.14   Feed-forward phase estimator

where k is associated with the position in the preamble, which will be the starting point for the likelihood function calculation.

Because of the difficulties in finding closed form expression for maximizing the above log-likelihood function, other mathematical expressions must be explored. It can be shown [Viterbi83] that averaging operation can lead to a straightforward determination of the phase estimate. The time average is given by the following expression:

$$E(y_{F,k}) \approx \frac{1}{N_P} \sum_{k=K}^{K+N_P-1} y_{F,k} = C \frac{1}{N_P} \sum_{k=K}^{K+N_P-1} \exp\left[j(k 2\pi \Delta f T_S + \varphi)\right]$$

$$= \frac{C}{N_P} \frac{\sin\left(N_P \Delta f /2\right)}{\sin\left(\Delta f /2\right)} \exp j\left[\varphi + 2\pi \Delta f T_S \left(K + \frac{N_P-1}{2}\right)\right] \qquad (5.39)$$

which finally leads to the phase estimate:

$$\hat{\varphi} = \tan^{-1}\left(\frac{\mathrm{Im}\left[\sum_{k=K}^{K+N_P-1} y_{F,k}\right]}{\mathrm{Re}\left[\sum_{k=K}^{K+N_P-1} y_{F,k}\right]}\right) = \varphi + 2\pi \Delta f T_S \cdot \left(K + \frac{N_P-1}{2}\right) \qquad (5.40)$$

Figure 5.14 shows the simple implementation of this phase estimator.

The above equation indicates that the estimated phase is a linear function of Δf which will always have a small but nonzero value. However, choosing $K = -(N_P-1)/2$ the averaging is centred at the middle of the preamble and the phase estimate is no longer a function of Δf.

5.1.5 Feedback Timing Estimation Algorithms

In high-frequency communications timing recovery involves both frame synchronization and symbol timing recovery. Frame synchronization has to do with finding the correct timing instant at which a useful packet begins. On the other hand the purpose of symbol timing recovery is to find the optimum time instant $nT_S + \hat{\tau}$ at which to sample data symbols.

5.1.5.1 Decision-directed Techniques

Neglecting phase offset the baseband PAM signal at the receiver is given by Equation (5.5):

$$r(t) = \sum_n a_n g(t - nT_S - \tau) + n(t) \qquad (5.41)$$

If we now use Equation (5.3) and eliminate φ we result in a log-likelihood function:

$$\Lambda_L(\tau) = \sum_n a_n \int_{T_O} r(t)g(t - nT_S - \hat{\tau})dt = \sum_n a_n y_n(\tau) \tag{5.42}$$

where in this case the output of the matched filter should include the time delay:

$$y_n(t) = \int_{T_O} r(t)g(t - nT_S - \hat{\tau})dt \tag{5.43}$$

We can get the ML timing estimate by setting the derivative to zero

$$\frac{d\Lambda_L(\tau)}{d\tau} = \sum_n a_n \frac{d}{d\tau}\left[\int_{T_O} r(t)g(t - nT_S - \hat{\tau})dt \right] = 0 \tag{5.44}$$

The above expression for timing synchronization can be implemented using a feedback loop such as the one shown in Figure 5.15, where the differentiated output from the matched filter enters a sampler which is part of a feedback loop. This loop has a filter in the form of a finite summation block. The length of the summation determines the bandwidth of the filter. The voltage-controlled oscillator following the filter produces a clock with variable phase depending on the output of the filter. In this way we control the time instant at which the sampler will be activated.

5.1.5.2 Nondecision-directed Techniques

Similarly to frequency estimation NDA techniques, the likelihood function must be averaged over the p.d.f. of the transmitted data symbols. For a PAM-modulated signal we get [Proakis01]:

$$\overline{\Lambda}_L(\tau) = \sum_n \ln \cosh [Cy_n(\tau)] \approx \frac{C^2}{2}\sum_n y_n^2(\tau) \tag{5.45}$$

The approximation at the right-hand side of the above equation holds for small x for which $\ln [\cosh x] \approx x^2/2$. C is a constant proportional to the received signal power. Taking the derivative we get:

$$\frac{d}{d\tau}\sum_n y_n^2(\tau) = 2\sum_n y_n(\tau) \cdot \frac{dy_n(\tau)}{d\tau} = 0 \tag{5.46}$$

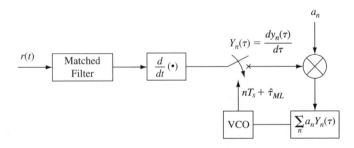

Figure 5.15 ML DD timing estimator for PAM signal

Figure 5.16 illustrates a loop which implements the tracking process described by Equation (5.45). We must note again here that the summation block functions as a loop filter. Equation (5.46) can be implemented using a feedback loop with two branches which strongly resembles the Costas loop.

Another way to implement nondata aided (NDA) timing synchronizers is by using early–late gate loops. This architecture is based on the fact that the output of the matched filter is a symmetrical signal. If the basic unit pulse is a rectangular pulse, the output of the matched filter is a triangle as in Figure 5.17. When the signal is sampled at time instants $T_S - \delta$ and $T_S + \delta$, in order to have a good estimate of T_S, the values $y(k(T_S - \delta))$ and $y(k(T_S + \delta))$ must be equal.

A dual loop in the form of a Costas loop as shown in Figure 5.18 can be used to identify the advance and delay value δ at which the outputs of the matched filter become equal. This will give the estimate $\hat{\tau}$ of the time delay. It can be shown [Proakis01], that this loop implements the differentiation of the likelihood function $\overline{\Lambda}_L(\tau)$:

$$\frac{d\overline{\Lambda}_L(\tau)}{d\tau} \approx \frac{\overline{\Lambda}_L(\tau + \delta) - \overline{\Lambda}_L(\tau - \delta)}{2\delta} \tag{5.47}$$

5.1.6 Feed-forward Timing Estimation Algorithms

To demonstrate the feed-forward method in timing estimation we follow [Morelli97] in presenting a NDA feed-forward timing estimation algorithm for M-PSK modulated data. The likelihood function incorporating the time estimate $\hat{\tau}$ is:

$$\Lambda(\{a_i\}, \varphi, \tau) = \exp\left\{ \frac{2E_S}{N_0} \sum_{i=0}^{L-1} \mathrm{Re}[e^{-j\hat{\varphi}} \hat{a}_i^* y(iT_S + \hat{\tau})] \right\} \tag{5.48}$$

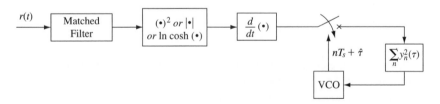

Figure 5.16   NDA timing estimator for PAM signal

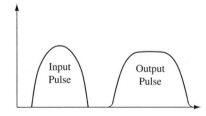

Figure 5.17   Typical input pulse and corresponding matched filter output

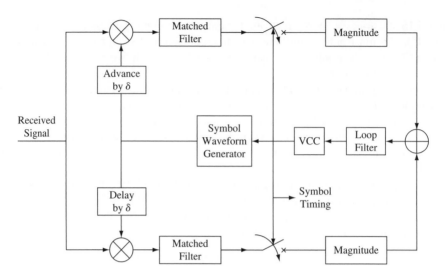

Figure 5.18 Early–late sample synchronizer [Proakis04]

To derive the NDA maximum likelihood $\Lambda(\tau)$ averaging of $\Lambda(\{a_i\}, \varphi, \tau)$ must be performed over the estimated data and carrier phase $\{a_i\}, \varphi$ respectively. Averaging is first performed over data $\{a_i\}$ considered as i.i.d. random variables:

$$\Lambda(\varphi, \tau) = \prod_{i=0}^{L_0-1} \frac{1}{M} \sum_{m=0}^{M-1} \exp\left(\frac{2}{N_0} \mathrm{Re}[y(iT_S + \tau) \cdot \exp[-j(\varphi + 2\pi m/\mathrm{M})]] \right) \qquad (5.49)$$

At low SNR values and for $M \geq 4$ the second exponent within the summation has a small argument which justifies the approximation $\exp(x) \approx 1 + x + x^2/2$. Taking this into account, the above equation gives the required likelihood function:

$$\Lambda(\tau) \approx \prod_{i=0}^{L_0-1} \left[1 + \frac{|x(iT_S + \hat{\tau})|^2}{4(N_0/2)^2} \right] \qquad (5.50)$$

It is easy to note that the argument φ within the ML function has been eliminated since there is no dependence of the right-hand side of Equation (5.50) on it. The logarithms of the above equation will give the log-likelihood function:

$$\Lambda_L(\tau) \approx \sum_{i=0}^{L_0-1} \ln\left[1 + |y(iT_S + \tau)|^2 \left(\frac{E_S}{N_0} \right)^2 \right] \qquad (5.51)$$

Maximization of $\Lambda_L(\tau)$ will produce the following estimate for τ [Morelli97]:

$$\hat{\tau} = -\frac{T_S}{2\pi} \arg\left\{ \frac{1}{N} \sum_{k=0}^{N-1} \Lambda_L\left(\frac{kT_S}{N} \right) \exp[-j2\pi k/N] \right\}$$

$$= \frac{T_S}{2\pi} \arg\left\{ \frac{1}{N} \sum_{k=0}^{NL_0-1} \ln\left[1 + \left| y\left(\frac{kT_S}{N} \right) \right|^2 \left(\frac{E_S}{N_0} \right)^2 \right] \exp[-j2\pi k/N] \right\} \qquad (5.52)$$

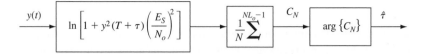

Figure 5.19   Feed-forward timing estimator

Figure 5.19 depicts the block diagram for the realization of this estimator. One can see that the nonlinearity has the form of $\ln(1 + Kx^2)$. Under low SNR, the nonlinearity can take the form of a squarer and the system resembles the one in Oerder and Meyr [Oerder88]. Although this algorithm requires knowledge of SNR, simulations, Morelli [Morelli97] have shown that the estimator is not very sensitive to differences between the actual and the used SNR.

5.2 Diversity

5.2.1 Diversity Techniques

Multipath fading induces considerable degradation in radio systems performance mainly due to large variations of received power. The basic principle of diversity is associated with the fact that multiple copies of the same signal arriving at the receiver will not go through the same deep fading simultaneously. Consequently, the receiver can combine all or part of these copies of the signal to achieve higher signal-to-noise ratio.

Regarding the copies of the transmitted signal, they may represent the signals detected by M different antennas at the receiver. This is called receiver diversity. On the other hand, they may arrive from different paths associated with different transmitting antennas. In this case we have transmitter diversity. Both configurations belong to the general category of antenna diversity which is the most frequently encountered form of diversity. For any configuration it is sufficient to say that we have multiple independent versions of the transmitted signal between the transmitter and the receiver.

In all cases diversity is associated with some form of losses in implementation. In receiver diversity, and depending on the kind of combining, it may be necessary to have multiple antenna subsystems. In transmitter diversity, apart from the multiple antennas, we may need more transmitted power or bandwidth.

Another form of diversity is polarization diversity based on the fact that, when a signal is transmitted simultaneously in different polarizations, the resulting signal components at the receiver are highly decorrelated. This helps to combine constructively the received signal components.

Frequency diversity is based on the principle of transmitting the signal simultaneously in several frequencies which differ by more than the coherence bandwidth. This results in independent fading of the resulting signals.

Time diversity is associated with transmission at different time instants such that the resulting signals are decorrelated at the receiver. An interesting and effective form of time diversity used frequently in modern transceivers is the RAKE receiver, which has the ability to detect a number of the multipath components produced by the same transmitted signal due to reflections. We will examine the RAKE receiver in some extent in Section 5.4, which concerns spread spectrum communications.

5.2.2 System Model

If we consider a radio link involving M_t transmitter antennas and M_r receiver antennas as shown in Figure 5.20, the received signal by the lth antenna at the kth time instant is given by the following summation of convolutions:

$$r_l(k) = \sqrt{\frac{P_t}{N}} \cdot \sum_{i=1}^{M_t} h_{li}(k) * s_i(k) + n_l(k) \tag{5.53}$$

where h_{li} represents the channel impulse response of the $l-i$ link (a $M_t \times M_r$ matrix), $s_i(t)$ is the transmitted signal and $n(t)$ is AWG noise with variance N_0. P_t is the total average energy at the transmitter within a symbol period.

The signal from every transmitting antenna (for example Tx_2 in Figure 5.20) reaches all receiving antennas, whereas every receiving antenna (for example Rx_1 in Figure 5.20) receives signals from all the transmitting antennas.

Such an $M_t \times M_r$ arrangement is called multiple-input multiple-output (MIMO) communication system in which the $(l-i)$ channels ($l=1,\ldots,M_r$ $i=1,\ldots,M_t$) comprise multiple independently fading channels. A transceiver with one antenna at the transmitter and one at the receiver is also called a single-input single-output (SISO) system.

At this point some considerations on the capacity of a multiantenna system are in order. The capacity of a transmission technique is the highest transmission rate at which the error probability of transmission can be kept below a given threshold of any value. The capacity of a SISO system for an AWGN channel is given by:

$$C = \log_2 \left(1 + \frac{P_t}{N_0} \right) \tag{5.54}$$

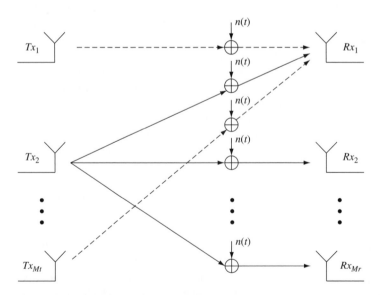

Figure 5.20   Model of a MIMO communication system

For a SISO system in a flat fading stationary and ergodic channel the capacity is:

$$C = E\left[\log_2\left(1 + \frac{P_t}{N_0}|h|^2\right)\right] \tag{5.55}$$

where expectation is performed with respect to the channel response $h(k)$. For a system with one transmitting and M receiving antennas we have:

$$C = E\left[\log_2\left(1 + \frac{P_t}{N_0}\|\mathbf{h}\|^2\right)\right] \tag{5.56}$$

where in this case $\mathbf{h}$ is a vector representing the channel response and $\|\bullet\|$ is the norm of the vector given by $\sum_i |h_i|^2$. It should be noted here that P_t/N_0 represents the average SNR at each receiver antenna.

For a general $M_t \times M_r$ MIMO arrangement the capacity is given by [Diggavi04]:

$$C = E_H\left[\log\left|\mathbf{I} + \frac{1}{\sigma^2}\frac{P}{M_t}\mathbf{H}\mathbf{H}^*\right|\right] \tag{5.57}$$

for the special case where the autocorrelations $\mathbf{R}_x$ and $\mathbf{R}_z$ for transmitted information signal and noise respectively are given as:

$$\mathbf{R}_x = (P/M_t)\mathbf{I}, \quad \mathbf{R}_z = \sigma^2\mathbf{I} \tag{5.58}$$

5.2.3 Diversity in the Receiver

If we have one Tx antenna and M_r receiving antennas the received signal for the ith antenna is:

$$r_i(t) = a_i(t)e^{j\theta_i(t)}s_i(t) + n_i(t) \tag{5.59}$$

taking into account that the channel introduces a gain $a_i(t)$ and a phase shift $\theta_i(t)$ for the ith antenna.

The ith antenna branch has an average SNR $\overline{\gamma}_i$ which is:

$$\overline{\gamma}_i = \frac{P_t}{N_0}E(a_i^2) \tag{5.60}$$

To benefit from diversity, a combining scheme must be employed at the output of the antennas. Usually a linear combining scheme is used according to which the combined output is:

$$R = \sum_{i=1}^{M_r} w_i a_i \tag{5.61}$$

Figure 5.21 illustrates the general block diagram of a diversity receiver configuration. For coherent combining, the weight factors must be complex numbers to eliminate phase shifts θ_i:

$$w_i = |w_i|e^{-j\theta_i} \tag{5.62}$$

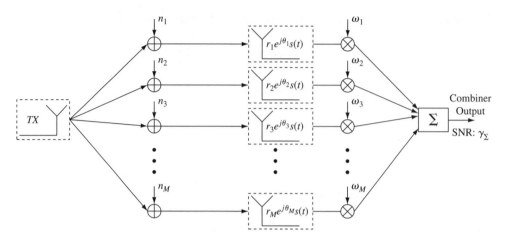

Figure 5.21 Diversity receiver configuration [Goldsmith05]

The combining action will produce a composite signal R at the output of the combiner which will have an SNR γ_C higher than the average SNR of all branches:

$$\gamma_C = \frac{\left(\sum_{i=1}^{M_r} w_i a_i\right)^2}{N_0 \sum_{i=1}^{M_r} w_i^2} \tag{5.63}$$

If we assume that the average SNR of all branches are equal ($\overline{\gamma}_i = \overline{\gamma}$), the ratio between the average of the combined signal SNR $\overline{\gamma}_C$ and $\overline{\gamma}$ is the SNR gain or array gain as it is frequently called. For all combining strategies presented below, it is necessary to calculate the average SNR $\overline{\gamma}_C$ at the combiner output as well as the cumulative distribution function (CDF) $P_{\gamma_C} = p(\gamma_C < \gamma)$ for γ_C.

To do that it is necessary to have expressions for the p.d.f. and CDF of SNR. It is easy to show that for the ith receiver branch the p.d.f. of the instantaneous SNR is [Rappaport02], [Goldsmith05]:

$$p(\gamma_i) = \frac{1}{\overline{\gamma}_i} e^{-\gamma_i/\overline{\gamma}_i} \tag{5.64}$$

and the CDF for a specific SNR threshold γ_0 is:

$$P(\gamma_i \leq \gamma_0) = \int_0^{\gamma_0} p(\gamma_i) \mathrm{d}\gamma_i = 1 - e^{\gamma_0/\overline{\gamma}_i} \tag{5.65}$$

The p.d.f. $p_{\gamma_C}(\gamma)$ of the combiner SNR γ_C can be used to calculate the average probability of error for the combining scheme as follows:

$$\overline{P}_{SC} = \int_0^\infty P_S(\gamma) p_{\gamma_C}(\gamma) \mathrm{d}\gamma \tag{5.66}$$

5.2.3.1 Selection Combining

Selection combining (SC) is the strategy according to which the combiner selects and delivers to the output the signal from the branch with the highest SNR. A threshold γ_S for SNR must be specified to apply selection combining. For this threshold the outage probability $P_{out}(\gamma_S)$ and the average SNR $\overline{\gamma}_C$ are calculated as follows.

The CDF for the combiner output being lower than γ is:

$$P_{\gamma_C}(\gamma) = p\,(\gamma_C < \gamma) = p\,(\max\,[\gamma_1, \gamma_2, \ldots, \gamma_M] < \gamma) = \prod_{i=1}^{M_r} p\,(\gamma_i < \gamma) \tag{5.67}$$

For a threshold γ_S we use the above expression and Equation (5.65) to obtain the outage probability $P_{out}(\gamma_S)$ given by:

$$P_{out}(\gamma_S) = p(\gamma_C < \gamma_S) = \prod_{i=1}^{M_r} p(\gamma_i < \gamma_S) = \prod_{i=1}^{M_r} \left(1 - e^{-\gamma_S/\overline{\gamma}_i}\right) \underset{\overline{\gamma}_i = \overline{\gamma}}{=} \left(1 - e^{-\gamma_S/\overline{\gamma}}\right)^{M_r} \tag{5.68}$$

where the last equality is true when all average SNRs are equal ($\overline{\gamma}_i = \overline{\gamma}$). Differentiation of the above equation with respect to γ_S will produce the p.d.f. $p_{\gamma_C}(\gamma)$, which can give the average SNR $\overline{\gamma}_C$ at the output of the combiner:

$$\overline{\gamma}_C = \int_0^\infty \gamma p_{\gamma_C}(\gamma)\mathrm{d}\gamma = \overline{\gamma}\sum_{i=1}^{M_r} 1/i \tag{5.69}$$

Figure 5.22 shows $P_{out}(\gamma_S)$ for selection combining as a function of $\overline{\gamma}/\gamma_S$ for Rayleigh fading and various values of M. Considerable gain is achieved by introducing diversity ($M \neq 1$) in SNR for a typical outage probability equal to 1%. This gain is 12 dB for a system with two receiving antennas ($M = 2$) when compared with a single antenna receiver. It is important to see that it falls drastically for higher values of M.

5.2.3.2 Switch-and-stay Combining

Switch and stay combining (SSC) also involves a selection strategy associated with an SNR threshold γ_{SSC}. At start-up, the receiver chooses the first branch in which the signal exceeds γ_{SSC}. After that, the combiner output remains at that branch until its signal SNR drops below the given threshold γ_{SSC}. At this point, the receiver switches to another branch according to a switching strategy. It can look for the branch with the highest SNR, which would result in the selection combining approach presented previously. Instead, it can switch to another receiver branch at random.

Since we have a switching strategy, γ_{SSC} is equal to γ_1 during the time the combiner uses the output of the first branch and equal to γ_2 when it is connected to the second branch. On the basis of that and using the same notation as in the previous section, the CDF for a two-antenna diversity is calculated as follows [Goldsmith05]:

$$P_{\gamma_C}(\gamma) = p(\gamma_C \leq \gamma) = \begin{cases} P[(\gamma_1 \leq \gamma_{SSC})\text{ AND }(\gamma_2 \leq \gamma)], & \gamma < \gamma_{SSC} \\ P[(\gamma_1 \leq \gamma_{SSC})\text{ AND }(\gamma_2 \leq \gamma)]OR \ldots P[\gamma_{SSC} \leq \gamma_2 \leq \gamma], & \gamma \geq \gamma_{SSC} \end{cases}$$
$$\tag{5.70}$$

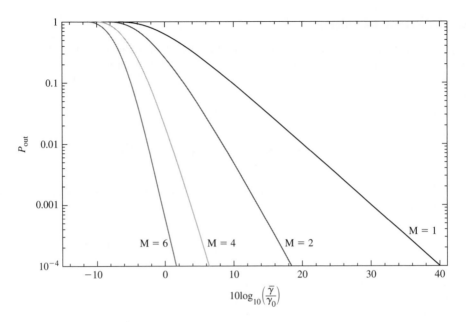

Figure 5.22 Outage probability for selection combining for $M = 1, 2, 4, 6$ and for switch and stay combining for $M = 2$

For Rayleigh fading and taking into account that both branches have the same average SNR ($\overline{\gamma}_i = \overline{\gamma}$), the outage probability for a specific $\gamma = \gamma_0$ is given as:

$$P_{\text{out}}(\gamma_0)\big|_{\gamma_{\text{SSC}}} = \begin{cases} 1 - e^{-\gamma_{\text{SSC}}/\overline{\gamma}} - e^{-\gamma_0/\overline{\gamma}} + e^{-(\gamma_{\text{SSC}}+\gamma_0)/\overline{\gamma}} & \gamma_0 < \gamma_{\text{SSC}} \\ 1 - 2e^{-\gamma_0/\overline{\gamma}} + e^{-(\gamma_{\text{SSC}}+\gamma_0)/\overline{\gamma}} & \gamma_0 \geq \gamma_{\text{SSC}} \end{cases} \tag{5.71}$$

It is noted from the above equation that the outage probability has as a parameter the threshold γ_{SSC}, which can be chosen such that $P_{\text{out}}(\gamma_0)$ is optimum. In that case it can be shown that it is equal to the corresponding outage probability for the selection combining case. For example for $\gamma_{\text{SSC}} = \frac{1}{2}\gamma_0$ we get the performance illustrated in Figure 5.22, where the performance for selection combining is also given.

5.2.3.3 Maximum Ratio Combining

In Maximum Ratio Combining (MRC), the outputs of all branches are coherently combined in an optimum manner. This is done by employing a complex weight $w_i = |w_i|e^{-j\theta_i}$ for each branch to eliminate phase shifts θ_i of the corresponding received signals and at the same time 'match' the SNR of each individual branch. This means that branches with high SNR will have higher weights. As shown below, the optimal weights $|w_i|$ are proportional to the corresponding branch amplitude a_i of the received signal.

The optimization criterion is to maximize γ_C, the expression of which is given in Equation (5.63) and repeated below for convenience:

$$\gamma_C = \frac{\left(\sum_{l=1}^{M} w_l a_l\right)^2}{N_0 \sum_{l=1}^{M} w_l^2} \tag{5.63}$$

Optimization is achieved by employing partial derivatives of Equation (5.63), which results in the optimum weights given as follows:

$$w_i^2 = \frac{a_i^2}{N_0} \tag{5.72}$$

Inserting these values in Equation (5.63) we obtain the resulting combiner SNR which is

$$\gamma_C = \sum_{i=1}^{M_r} \frac{a_i^2}{N_0} = \sum_{i=1}^{M_r} \gamma_i \tag{5.73}$$

Assuming Rayleigh fading on the M_r branches, γ_C is chi-square distributed with a resulting p.d.f. as follows [Rappaport02]:

$$p_{\gamma_C}(\gamma) = \frac{\gamma_C^{M_r-1} e^{-\gamma/\overline{\gamma}}}{(\overline{\gamma})^{M_r} (M_r - 1)!} \tag{5.74}$$

The outage probability is given as [Goldsmith05]:

$$P_{\text{out}} = p\left(\gamma_C < \gamma_0\right) = 1 - e^{-\gamma_0/\overline{\gamma}} \sum_{k=1}^{M_r} \frac{(\gamma_0/\overline{\gamma})^{k-1}}{(k-1)!} \tag{5.75}$$

Appling Equation (5.66) we get for the average probability of error $\overline{P}_{\text{MRC}}$:

$$\overline{P}_{\text{MRC}} = \int_0^\infty Q\left(\sqrt{2\gamma}\right) p_{\gamma_C}(\gamma) d\gamma = \left(\frac{1 - \sqrt{\overline{\gamma}/(1+\overline{\gamma})}}{2}\right)^{M_r} \sum_{m=0}^{M_r-1} \binom{M_r - 1 + m}{m}$$

$$\times \left(\frac{1 + \sqrt{\overline{\gamma}/(1+\overline{\gamma})}}{2}\right)^m \tag{5.76}$$

Figure 5.23 gives P_{out} for MRC for $M = 1, 2, 4, 6$.

5.2.3.4 Equal Gain Combining

In equal gain combining (EGC) a simpler (compared with maximum gain combining, MGC) combining method is employed using weights of equal (unit) amplitude for all branches and matching the phase of the corresponding branch. This is done by setting $w_i = e^{-j\theta_i}$.

In this case, the resulting combiner SNR γ_C can be shown to be [Goldsmith05]:

$$\gamma_C = \frac{1}{M_r N_0} \left(\sum_{i=1}^{M_r} a_i\right)^2 \tag{5.77}$$

It is not possible to obtain a closed-form expression for the p.d.f. and CDF of γ_C. Only in the case of two-branch receiver diversity in a Rayleigh fading channel are the expressions for the CDF and P_{out} respectively [Goldsmith05]:

$$P_{\gamma_C}(\gamma) = 1 - e^{-2\gamma/\overline{\gamma}} \sqrt{\frac{\pi\gamma}{\overline{\gamma}}} e^{-\gamma/\overline{\gamma}} \left(1 - 2Q\left(\sqrt{2\gamma/\overline{\gamma}}\right)\right) \underset{\gamma=\gamma_0}{=} P_{\text{out}}(\gamma_0)$$

$$= 1 - e^{-2\gamma_0/\overline{\gamma}} \sqrt{\frac{\pi\gamma_0}{\overline{\gamma}}} e^{-\gamma_0/\overline{\gamma}} \left(1 - 2Q\left(\sqrt{2\gamma_0/\overline{\gamma}}\right)\right) \tag{5.78}$$

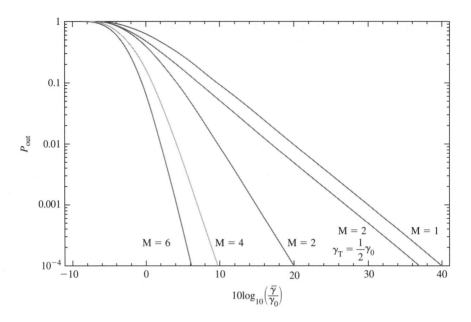

Figure 5.23 Outage probability for maximum ratio combining for $M = 1, 2, 4, 6$

The average error probability of the simple case of BPSK can be determined by first evaluating the p.d.f. $p_{\gamma c}(\gamma)$ by differentiation of Equation (5.78). After that, Equation (5.66) is used to obtain:

$$\overline{P}_{b-\text{BPSK}} = \int_0^\infty Q\left(\sqrt{2\gamma}\right) p_{\gamma c}(\gamma) d\gamma = \frac{1}{2}\left[1 - \sqrt{1 - \left(\frac{1}{1+\overline{\gamma}}\right)^2}\right] \qquad (5.79)$$

In Figure 5.23, the outage probability for EGC is given for $M = 2$ along with the performance for MRC presented in the previous subsection.

5.2.4 Implementation Issues

As mentioned before, antenna diversity is the most frequently used form of diversity and, from the implementation point of view, requires minimum additional hardware to get in return a diversity gain in BER performance in the order of 5–15 dB. The extra hardware needed is antennas and a combining device, the complexity of which is determined by the combining approach. Referring to the combining techniques presented above we have:

(1) *Selection combining.* This type of combining is the simplest and is implemented by attaching a signal-strength measuring device (like an RSSI indicator) in each antenna of the receiver as shown in Figure 5.24. All the antennas are connected to a switch, controlled to connect the strongest antenna signal to the input of the receiver. An important requirement for the RSSI is to be fast enough with tuning time comparable to the duration of fades. It is clear that in this case we need just one receiver, which at each instant is connected to the antenna with the highest SNR.

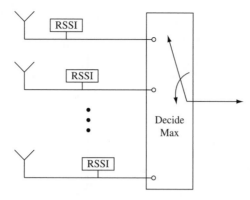

Figure 5.24   Selection combining configuration

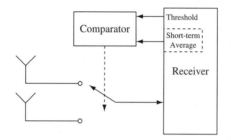

Figure 5.25   Switch-and-stay combining realization

(2) *Switch and stay combining.* The implementation approach for this technique is very close to selection combining because it also chooses an antenna branch to connect to the receiver/demodulator. The criterion for this choice is different and consists of selecting the antenna branch in which the received signal strength is above a preset threshold. To have a reliable estimate for the signal strength, a device is needed to provide a short-term average of the measured signal strength [Rappaport02]. Hence, the extra hardware needed is a comparator, an averaging circuit and a control switch, as shown in Figure 5.25. The procedure is simple. The system scans the antennas until it finds one whose signal exceeds the threshold. It remains there until the measured strength falls below this threshold.

(3) *Maximum ratio combining.* Here, in terms of implementation we have a considerably complicated structure because the system consists of M_r different front-end receivers (as many as the receiving antennas). This is shown in Figure 5.26, where each branch must be aligned in phase with the incident signal. This can be achieved by employing phase estimators in each antenna branch. Furthermore, the signal level of each branch is adjusted such that the particular branch weight is proportional to the corresponding branch SNR [see Equation (5.64)]. After the gains and phases of all antenna branches have been adjusted, the signals of all branches go through a summation device which produces the combined output according to the MRC criterion. Although it is a complex technique in both hardware and signal processing, this approach gives the highest diversity gain since it reduces the effect of fading in the most effective way. Because modern signal processors can perform very complex processing very fast, MRC becomes very attractive.

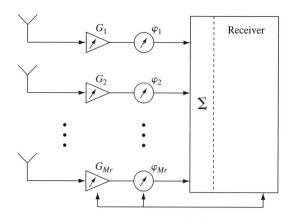

Figure 5.26   Maximum ratio combining configuration

(4) *Equal gain combining.* It is much simpler to implement EGC since it only requires phase cancellation in each branch. This is achieved by multiplying the input signal of the ith branch by $w_i = e^{-j\theta_i}$. Although much simpler in implementation than MRC, its performance is very close to that of MRC.

5.2.5 Transmitter Diversity

In transmitter diversity the modulated signal is divided into M_t different branches which feed an equal number of antennas. In this way, the signal is transmitted through M_t different channels whereas M_t signal components are received at each antenna in the receiver side. Because the diversity is due to the different channel paths received in one antenna, knowledge of the gains of these paths (i.e. the channel gains) is, in most cases, necessary to exploit diversity and obtain diversity gain. This is the case for delay diversity and diversity due to Trellis Space–Time Codes (TSTC). However, in other approaches (like in block space–time codes – BSTC) the transmitted signals can be arranged in such a way that diversity gain can be achieved without knowledge of channel state information (CSI).

The signal received by the ith antenna at the kth time instant is given as:

$$r_l(k) = \sum_{i=1}^{M_t} \sqrt{E_s} a_{il} s_i(k) + n_l(k) \qquad (5.80)$$

where a_{il} is the channel gain from the ith transmitting antenna to the lth antenna at the receiver.

Delay diversity is the simplest form of transmitter diversity and comprises of transmitting the same block of symbols with a delay between transmitting antennas. For example, for a block of four symbols the following matrix can represent the transmission in time. The ith row of the matrix represents the ith time instant whereas the jth column corresponds to the jth receiving antenna:

$$S = \begin{bmatrix} s_1 & s_2 & s_3 & s_4 \\ 0 & s_1 & s_2 & s_3 \\ 0 & 0 & s_1 & s_2 \\ 0 & 0 & 0 & s_1 \end{bmatrix} \begin{matrix} \\ \\ \leftarrow t_2 \\ \\ \end{matrix} \qquad (5.81)$$

$$\underset{\text{3rd antenna}}{\uparrow}$$

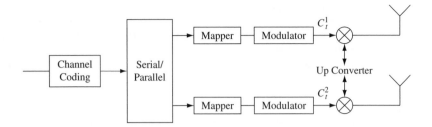

Figure 5.27 Trellis space–time code diversity transmitter

Trellis space–time code diversity is realized by forming M_t parallel streams from the single stream produced at the output of the channel encoder. This technique is depicted in general form in Figure 5.27. The M_t signals are transmitted simultaneously and have the same symbol rate. The design goal of TSTC diversity is minimization of the pairwise error probability (PEP) [Tarokh98], [Diggavi04]:

$$P(\mathbf{s} \to \mathbf{e}|h_{i,j}, i = 1, 2, \ldots, n,\ j = 1, 2, \ldots, m) \leq \exp\left[-d^2(\mathbf{s},\mathbf{e})\frac{E_S}{4N_0}\right]$$

$$d^2(\mathbf{s},\mathbf{e}) = \sum_{j=1}^{m}\sum_{t=1}^{l}\left|\sum_{i=1}^{n} h_{i,j}(s_t^i - e_t^i)\right|^2 \tag{5.82}$$

This represents the probability that the ML receiver is in error because it decides that the coded stream **e** is transmitted instead of the actual transmitted sequence **s**:

$$\left. \begin{aligned} \mathbf{e} &= [e_1(1), e_2(1), \ldots, e_{M_t}(1), e_1(2), e_2(2), \ldots, e_{M_t}(2), \ldots \ldots, e_1(k), e_2(k), \ldots, e_{M_t}(k)] \\ \mathbf{s} &= [s_1(1), s_2(1), \ldots, s_{M_t}(1), s_1(2), s_2(2), \ldots, s_{M_t}(2), \ldots \ldots s_1(k), s_2(k), \ldots, s_{M_t}(k)] \end{aligned} \right\}$$
$$\tag{5.83}$$

5.2.5.1 The Alamouti Technique

Because of the complexity of TSTC, which exhibits exponential increase with diversity order and transmission, rate a block space–time coding scheme was presented by [Alamouti98], according to which two successive (in time) symbols that are to be transmitted are grouped in pairs. At time t the two antennas transmit symbols s_1 and s_2, whereas at the next interval $(t + T_S)$ the two antennas transmit $-s_2^*$ and s_1^*, respectively. We assume a MIMO transceiver system with two transmitting antennas and two receiving antennas employing Alamouti scheme. Table 5.1 shows the diversity scheme in the transmitter, the channel gains h_1, h_2, h_3 and h_4 for the four possible paths between the antennas of the transmitter and the receiver. The same table also gives the notation of the received signals at the two receiving antennas during the successive time instants t and $t + T$.

We must note that channel gains correspond to complex distortion terms of the form $h_i = a_i \exp(j\theta_i)$ affecting the transmitted signals in the way illustrated in Figure 5.28, where the transmitting and receiving antennas are shown. In addition, Figure 5.28 depicts the basic receiver structure comprising channel estimation, combining device and detection scheme.

Table 5.1 Alamouti diversity technique

	Tx antenna 1	Tx antenna 1	Rx antenna 1	Rx antenna 2
Time instant t	s_1	s_2	r_1	r_3
Time instant $t+T$	$-s_2^*$	s_1^*	r_2	r_4
Rx antenna 1	h_1	h_2		
Rx antenna 2	h_3	h_4		

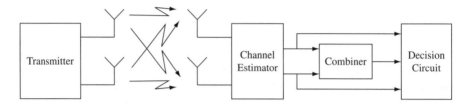

Figure 5.28 Space–time coding using Alamouti technique

From the above, the received signals r_1, r_2, r_3 and r_4 are given as:

$$r_1 = h_1 s_1 + h_2 s_2 + n_1$$
$$r_2 = -h_1 s_2^* + h_2 s_1 + n_2$$
$$r_3 = h_3 s_1 + h_4 s_2 + n_3$$
$$r_4 = -h_3 s_2^* + h_4 s_1^* + n_4 \tag{5.84}$$

where n_1, n_2, n_3, n_4 and represent complex additive noise.

The combiner output can be designed to provide the detector with signals of the general form [Alamouti98]:

$$y_i = \sum_{i=1}^{2} h_{ki} r_i + \sum_{j=3}^{4} h_{kj} r_j \tag{5.85}$$

where h_{ki} represents one of h_1 or h_2 and h_{kj} represents one of h_3, h_4. ML decoding of these signals corresponds to maximum ratio combining of order 4.

5.3 OFDM Transmission

5.3.1 Introduction

When the available bandwidth is divided into a number of subchannels, each of which is modulated independently from the others then we have Frequency Division Multiplexing (FDM). This results in much lower data rate in each subchannel compared with the overall system data rate. Consequently, each subchannel is characterized by a narrow bandwidth much lower than the coherence bandwidth of the transmission channel. This in turn reduces the effect of impulse response spread of the channel and transforms a frequency selective channel into many narrow flat-fading subchannels. In order to have a correct application of FDM principle, inter-channel interference (ICI) must be effectively reduced. One way to do that and at the same time keep spectral efficiency high is to have orthogonal overlapping subchannels. To do this, we should guarantee that no subcarrier is a linear combination of all others. This is similar to the basic principle of orthogonality in vectors. Practically this is achieved by making all subcarriers be an integer multiple of a fundamental frequency.

In order to show the implementation of the above principle in a simple way, we depict an N-subcarrier OFDM system in Figure 5.29. It is an example of an analog OFDM implementation where a serial data stream with data rate equal to R is first converted to N parallel streams using a serial-to-parallel (S/P) converter. Each parallel stream (of rate R/N) is subsequently applied to the input of a respective mapper, transforming the data stream into complex data symbols a_k from a constellation diagram. These data symbols are in turn modulated by complex sinusoids with frequencies that are multiples of a fundamental frequency. As a result, the complex data symbol $a_k = (a_{k,r}, \; a_{k,i})$ is multiplied by parts by the complex sinusoid $[\sin 2\pi f_k t, \; \cos(2\pi f_k t)]$ as depicted in the figure. It will be shown later that this operation can be performed in a digital manner by employing inverse fast Fourier transform (IFFT) (or inverse discrete Fourier transform, IDFT, in the digital domain) on the data symbols a_k.

As explained later in this section, the result of IDFT is a time sequence $s[n]$, $n = 0, 1, \ldots, N - 1$, representing the lth OFDM symbol $s_l(t)$. However, a guard interval (GI) must be inserted in front of the resulting OFDM time sequence in order to eliminate ISI due to the dispersive channel. This interval is a time slot, which can be empty or it can include a specific signal. Where there is no signal during the guard interval, the orthogonality property between subcarriers is violated and the result is ICI. If the resulting (after IDFT) OFDM symbol is cyclically extended by placing in front of it the last N_g samples we call the inserted interval the cyclic prefix (CP). In this way, it can be shown that ICI can be eliminated. Figure 5.30 shows the way cyclic prefix is appended on the IDFT output. T_{FFT} represents the FFT integration time whereas T_g represents the cyclic prefix (guard time) corresponding to the last N_g samples of the IDFT output. From the same figure it can be seen that T_g must be at least equal to the duration of the channel impulse response (CIR) in order to eliminate ISI. The maximum excess delay τ_{max} of the channel is the typical figure chosen to determine the value of T_g.

Figure 5.30 also gives the OFDM symbol indicating the way time sequence $s[0]$, $s[1], \ldots, s[N-1]$ is arranged to produce the cyclic prefix. Furthermore this figure, (adapted from [VanNee00]) gives an example of an OFDM signal containing only three subcarriers. For each subcarrier there is a direct component and a reflected one (corresponding to some delay) reaching the receiver. The subcarriers are BPSK modulated and one can see the phase discontinuities, which occur at the boundaries (beginning and end) of the OFDM symbol. Because of cyclic extension there are no phase discontinuities during the FFT integration, as can be seen from the same figure. As a result there is no ISI because the OFDM receiver detects the summation of three sinewaves and their reflected components with different amplitudes and phases but with no discontinuities. This does not affect orthogonality and does not inflict ICI between the subcarriers. In general if the duration of cyclic prefix is longer than the duration of CIR (for a linear dispersive channel) then the linear convolution of the transmitted data with CIR becomes a circular convolution. This results in multiplication between the detected constellation symbols and the channel transfer function, which means no ICI between the received subcarriers. These will be mathematically demonstrated later in this section.

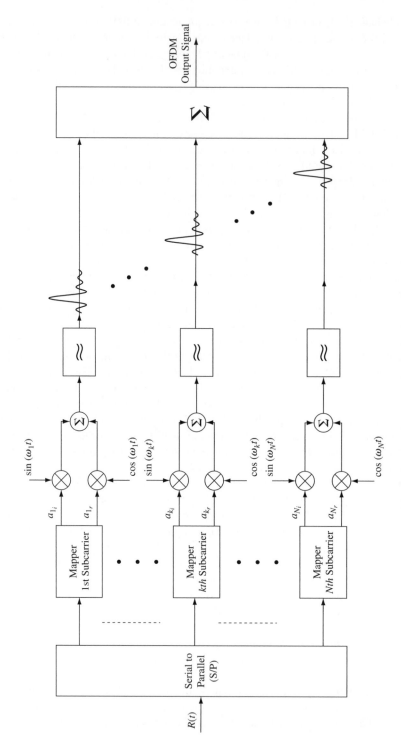

Figure 5.29 OFDM transmitter principle of operation

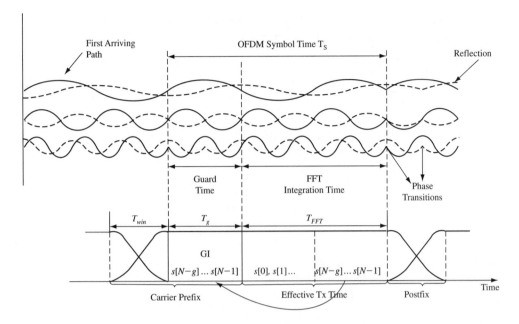

Figure 5.30 OFDM signalling with three subcarriers, representation of OFDM symbol [VanNee00]

Before transmission, the formulation of the OFDM symbol must be suitable to satisfy the requirements for low spectral density at the end of the spectrum. For this reason the time-domain signal after IFFT must go through a windowing operation which will result in an acceptable level of the spectrum of the OFDM signal at the edges of the transmitted spectrum. We will elaborate on windowing in a subsequent subsection.

5.3.2 Transceiver Model

Taking into account all the above considerations we can proceed with the mathematical formulation of the transmitter and receiver operations. The lth OFDM modulated symbol $s_l(t)$, which begins at $t = t_S$ and ends at $t = t_S + T_u$ can be represented in the standard baseband notation as follows [Proakis04], [Speth99]:

$$s_l(t) = \frac{1}{\sqrt{T}} \sum_{k=-N/2}^{(N/2)-1} a_{l,k} \exp\left[j2\pi f_k(t - T_g - lT_s)\right] u(t - lT_s) \tag{5.86}$$

where $u(t)$ is the unit step function:

$$u(t) = \begin{cases} 1, & 0 \leq t < T_s \\ 0, & t < 0, t \geq T_s \end{cases} \tag{5.87}$$

In the above equation, $a_{l,k}$ represents the data symbol whereas $f_k = k/T_u$ represents the corresponding subcarrier baseband frequency.

Employing discrete time t_n and splitting the exponential into two terms (expressing the current instant and the past $l - 1$ OFDM symbols) we have:

$$s_l(t_n) = \frac{1}{\sqrt{T}} \sum_{k=-N/2}^{(N/2)-1} a_{l,k} \exp\{ j(2\pi f_k t_n) \cdot \exp[j2\pi f_k(-T_g - lT_s)]\} u(t_n - lT_s)$$

$$= \frac{1}{\sqrt{T}} \sum_{k=-N/2}^{(N/2)-1} a_{l,k} \exp\left(j2\pi \frac{k \cdot n}{N} \right) \cdot \exp[j2\pi f_k(-T_g - lT_s)] \cdot u(t_n - lT_s) \qquad (5.88)$$

This is because:

$$\frac{1}{T} = \frac{N}{T_u} = \frac{N_S}{T_S}$$

and $t_n = nT$ becomes equal to nT_u/N.

It can now be seen that the first exponential

$$\exp\left(j2\pi \frac{k \cdot n}{N} \right)$$

in the above equation represents the inverse Fourier transform (IFT) of the data sequence $a_{l,k}$.

Figure 5.31 shows pictorially the loading of data symbols $a_{l,k}$ on OFDM symbols, each of which corresponds to a particular column in this figure. Each row represents a subcarrier

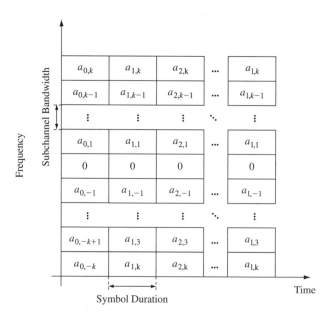

Figure 5.31 Data symbols $a_{l,k}$ within the time–frequency grid

frequency. In the receiver, demodulation along the ith path is achieved by passing the complex modulated data of the ith subcarrier through an integrator:

$$\int_{t_s}^{t_s+T_u} \left\{ \exp\left(-j2\pi\frac{i}{T_u}(t-t_s)\right) \sum_{m=-\frac{N}{2}}^{\frac{N}{2}-1} \left[a_{m+N/2} \exp\left(j2\pi\frac{m}{T_u}(t-t_s)\right)\right]\right\} dt$$

$$= \underbrace{\sum_{\substack{m=-\frac{N}{2}\\m\neq i}}^{\frac{N}{2}-1} \left[a_{m+N/2} \int_{t_s}^{t_s+T_u} \exp\left(j2\pi\frac{m-i}{T_u}(t-t_s)\right)dt\right]}_{0}$$

$$+ a_{i+N/2} \cdot \int \exp\left(j2\pi\frac{i-i}{T_u}(t-t_s)\right)dt = a_{i+N/2}T_u \qquad (5.89)$$

Figure 5.32 illustrates detection [multiply by $\exp(j\omega_i t)$ and integrate] and demonstrates its equivalence with DFT operation. In addition, it shows the orthogonality property of OFDM modulation, since from all N components of the transmitted signal $s_l(t)$, in the ith receiver path only the ith data symbol is demodulated.

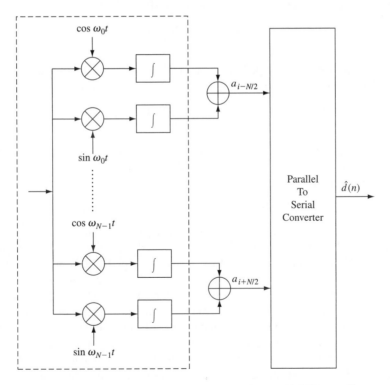

Figure 5.32 OFDM detection showing equivalence with FFT operation

5.3.3 OFDM Distinct Characteristics

OFDM transmission has two distinct characteristics that make it different from every other transmission system. The first is the use of IDFT/DFT in the transmitter/receiver for modulation/demodulation respectively. The second refers to the existence of a guard interval or cyclic prefix used to leave data blocks unaffected from ISI. Both concepts were qualitatively presented above. Here we briefly elaborate on the mathematics of DFT/IDFT. Furthermore, we demonstrate how the cyclic prefix guarantees ISI-free reception in multipath fading channels.

If the IDFT of a sequence $X[i]$ is $x[n]$ then we have:

$$x[n] = \frac{1}{\sqrt{N}} \sum_{i=0}^{N-1} X[i] \exp \left[j \frac{2\pi n i}{N} \right], \quad 0 \le n \le N-1 \tag{5.90}$$

From the above and Equation (5.88) it turns out that the complex QAM data symbols $a_{l,k}$ correspond to the inputs $X[i]$ of the IDFT in the transmitter which will produce the transmitted time-domain signal $x[n]$ which is identical to the transmitted sampled signal $s[n]$.

In similar fashion we have

$$DFT\{s(n)\} = DFT\{x(n)\} = X[i] \triangleq \frac{1}{\sqrt{N}} \sum_{n=0}^{N-1} x(n) \exp \left[-j \frac{2\pi n i}{N} \right], \quad 0 \le n \le N-1 \tag{5.91}$$

If we also take into account the linear time-invariant channel $h[n]$ the received signal is the linear convolution of it and the transmitted sequence $s[n]$:

$$r[n] = h[n] * s[n] = \sum_k h[k]s[n-k] \tag{5.92}$$

If $s[n-k]$ can take the form of a periodic function the linear convolution will be transformed into a circular convolution [Goldsmith05]:

$$r[n] = h[n] \otimes s[n] = \sum_k h[k]s[(n-k)_{\text{mod }N}] \tag{5.93}$$

where $(n-k)_{\text{mod }N}$ represents the number $[n-k]$-modulo N. It can be shown that since $s[(n-k)_{\text{mod }N}]$ is a periodic function with period N, the same holds for $r[n]$. From the basic properties of DFT we have:

$$DFT\{r[n]\} = z[n] = S[n].H[n] = a_n H[n] \tag{5.94}$$

Since $s[n-k]$ is a time sequence, to transform it into a periodic sequence, a block of values is repeated at the end or in the beginning of the transmitted symbol when its length exceeds N. As we see immediately, this is achieved by appending a prefix in front of the transmitted sequence $s[n-k]$ constituting from the last C_P values of the time sequence.

Considering again Figure 5.30(a), we can see that the transmitted time sequence has the form $s[N-g]$, $s[N-g+1]$, ..., $s[N-1]$, $s[0]$, $s[1]$, ..., $s[N-1]$. This, as a block, constitutes the OFDM symbol. The sampled signal values before $s[0]$ are the last $(g-1)$ values of the time sequence corresponding to the length g of the CIR, $g = (\tau_{rms}/T_S) - 1$. They represent the cyclic prefix. That is how the linear convolution is transformed to a circular one and consequently Equation (5.94) is valid and can be readily used to estimate transmitted data symbols a_n by dividing the DFT of the received data $r[n]$ by the estimated channel transfer function $H[n]$.

This will be shown further below in this section. The first $(g-1)$ samples of the OFDM symbol are corrupted by ISI due to CIR of the radio channel. However, these samples are not needed for the estimation of the transmitted data symbols a_n and are discarded at the input of the digital receiver. Finally we must note that, by adding the cyclic prefix at the beginning of the transmitted symbol, the effective data rate is reduced by $N/(N+g)$.

5.3.4 OFDM Demodulation

For a multipath channel characterized by $h(\tau, t)$ the baseband received signal is given by $r(t)$:

$$h(\tau, t) = \sum_i h_i(t)\delta(\tau - \tau_i) \qquad (5.95)$$

$$r(t) = \sum_i h_i(t) \cdot s(t - \tau_i) + n(t) \qquad (5.96)$$

where τ_i is the delay of the ith ray and $h_i(t)$ is the corresponding complex number that gives the measure and phase of the ith ray in the channel impulse response.

To simplify derivations we drop the lth index representing the current OFDM symbol and modify Equation (5.86) as follows:

$$s_l(t) = \frac{1}{\sqrt{T_u}} \cdot \sum_{k=-K/2}^{K/2-1} a_k \exp\left(2\pi f_k t\right) \cdot u(t) \qquad (5.97)$$

The sampled signal $s(t_n)$ at time instants $t_n = nT_S$ is given as:

$$s(t_n) = \frac{1}{\sqrt{T_u}} \cdot \sum_{k=-K/2}^{K/2-1} a_k \exp\left(j2\pi \frac{kn}{N}\right) \qquad (5.98)$$

The sampled signal at the receiver corresponds to the sampled version of Equation (5.96):

$$r(t_n) = \sum_i h_i(nT) \cdot s(nT - \tau_i) + n(nT) = s[n] \otimes h[n] \qquad (5.99)$$

where $\otimes$ represents circular convolution.

After DFT we get the demodulated signal at the receiver:

$$z_{l,k} = DFT\{r(t_n)\} = \sum_{n=0}^{N-1} r_k \exp\left(-j2\pi \frac{n}{N}k\right) \qquad (5.100)$$

This gives:

$$z_k = DFT\{s[k] \otimes h[k]\} = DFT\{s[k]\} \cdot DFT\{h[k]\} = a_k \cdot H[k] = a_k \cdot H_k \qquad (5.101)$$

The above is due to the property of the DFT of circular convolution, which states that the DFT of two circularly convolved signals is equal to the product of the DFTs of the individual signals. In addition we note that $H[k] = H_k$ represents the frequency response of the channel at subcarrier k.

Taking into account complex white Gaussian noise $n(t)$ added by the channel, we get for the demodulated data symbols at the receiver:

$$z_k = a_k \cdot H_k + n_k \tag{5.102}$$

Equation (5.102) demonstrates that by dividing the received data symbols z_k by the channel response H_k at subcarrier k we can have the estimate of the transmitted data a_k:

$$\hat{a}_{l,k} = \frac{z_{l,k}}{\hat{H}_{l,k}} \tag{5.103}$$

5.3.5 Windowing and Transmitted Signal

Because the composite OFDM signal within an OFDM symbol consists of a summation of QAM subcarriers convolved with a square pulse of dulation equal to that of the OFDM symbol, the unfiltered out-of-band spectrum decreases according to a *sinc* function. However, this degradation is unsatisfactory, resulting in high $-40\,\text{dB}$ bandwidth which frequently characterizes the compactness of the transmitted spectrum. Furthermore, it is easy to see that, as the number of subcarriers in an OFDM system decreases, the $-40\,\text{dB}$ bandwidth increases.

By applying windowing, the out-of-band spectrum can be drastically reduced in a smooth fashion. Raised cosine is the most frequently applied window and is given by the following expressions:

$$w_{RS}(t) = \begin{cases} \sin^2(\pi t/2\beta T_S), & 0 \leq t \leq \beta T_S \\ 1, & \beta T_S \leq t \leq T_S \\ \cos^2(\pi(t - T_S)/2\beta T_S), & T_S \leq t \leq (\beta + 1)T_S \end{cases} \tag{5.104}$$

where β is the roll-off factor and T_S is the OFDM symbol duration.

Figure 5.33(a) shows how windowing is applied in an OFDM symbol before transmission. At the beginning of the OFDM symbol the last T_{CP} samples (cyclic prefix) of the IFFT are appended. Similarly a postfix can be placed at the end of the symbol. After that, the OFDM symbol is multiplied by the window function. The window position can be adjusted such that we have one section at the beginning and one at the end of the OFDM symbol of duration βT_S each. In this way there is no need for any postfix section at the end of the symbol. Finally, the next OFDM symbol must be delayed such that the βT_S part in its start overlaps with the βT_S part of the last one as depicted in Figure 5.33(a).

After windowing the signal to be transmitted is:

$$s_l(t) = \frac{1}{\sqrt{T}} \sum_{k=-N/2}^{(N/2)-1} a_{l,k} \cdot q_n(t) \tag{5.105}$$

$$q_n(t) = \begin{cases} \sin^2\left(\dfrac{\pi t}{2\beta T_S}\right) \cdot \exp[j2\pi nB(t - T_{CP})], & t \in [0, \beta T_S] \\ \exp[j2\pi nB(t - T_{CP})], & t \in [\beta T_S, T_S] \\ \cos^2\left(\dfrac{\pi t(t - T_S)}{2\beta T_S}\right) \cdot \exp[j2\pi nB(t - T_{CP})], & t \in [T_S, (1 + \beta)T_S] \end{cases} \tag{5.106}$$

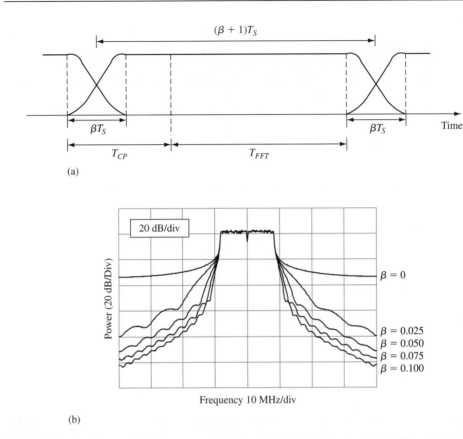

Figure 5.33 (a) Effect of windowing in OFDM symbol; (b) transmitted spectrum for various roll-off factors β

where $B = 1/(T_S - T_{CP})$. After some manipulations, the PSD of $s_l(t)$ is obtained [Rodrigues02]:

$$\Phi_{S_l}(f) = \frac{1}{T_S} \sum \sigma_a^2 \cdot \Phi_{qn}(f) \tag{5.107}$$

where σ_a^2 is the variance of the data symbols $a_{l,k}$ and $\Phi_{qn}(f)$ is the power spectral density of $q_n(t)$.

Depending on the roll-off factor the transmitted spectrum can decrease rapidly. Figure 5.33(b) shows the spectrum of a raised cosine window for $\beta = 0$ (no window) and various values of β.

5.3.6 Sensitivities and Shortcomings of OFDM

Sensitivities of OFDM can be categorized in two groups: one is associated with the typical digital modem problems of synchronization, and phase-frequency offset correction. The other type of problems is closely related to IF and RF imperfections associated with up- and down-conversion subsystems. These include the impact of phase noise, I–Q imbalances in frequency translation subsystems and power amplifier nonlinearities. A different type of shortcoming found in OFDM systems is the peak-to average power ratio (PAP), which is associated with the property of OFDM to have a variable instantaneous power due to the fact that, at each time instant, we have a superposition of sinusoids with different amplitudes and phases. This results in high and low values for the composite OFDM waveform, which in turn gives the PAP ratio. From the above it can be easily deduced that PAP ratio combined with PA nonlinearities have a considerable impact on the OFDM system performance.

The general equation expressing carrier frequency offset (CFO) and sampling frequency (clock) offset (SFO) is as follows [Speth99]:

$$r(t_n) = \exp\left(j2\pi\Theta_0(n'T')\right) \sum_i h_{\varepsilon,i}(n'T') \cdot s(n'T' - \tau_i) + n(n'T') \tag{5.108}$$

where the phase offset is

$$\Theta_0(n'T') = (1 + \zeta)\Delta f_C n'T \tag{5.109}$$

and the time scale n' is given by

$$n' = n + N_g + lN_S + n_\varepsilon \tag{5.110}$$

In addition, corresponding to a symbol timing offset of $\varepsilon = n_\varepsilon T$, the normalized clock frequency offset is:

$$\zeta = \frac{T' - T}{T} \tag{5.111}$$

Depending on the type of misalignment between transmitter and receiver parameters (symbol time, carrier frequency or sampling period), we have the corresponding cases, as shown in Table 5.2.

5.3.6.1 Sampling Clock Frequency Offset [Speth99]

This is the case for $\Delta f_C = 0$, $n_\varepsilon = 0$ and $\zeta \neq 0$. Using Equation (5.108) the received samples are:

$$r_{l,n} = \sum_i h_i \sum_l \sum_k a_{l,k} \psi\left(n'T' - \tau_i\right) + n_{l,n} \tag{5.112}$$

where

$$\psi(n'T' - \tau_i) = \exp\left(j2\pi \frac{k}{N}\left(n'(1 + \zeta) - N_g - lN_s - \frac{\tau_i}{T}\right)\right) \cdot u(n'(1 + \zeta)T - lT_s - \tau_i) \tag{5.113}$$

Table 5.2 Categorization of offsets in OFDM receivers

Carrier frequency offset	$\Delta f_C \neq 0, \zeta = 0, n_\varepsilon = 0$
Symbol (timing) offset	$n_\varepsilon \neq 0, \Delta f_C = 0, \zeta = 0$
Clock frequency offset	$\zeta \neq 0, \Delta f_C = 0, n_\varepsilon = 0$

Interpretation of the above equation leads to the observation that the lth OFDM symbol is shifted in time during reception and its position is the following interval:

$$n' \in [(lN_s - \zeta lN_s), ((l+1)N_s - \zeta(l+1)N_s))] \tag{5.114}$$

Demodulation of the lth OFDM symbol by applying DFT on $r_{l,n}$ and assuming there is no ISI (perfect synchronization) gives:

$$z_{l,k} = \left(\exp(j\pi\varphi_{k,\text{SFO}}) \exp\left(j2\pi \frac{lN_s + N_g}{N} \varphi_{k,\text{SFO}} \right) \right) \sin c\,(\pi\varphi_{k,\text{SFO}}) a_{l,k} H_k$$

$$+ \sum_{i;i\neq k} \left(\exp(j\pi\varphi_{i\neq k,\text{SFO}}) \exp\left(j2\pi \frac{lN_s + N_g}{N} \varphi_{i\neq k,\text{SFO}} \right) \right) \sin c\,(\pi\varphi_{i\neq k,\text{SFO}}) a_{l,i} H_i + n_{l,k}$$

$$\underbrace{\phantom{+ \sum_{i;i\neq k} \left(\exp(j\pi\varphi_{i\neq k,\text{SFO}}) \exp\left(j2\pi \frac{lN_s + N_g}{N} \varphi_{i\neq k,\text{SFO}} \right) \right) \sin c\,(\pi\varphi_{i\neq k,\text{SFO}}) a_{l,i} H_i}}_{ICI}$$

$$\tag{5.115}$$

where phase offsets for subcarrier k and for subcarriers i with respect to subcarrier $k(i \neq k)$ are as follows:

$$\varphi_{k,\text{SFO}} \approx \zeta k$$

$$\varphi_{i\neq k,\text{SFO}} = (1+\zeta)i - k \tag{5.116}$$

Expressing the useful part of the demodulated signal $z_{l,k}$ as a function of an equivalent channel transfer function $H'_{k,\text{SFO}}$, the above equation becomes:

$$z_{l,k} = \left(\exp\left(j2\pi \frac{lN_s + N_g}{N} \varphi_{k,\text{SFO}} \right) \right) a_{l,k} H'_{k,\text{SFO}} + ICI + n_{l,k}$$

$$H'_{k,\text{SFO}} = H_k \cdot \exp(j\pi\varphi_{k,\text{SFO}}) \cdot \sin c\,(\varphi_{k,\text{SFO}})$$

$$ICI = \sum_{i;i\neq k} \left(\exp(j\pi\varphi_{i\neq k,\text{SFO}}) \exp\left(j2\pi \frac{lN_s + N_g}{N} \varphi_{i\neq k,\text{SFO}} \right) \right) \sin c\,(\pi\varphi_{i\neq k,\text{SFO}}) a_{l,i} H_i$$

$$\tag{5.117}$$

Because the terms $\exp(j\pi\varphi_{k,\text{SFO}})$ and $\sin c\,(\varphi_{k,\text{SFO}})$ are time-invariant, they can be considered as scaling factors on H_k. In addition, the second term is very close to 1 because $\varphi_{k,\text{SFO}}$ is very small.

In conclusion, the impact of the sampling clock frequency offset is twofold:

- the variation of the phase of each subcarrier k by the amount

$$\theta_{k,SFO} = 2\pi \frac{lN_s + N_g}{N} \zeta k$$

- the introduction of a time-variant ICI.

5.3.6.2 Symbol (Timing) Offset [Speth99]

We assume there is a synchronization offset of n_e samples with corresponding timing offset $\varepsilon = n_e T$. Figure 5.34 shows three successive OFDM symbols indicating also the cyclic prefix and impulse response length.

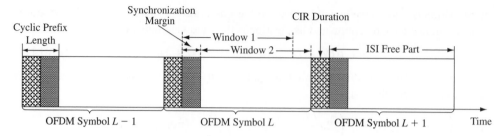

Figure 5.34 Time relation of cyclic prefix and impulse response duration in successive OFDM symbols

Equations (5.112) and (5.113) for $\Delta f_C = 0$, $\zeta = 0$ and $n' = n + n_\varepsilon + N_g + lN_S$ give [Speth99]:

$$r_{l,n} = \sum_i h_i(t) \cdot \sum_l \sum_k a_{l,k} \psi(n'T - \tau_i) + n_{l,k}$$

$$\psi(n'T - \tau_i) = \exp\left(j2\pi \frac{k}{T_u}(n'T - T_g - lT_s - \tau_i)\right) u(n'T - lT_s - \tau_i) \qquad (5.118)$$

The receiver signal (in vector form) r_l below should be such that it is affected only by the lth transmitted OFDM symbol and not by the $(l-1)$th or $(l+1)$th symbol.

$$\mathbf{r_l} = \{r_{l,n_e}, r_{l,n_e+1}, \ldots, r_{l,N-1-n_e}, r_{l+1,0}, r_{l+1,1}, \ldots, r_{l+1,n_e-1}\} \qquad (5.119)$$

The above consideration led to the following inequality for the range of timing error n_ε that can be permitted [Speth99]:

$$\left[\frac{\tau_{max}}{T}\right] - N_g \leq n_\varepsilon \leq 0 \qquad (5.120)$$

The above inequality suggests that there is more than one acceptable value for the synchronization offset n_ε depending on the value of the excess delay τ_{max}.

To investigate how ISI and ICI are introduced when inequality (5.120) is violated, we assume a gaussian channel with no dispersion ($\tau_{max} = 0$) and we also eliminate the guard interval ($N_g = 0$). This leads to $n_\varepsilon = 0$. However we impose a positive $n_\varepsilon > 0$, which means that the received signal vector $\mathbf{r}_l$ has the form of Equation (5.119). After applying FFT we obtain the demodulated signal:

$$z_{l,k} = \frac{N - n_\varepsilon}{N} a_{l,k} \exp\left[j\frac{2\pi k}{N} n_\varepsilon\right] + \text{ICI} + \text{ISI} + n_{l,k} \qquad (5.121)$$

$$\text{ICI} = \frac{1}{N} \sum_{n=0}^{N-1-n_\varepsilon} \sum_{\substack{i=-K/2 \\ i \neq k}}^{K/2-1} a_{l,i} \exp\left[j\frac{2\pi i}{N}(n + n_\varepsilon)\right] \exp\left[-j\frac{2\pi n}{N}k\right] \qquad (5.122\text{a})$$

$$\text{ISI} = \frac{1}{N} \sum_{n=N-n_\varepsilon}^{N-1} \sum_{i=-K/2}^{K/2-1} a_{l+1,i} \exp\left[j\frac{2\pi i}{N}(n - N + n_\varepsilon)\right] \exp\left[-j\frac{2\pi n}{N}k\right] \qquad (5.122\text{b})$$

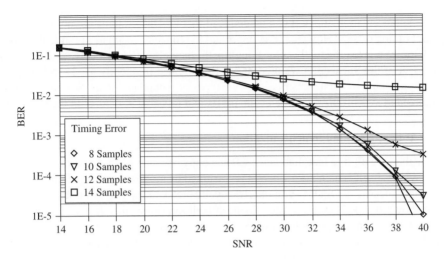

Figure 5.35 Effect of synchronization error in Hiperlan/2 BER performance (QAM-64 BRAN-B channel, no coding)

Apart from ISI and ICI the transmitted data symbols $a_{l,k}$ are distorted in amplitude and phase. Attenuation and phase change are constant and depend on $n_\varepsilon = 0$.

In multipath transmission we have two cases. The FFT window can be within or outside of the synchronization margin. Where the FFT starts within this margin we have [Speth97]:

$$z_{l,k} = a_{l,k} H_{l,k} \exp\left[j2\pi \frac{k}{N} n_\varepsilon \right] + n_{l,k} \tag{5.123}$$

Where demodulation (FFT window) starts outside of the synchronization margin we have:

$$z_{l,k} = \frac{N - n_\varepsilon}{N} a_{l,k} H_{l,k} \exp\left[j2\pi \frac{k}{N} n_\varepsilon \right] + n_{l,k} + n_{n\varepsilon}(l, k) \tag{5.124}$$

In this case additive noise $n_{n\varepsilon}(l, k)$ has the highest impact on the system and includes ISI and ICI disturbances. Its nature can be considered Gaussian while its power is approximated as [Speth99]:

$$\sigma_\varepsilon^2 = \sum_i |h_i(t)|^2 \left[2 \frac{\Delta\varepsilon_i}{N} - \left(\frac{\Delta\varepsilon_i}{N} \right)^2 \right] \tag{5.125}$$

$$\Delta\varepsilon_i = \begin{cases} n_\varepsilon - \dfrac{\tau_i}{T}, & n_\varepsilon T > \tau_i \\[2mm] -\dfrac{T_g}{T} - \left(n_\varepsilon - \dfrac{\tau_i}{T} \right), & 0 < n_\varepsilon T < -(T_g - \tau_i) \\[2mm] 0, & \text{elsewhere} \end{cases} \tag{5.126}$$

The amount of the symbol offset n_ε cannot be effectively used to determine symbol timing accuracy but deterioration in system performance due to n_ε must be evaluated. An example of that can be seen at Figure 5.35, where the effect of offsets $n\varepsilon$ ranging from eight to 14 samples is seen for Hiperlan/2 64-QAM without coding in multipath channel BRAN-B.

From the presentation above it can be deduced that OFDM systems are not very sensitive in symbol timing offsets. However, it is important to reduce as much as possible this offset in order for the system to effectively combat multipath channels with long delay response profiles.

In the rest of this subsection we present briefly the most frequently used symbol synchronization approaches in burst transmission. Two types of redundancies are exploited in order to achieve symbol synchronization: the cyclic prefix and repeated structures in the preamble.

5.3.6.3 CP-based Symbol Synchronization

In this method we use the fact that, since CP is a copy of a part of the symbol, its correlation with the symbol itself is high even where it is corrupted by multipath fading. Figure 5.36 below depicts three successful OFDM symbols. The cyclic prefix of the two last ones [lth, $(l+1$th)] are also shown (shaded) in the figure which, in addition shows the observation interval and the correct timing instant θ. Following [Sandell95], we define two sets of indices I and I' for the CP parts and the observation vector $\mathbf{r}$ containing $2N + N_g + 1$ consecutive samples:

$$I \equiv \{\theta, \ldots, \theta + N_g - 1\}$$

$$I' \equiv \{\theta + N, \ldots, \theta + N + N_g - 1\} \tag{5.127}$$

$$\mathbf{r} \equiv [r_1 \ldots r_{2N+Ng+1}]^{\mathrm{T}} \tag{5.128}$$

The log-likelihood function for θ, given $\mathbf{r}$ is:

$$\Lambda_r(\theta) = \sum_{k=\theta}^{\theta=N_g-1} 2\mathrm{Re}\{r_k r_{k-N}^*\} - \rho(|r_k|^2 + |r_{k-N}^*|^2) \tag{5.129}$$

where ρ is a correlation coefficient indicating the correlation between samples of the CP and the corresponding ones from the data symbol

$$\rho = \frac{E\{|s_k|^2\}}{E\{|s_k|^2\} + E\{|n_k|^2\}} = \frac{\mathrm{SNR}}{\mathrm{SNR} + 1} \tag{5.130}$$

Taking the derivative we obtain the estimate of the symbol timing $\hat{\theta}_r$:

$$\hat{\theta}_r = \arg\max_{\theta}\{\lambda(\theta)\}$$

$$\lambda(\theta) = 2\left|\sum_{k=\theta}^{\theta+N_g-1} r_k r_{k-N}^*\right| - \rho \sum_{k=\theta}^{\theta+N_g-1} (|r_k|^2 + |r_{k-N}^*|^2) \tag{5.131}$$

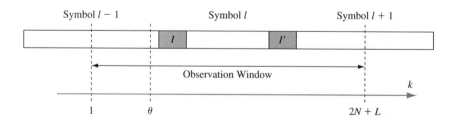

Figure 5.36 Position of observation window with respect to successive OFDM symbols

Figure 5.37(a) illustrates the functional diagram of the timing algorithm and Figure 5.37(b) presents simulation results for the timing metric $\lambda(\theta)$ for an OFDM system with 256 subcarriers, a CP equal to 64 (25%) and SNR $= 15$ dB. Reducing the number of subcarriers (for example for $N = 64$, $N_g = 16$ as in Hiperlan2) results in less clear peaks because the moving average [the first term in $\lambda(\theta)$] is calculated using fewer samples.

In cases of OFDM systems with a large number of subcarriers, an approximate estimator, which can give satisfactory results, is:

$$\hat{\theta} = \arg \max_{\theta} \left\{ \left| \sum_{k=\theta}^{\theta+N_g-1} r_n \cdot r_{n+N}^* \right| \right\} \tag{5.132}$$

5.3.6.4 Preamble-based Symbol Synchronization

Many OFDM systems have a preamble which includes a structure (such as a short symbol) which is repeated one or more times. For, example the Broadcast Physical burst of Hiperlan2 (Figure 5.38) contains section 3, which has two similar short symbols preceded by a CP.

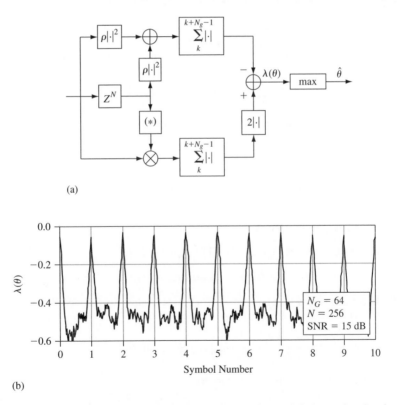

(a)

(b)

Figure 5.37 (a) Block diagram of timing system and (b) magnitude of timing estimation function for OFDM system with 256 subcarriers and 25% cyclic prefix in AWGN channel

Figure 5.38 The broadcast physical burst of Hiperlan2

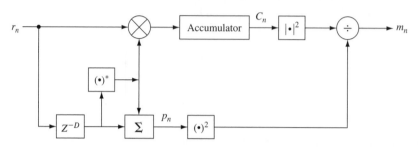

Figure 5.39 Block diagram for delay-and-correlate technique for symbol synchronization using preamble

A delay-and-correlate method suggested by Schmidl and Cox [Schmidl97] performs correlation and power estimation calculating c_n and p_n as follows:

$$c_n = \sum_{k=0}^{L-1} r_{n+k} r_{n+k+D}^*$$

$$p_n = \sum_{k=0}^{L-1} r_{n+k+D} r_{n+k+D}^* = \sum_{k=0}^{L-1} |r_{n+k+D}|^2 \qquad (5.133)$$

The decision metric is the following:

$$m_n = \frac{|c_n|^2}{p_n^2} \qquad (5.134)$$

Figure 5.39 gives the block diagram of this algorithm.

5.3.6.5 Carrier Frequency Offset

For $\Delta f_C \neq 0$, $\zeta = 0$, $n_\varepsilon = 0$ we have only CFO expressed as:

$$\varphi_{k,CFO} \approx \Delta f_C T_u, \quad \varphi_{i \neq k,CFO} = \Delta f_C T_u + i - k \qquad (5.135)$$

By substituting $\varphi_{k,SFO}$ and $\varphi_{i \neq k,SFO}$ from above we take the following equation which is similar to Equation (5.117) if $\varphi_{k,SFO}$ and $\varphi_{i \neq k,SFO}$ are replaced by $\varphi_{k,CFO}$ and $\varphi_{i \neq k,CFO}$

$$z_{l,k} = \left(\exp\left(j2\pi \frac{lN_s + N_g}{N} \varphi_{k,CFO} \right) \right) a_{l,k} H'_{k,CFO} + ICI + n_{l,k}$$

$$H'_{k,SFO} = H_k \cdot \exp(j\pi\varphi_{k,CFO}) \cdot \sin c(\varphi_{k,CFO}) \qquad (5.136)$$

$$ICI = \sum_{i;i \neq k} \left(\exp(j\pi\varphi_{i \neq k,CFO}) \exp\left(j2\pi \frac{lN_s + N_g}{N} \varphi_{i \neq k,CFO} \right) \right) \sin c(\pi\varphi_{i \neq k,CFO}) a_{l,i} H_i$$

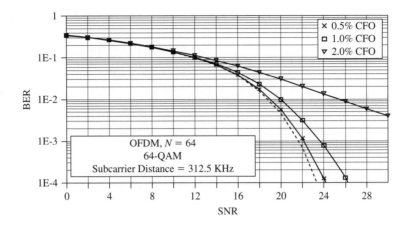

Figure 5.40 Effect of CFO in system performance in AWGN channel without coding

The major disturbances in the above equation are the ICI introduced by the CFO and the distortion of the transmitted data symbols $a_{l,k}$. The later constitutes both amplitude and phase distortion. The amplitude disturbance is associated with modified channel response $H'_{k,\text{SFO}}$, which is time-invariant and can be determined just in the same way channel estimation is performed in OFDM. The phase displacement of the data symbols is:

$$\Delta\theta_{l,k,\text{CFO}} = 2\pi \frac{lN_S + N_g}{N} \Delta f_C T_u \tag{5.137}$$

and it varies with time. Consequently it is not straightforward to remove when performing channel equalization. However, this phase change is common for all channels at a particular OFDM symbol, i.e. it is not a function of k.

The ICI term of the above equation can be greatly simplified by assuming that all subcarrier offsets are small and have the same value, $\varphi_{k,\text{CFO}} \approx \Delta f \cdot T_u$, and the channel transfer function amplitude is close to 1 ($|H_i|^2 \approx 1$). In that case it can be shown [Pollet95], [Classen94] that the noise power due to ICI is as follows:

$$\sigma^2_{\text{ICI}} = K \cdot \frac{\pi^2}{3} \cdot (\Delta f \cdot T_u)^2, \text{ with } K \approx 1 \tag{5.138}$$

Figure 5.40 gives the simulated performance of H/2 wireless system as affected by CFO in AWGN for three different CFO given as a fraction of subcarrier distance in H/2, which is 312.5 kHz. The overall CFO can be modelled as [Li03]:

$$x_{l,k} = \dots |H_{l,k-n_I}|^2 \exp\left(j2\pi(n_I + \varepsilon) \cdot N_S/N\right) \tag{5.139}$$

where n_I and ε represent the integer and fractional CFO respectively.

The overall normalized (with respect to subcarrier frequency) relative CFO:

$$e_T = \Delta f_C \cdot N \cdot T_S = \Delta f_C \cdot T_u \tag{5.140}$$

can take values higher than one when it is greater to the neighbouring subcarrier distance. In this case we have

$$e_T = e_I + e_F \tag{5.141}$$

where e_I represents the integer part and e_F the fractional part ($|e_F| < 1$).

Estimation of the integer frequency offset takes place using pilots after demodulation (at the output of the FFT) by employing [Li03], [Dlugaszewski02], [Zou01]:

$$\hat{n}_I = \arg \max \left| \sum x_{l,k} \right| \tag{5.142}$$

The effect of the integer part of CFO is different from that of the fractional part. It can be shown [Kim01] that if there was only the integer part, the demodulated signal would be given by:

$$z_n = H_{k-e_I} a_{k-e_I} + N_{k-2} \tag{5.143}$$

which indicates that the effect of e_I is the shifting of the subcarriers by e_I samples at the output of the FFT. The estimation of e_I can be done using a training symbol of double periodicity [Li03], [Schmidl97]. The correction of e_I effect on CFO can be achieved easily by shifting cyclically the samples at the output of the FFT.

Regarding fractional CFO, it can be estimated either at the output of FFT [Li03], or by using received pilots before demodulation (pre-FFT). Assuming that the integer part has been estimated and removed in the beginning, in this section we focus on techniques for fractional CFO estimation and correction. Fractional CFO estimation techniques are very important because they determine the convergence behavior and accuracy of estimation and correction.

CFO correction techniques in OFDM can be categorized using different criteria such as the nature of the technique (data-aided vs nondata aided), target application (continuous vs packet transmission) and operational characteristics (pre-FFT, post-FFT). An effective way to present CFO correction techniques is based on their properties and can be divided into two categories: those applied at the frequency domain and those using time domain estimators.

Frequency Domain Methods

As demonstrated above, CFO translates to phase revolution in every subcarrier from one symbol to the next. The fact that this revolution is common to all subcarriers can be used to estimate the phase displacement of the constellation diagram of known pilots. Various methods exist [Moose94], [McNair99] that apply the above principle on the demodulated OFDM symbol (at the output of the FFT block)

Time-domain Methods

These methods exploit the autocorrelation properties of the OFDM symbol in the time domain by employing a part of the symbol, which is repeated in time. Such a part is the cyclic prefix part of the OFDM symbol. The technique can also be applied in special training symbols that may exist in the format of the transmitted symbol or frame.

A variety of time-domain methods ([Sandell95], [Mochizuki98], [Kim01]) are reported in the literature which use the received OFDM symbols just before they are demodulated through the FFT. Hence these belong to the pre-FFT CFO estimation techniques.

As it is clear from the above discussion, in the context of OFDM, frequency- and time-domain methods correspond to post-FFT and pre-FFT techniques. Pre-FFT methods are faster to converge but post-FFT give more accurate estimates [Zou01].

In this section we briefly present CFO estimators for packet transmission systems, which usually exploit preamble and pilot structures to achieve fast feed-forward (open-loop) convergence [Chiavaccini04].

CFO Estimation in the Frequency Domain $\Delta \hat{f}_{C,FD}$

The phase change between pilots from one OFDM symbol to the next is connected to the CFO by:

$$\Delta\theta_{\text{CFO}} = 2\pi \frac{N_s}{N} \Delta f_c T_u = 2\pi \Delta f_c T_s \tag{5.144}$$

Therefore calculation of the phase of the correlation between two pilots of the same value belonging to successive OFDM symbols can be used to estimate $\Delta\theta_{\text{CFO}}$:

$$\Delta\hat{\theta}_{\text{CFO}} = \tan^{-1}(z_{l,k} z_{l-1,k}^*) \tag{5.145}$$

where $z_{l,k}$ is the demodulated (after the FFT) data symbol of the lth OFDM symbol at the kth subcarrier: $z_{l,k} = \hat{a}_{l,k}$.

Using Equation (5.136) we get:

$$z_{l,k} z_{l-1,k}^* = \exp\left(j2\pi \frac{N_s}{N} \Delta f_c T_u\right) \text{si}^2(\pi\Delta f_c)|H_{l,k}|^2 + n'_{ICI} + n' \tag{5.146}$$

where n'_{ICI} represents the ICI due to CFO, n' is the additive noise. In addition, it is assumed that the radio channel does not change during the two recent, $(l-1\text{th})$ and lth OFDM symbols.

The three last equations will give:

$$\Delta\hat{f}_c = \frac{\tan^{-1}(z_{l,k} z_{l-1,k}^*)}{2\pi T_s} = \Delta f_c + N'_{ICI} + N' \tag{5.147}$$

Both the ICI term N'_{ICI} and the additive noise reduce when the number of pilots within data symbols increases. If the two successive symbols $l-1$ and l contain only pilots having the same value in respective subcarriers $(a_{l-1,k} = a_{l,k})$, ICI reduces considerably and it can be shown [Moose94] that, in this case, the maximum likelihood CFO estimator using symbols $l-1$ and l is:

$$\Delta\hat{f}_{C,\text{FD}}(l) = \frac{1}{2\pi T_u} \tan^{-1}\left[\frac{\sum_{k=-\frac{N}{2}}^{\frac{N}{2}+1} \text{Im}\{z_{l,k} z_{l-1,k}^*\}}{\sum_{k=-\frac{N}{2}}^{\frac{N}{2}+1} \text{Re}\{z_{l,k} z_{l-1,k}^*\}}\right] \tag{5.148}$$

The estimation error for the above estimator has a mean value equal to zero. In addition, the variance is:

$$\text{Var}[\Delta\hat{f}_{C,\text{FD}}(l)|\Delta f_c, z_k] = \frac{N_0}{(2\pi)^2 \frac{T}{N} \sum_{n=0}^{N-1} |r_n|^2} \tag{5.149}$$

The denominator indicates that better estimation is achieved if we increase the number of subcarriers.

Because of $\tan^{-1}$ discontinuity, the region of operation of the above estimator is equal to half the frequency difference between successive subcarriers. To apply the above estimator we use two similar symbols, as shown in Figure 5.41. They have the same cyclic prefix and

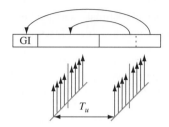

Figure 5.41 Structure of training symbols for CFO estimation in the frequency domain

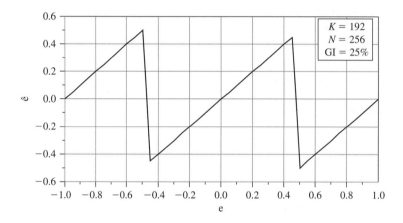

Figure 5.42 ML CFO estimator in the frequency domain

consequently their distance in time is T_u. For this example the FFT size is 256. The OFDM symbol contains 192 subcarriers (75%) whereas 25% of the symbol is used as cyclic prefix.

Figure 5.42 shows simulation results using Equation (5.148) for a system without noise or multipath reflections. The region of operation is $\pm 0.5 \cdot f_d$ where f_d represents the frequency separation between neighbouring subcarriers. It is interesting to note that, because CFO affects all subcarriers in the same way, the linearity of the estimator is not affected even at the limits of the operation region.

Figure 5.43(a) and (b) gives the RMS error of the estimator as a function of SNR and DFT size for SNR = 10 dB. As suggested by Equation (5.149), the estimation variance reduces when the number of subcarriers (or the size of DFT) increases.

Finally, Figure 5.44 presents the root mean square error for channels A and E of ETSI-BRAN. Because channel E impulse response extends beyond the cyclic extension of the OFDM symbol, RMSE assumes a constant value for SNR > 15 dB, indicating that, no matter how low the additive noise is, CFO cannot be further reduced due to severe channel impairment.

CFO Estimation in the Time Domain

While, in the previous subsection, CFO was achieved using the post-FFT samples $z_{l,k}$, the properties of the autocorrelation of the received signal samples, r_k, in the time domain can be used to produce a CFO estimator. For such an estimator to be feasible, the autocorrelation of the received samples r_k must contain information on the CFO.

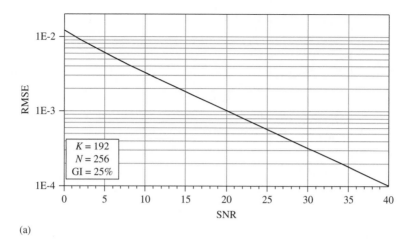

(a)

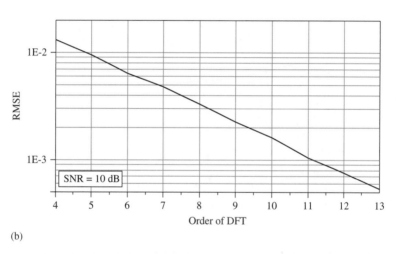

(b)

Figure 5.43 (a) Root mean square error of ML estimator as a function of SNR; (b) root mean square error of ML estimator as a function of the order of DFT

CFO Estimation Using the Cyclic Prefix $\Delta \hat{f}_{C,CP}$

Using $2N + N_g + 1$ consecutive received samples r_k (as in Section 5.3.6.3) that will contain a complete OFDM symbol, we can derive a CFO estimator based on the fact that in all samples there is a phase shift equal to $2\pi \Delta f_c T_u$.

It can be shown that the autocorrelation of the received samples is:

$$E\{r_k r_{k+m}^*\} = \begin{cases} \sigma_s^2 + \sigma_n^2, & m = 0 \\ \sigma_r^2 \exp(j2\pi \Delta f_c T_u), & m = N \\ 0, & m \neq 0 \cup m \neq N \end{cases} \tag{5.150}$$

Taking this into account it can be shown that the maximum likelihood CFO estimator is given by:

$$\Delta \hat{f}_{C,CP} = \frac{1}{2\pi T_u} \tan^{-1} \left[\frac{\sum_{k=\theta}^{\theta+N_g-1} \text{Im}\{r_k \cdot r_{k+N}^*\}}{\sum_{k=\theta}^{\theta+N_g-1} \text{Re}\{r_k \cdot r_{k+N}^*\}} \right] \tag{5.151}$$

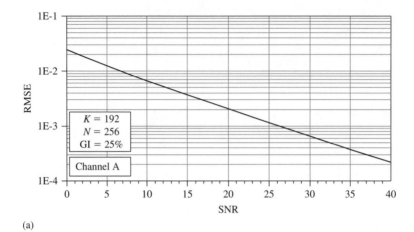

(a)

(b)

Figure 5.44 RMS estimation error for frequency-domain CFO estimation with (a) AWGN and channel BRAN-A; (b) AWGN and channel BRAN-E

where θ represents the estimated beginning of the OFDM symbol. A mistaken estimation of the symbol timing will introduce ISI in the CFO estimator above.

This method can be implemented in conjunction with the subsystem used for symbol timing estimation because it includes common functional blocks.

CFO Estimation Using Special Training Symbols in the Time Domain $\Delta \hat{f}_{C,TS}$

To achieve CFO estimation in this way, the same symbol must be repeated two or more times. In the case of two similar training symbols, the autocorrelation of two received samples coming from successive OFDM symbols l and $l+1$ is given as follows:

$$E\{r_{l,k} r_{l+1,k}^*\} = \sigma_r^2 \exp(j2\pi\Delta f_c T_s) \tag{5.152}$$

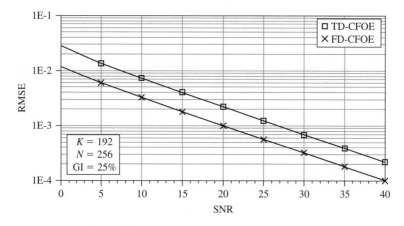

Figure 5.45   Comparison of time-domain and frequency-domain CFO estimation techniques

Because of the similarity of this autocorrelation and the one given for the case of cyclic prefix, it can be easily shown that the CFO estimator in this case is:

$$\Delta \hat{f}_{C,\text{TS}} \frac{1}{2\pi T_s} \tan^{-1} \left[\frac{\sum_{k=0}^{N+N_g-1} \text{Im}\{r_{l,k} r_{l+1,k}^*\}}{\sum_{k=0}^{N+N_g-1} \text{Re}\{r_{l,k} r_{l+1,k}^*\}} \right] \qquad (5.153)$$

Below, we give simulation results for CFO estimation in the time domain using the cyclic extension and special training symbols. The example system is similar to the one given above. Some 192 subcarriers are active out of a total of 256. Simulation results show that the time domain estimators have the same region of operation ($\pm 0.5 \cdot f_d$). Figure 5.45 give RMSE results for $\Delta \hat{f}_{C,CP}$ comparing it with the frequency domain estimator from above. One can see a deterioration exceeding 5 dB, which is due to the fact that the frequency domain estimator uses 192 samples whereas the TD-CP estimator uses only 64 samples. Simulation results give RMSE in Figure 5.46 for the TD-CP estimator for multipath channels BRAN-A and BRAN-E. It can be seen that TD-CP is a poor estimator compared with FD.

5.3.6.6 Effect of Phase Noise

The phase noise in any transceiver system is associated with the nature of the local oscillator signal and is generally divided in two categories. One is modelled as random Brownian motion process, which in the frequency domain is represented by a Lorentzian PSD. This model corresponds to free running tunable oscillators (such as VCOs). The second approach refers to frequency generators using phase locked loops to improve the noise characteristics of the oscillator and at the same time make it tunable across a frequency band while simultaneously possessing satisfactory frequency stability.

We will address the first category, which is most frequently analysed in recent literature ([Pollet95], [Tomba98], [El-Tanany01], [Wu04], [Costa02]). Its Lorentzian spectrum is represented by the transfer function of an LPF as follows:

$$S_s(f) = \frac{2}{\pi f_{\text{ln}}} \cdot \frac{1}{1 + \left(\frac{f}{f_{\text{ln}}}\right)^2} \qquad (5.154)$$

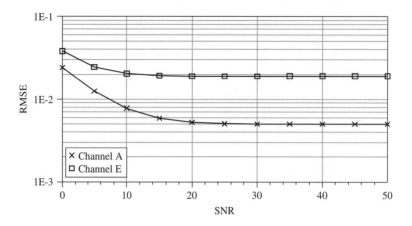

Figure 5.46 RMS estimator for TD-CP estimator

If we represent the cumulative (transmitter and receiver) effect of phase noise as $\theta(t)$, the transmitted OFDM signal is then given by:

$$s_l(t) = \frac{1}{\sqrt{T_u}} \cdot \left(\sum_{k=-K/2}^{K/2-1} a_k \exp(2\pi f_k t) \right) \cdot \exp(j\phi(t)) \tag{5.155}$$

where the corresponding random process $\theta(t)$ has zero mean and variance

$$\sigma_\phi^2 = 2\pi\beta t \tag{5.156}$$

Here β is the one-sided 3 dB bandwidth of the PSD of the carrier signal.

The demodulated signal at the output of the FFT is given by:

$$r_k = a_k I_0 + \sum_{\substack{l=0 \\ l \neq k}}^{N-1} a_l \cdot I_{l-k} + N_k \tag{5.157}$$

where I_n is [Pollet95]:

$$I_n = \frac{1}{T} \int_0^T \exp\left(-j2\pi\frac{n}{T}t\right) \exp[j\phi(t)]\mathrm{d}t \tag{5.158}$$

From the above equations we realize that the effect of phase noise is twofold. It attenuates and rotates all subcarriers by a random phase variation which is common to all subcarriers. In addition to the distorted (in amplitude and phase) useful term $a_k I_0$, the second effect is that it introduces ICI expressed as I_{k-m} terms. As a result, subcarriers are not orthogonal any more. Furthermore, calculation of the power of ICI can give the degradation in SNR due to phase noise (PN) as follows [Pollet95]:

$$D_{\mathrm{PN}} \approx \frac{11}{6 \cdot \ln 10} \cdot \left(4\pi\beta\frac{N}{R}\right) \cdot \frac{E_S}{N_0} \tag{5.159}$$

The derivation of the above formula is done for AWGN and under the assumption that the phase noise is small. In the more general case where phase noise can take high values while the transmission is done in a multipath channel, the received signal is [Wu04]:

$$r_m(k) = a_m(k) \cdot H_m(k) \cdot I_m(0) + \sum_{\substack{l=0 \\ l \neq k}}^{N-1} a_m(l) \cdot H_m(l) \cdot I_m(l - k) + N_m(k) \qquad (5.160)$$

where the effects of phase noise to the received signal are

$$\text{CPE} = I_m(0) \quad \text{and} \quad \text{ICI} = \sum_{\substack{l=0 \\ l \neq k}}^{N-1} a_m(l) \cdot H_m(l) \cdot I_m(l - k) \qquad (5.161)$$

For this general case, SINR is given by:

$$\text{SINR} = \frac{2\frac{d_0^{N+1}-(N+1)d_0+N}{(d_0-1)^2} - N}{\sum_{p=1}^{N-1}\left[2\frac{d_p^{N+1}-(N+1)d_p+N}{(d_p-1)^2} - N\right] + \frac{N^2}{\gamma_s}} \qquad (5.162)$$

where d_p and γ_S (SNR per subcarrier) are given as

$$d_p = \exp\left[j\frac{2\pi p}{N} - \frac{\pi\beta}{R}\right], \quad \gamma_S = \frac{E_S}{\sigma^2} \qquad (5.163)$$

It must be noted that, for large subcarrier number N and small phase noise, the above equation reduces to Equation (5.159).

Some qualitative observations [Wu04] are as follows:

(1) Because of the inherent dependence of SINR on the phase noise linewidth β (because of d_p) it can be seen that larger β degrades SINR. Consequently high-purity oscillators are required.

(2) Increasing the number N of subcarriers deteriorates system performance. Apart from Equation (5.162), this can be deduced by intuition because higher N leads to smaller subcarrier spacing, resulting in higher sensitivity to phase noise.

(3) In general, when $N\beta/R$ is in the order of 10^{-5}, SINR degradation is negligible and the system behaves as if there were no phase noise. When $N\beta/R$ is in the order of 10^{-3} or 10^{-2}, SINR degradation is moderate and can be compensated. Finally for $N\beta/R > 10^{-2}$ ICI dominates the useful signal.

We now give simulated results of error rate performance in terms of SNR with phase noise as a parameter. Figure 5.47 give the effect of Lorenzian-type phase noise on an OFDM system using QPSK modulation. Parameter γ expresses the amount of phase noise ($\gamma = 2\pi\beta T$, T = integration time of integrate-and-dump filter).

Figure 5.48 shows BER results for an OFDM with $N = 128$ using QPSK modulation in AWGN for a phase noise spectrum shaped by a PLL with PSD as in Figure 5.49. Figure 5.50 gives results for 64-QAM and phase noise equal to -31 dBc. The spectrum is formulated by a PLL and has a PSD, as in Figure 5.51. The parameters of the simulated system are similar

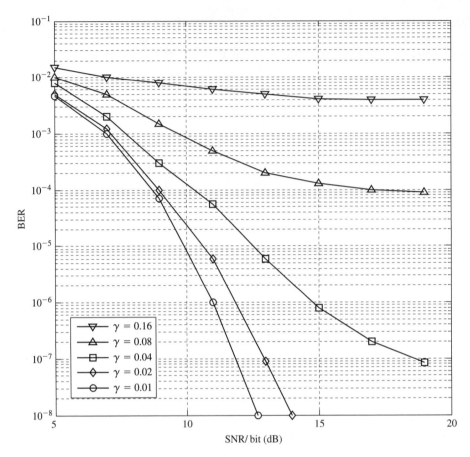

Figure 5.47 BER performance for QPSK modulated OFDM signal with phase noise of different levels (reprinted from L. Tomba, 'On the effect of Wiener phase noise in OFDM systems', IEEE Trans. Commun., vol. 46, pp. 580–583, ©1998 IEEE)

to those of OFDM-based WLANs. Finally, Figure 5.52 shows the phase noise effect for a Hiperlan2-type OFDM operating in mode 7 (64 subcarriers, 64-QAM, code rate ¾ using a PLL-shaped PSD.

In order to correct for phase noise disturbances, the average CPE is estimated and then cancelled out by rotation, as shown in Figure 5.53. This is done after equalization by multiplication of the data samples $a_m(k)$ of the mth OFDM symbol by $1/H_m(k)$ at the output of the FFT.

5.3.6.7 Effect of *I–Q* Imbalance

The *I–Q* imbalance at the receiver is generated by mismatching in the electronic manufacturing of the downconverting RF *I–Q* mixer. It can be modelled by introducing a $\Delta\varepsilon$ and a $\Delta\varphi$ term in the received signal as shown below:

$$y(t) = e^{-j\Delta\varphi} \cdot \text{Re}[r(t)] + (1 + \Delta\varepsilon) \cdot e^{j\Delta\varphi} \cdot \text{Im}[r(t)] \qquad (5.164)$$

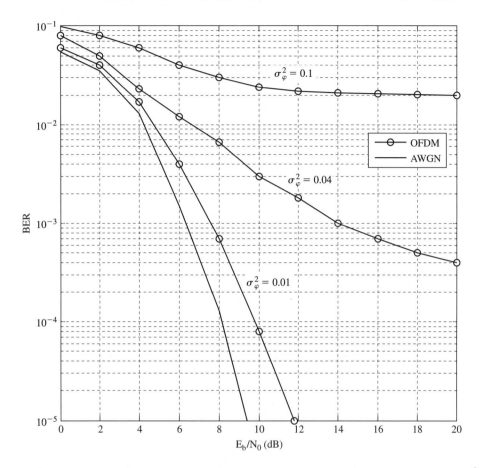

Figure 5.48 BER for QPSK modulated OFDM with $N = 128$ and different values of phase noise (σ_θ^2) (reprinted from [Shentu03], ©2003 IEEE)

In the case of a direct-conversion receiver (see Chapter 3) the effect of the IQ imbalance can be seen more clearly and the demodulated data symbol corresponding to the kth subcarrier of the mth OFDM symbol is [Windisch04]:

$$Z_k(m) = C_1 \cdot Y_k(m) + C_2 \cdot Y_{-k}^*(m) = \frac{1 + \Delta\varepsilon \exp(-j\Delta\varphi)}{2} Y_k(m)$$

$$+ \frac{1 - \Delta\varepsilon \exp(j\Delta\varphi)}{2} Y_{-k}^*(m) \qquad (5.165)$$

where Y^* represents the DFT of the image $y^*(t)$ of the received signal $r(t)$ after downconversion.

Interpreting the above equation, the downconverted OFDM signal contains not only the useful part $Y_k(m)$ but a portion of the image $y^*(t)$ as well. The impact of that is deterioration of the BER performance in the presence of I–Q imbalances. Figure 5.54(a, b) shows the effect of phase and amplitude imbalances respectively. From these one can note that the system is more sensitive to amplitude than phase imbalance.

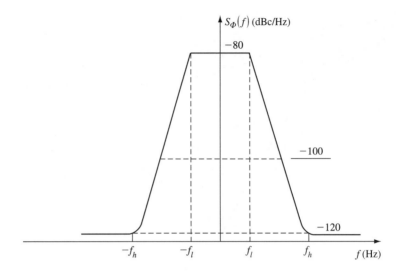

Figure 5.49 Power spectral density of phase noise shaped by phase-locked loop

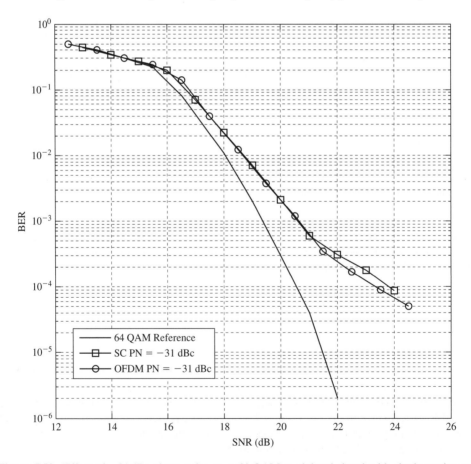

Figure 5.50 Effect of −31 dBc phase noise on a 64 QAM modulated signal with single-carrier and OFDM transmission (reprinted from [Tubbax01], ©2001 IEEE)

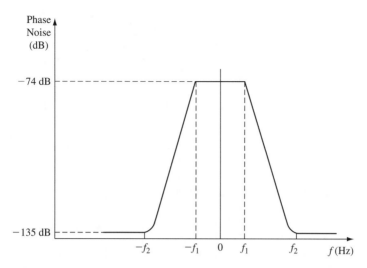

Figure 5.51 Phase noise spectrum shaped with PLL with $f_1 = 10\,\text{kHz}$ and $f_2 = 1\,\text{MHz}$ (reprinted from [Tubbax01], ©2001 IEEE)

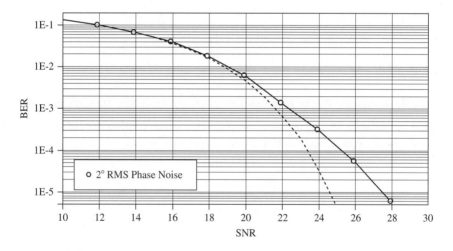

Figure 5.52 Effect of local oscillator phase noise on BER performance

To compensate for I–Q imbalances $\Delta\varepsilon$ and $\Delta\varphi$ must first be evaluated. One way to do it is to use cross-correlation and power statistics of the demodulated data symbols C_1, C_2 to estimate C_1, C_2 [Windisch04]:

$$\frac{E\{Z_k \cdot Z_{-k}\}}{E\{|Z_k + Z_{-k}^*|^2\}} = C_1 \cdot C_2 \tag{5.166}$$

$\Delta\varepsilon$ and $\Delta\varphi$ can be obtained from the expressions of C_1, C_2 as functions of $\Delta\varepsilon$, $\Delta\varphi$ [see Equation (5.165)] and the value of the product $C_1 \cdot C_2$.

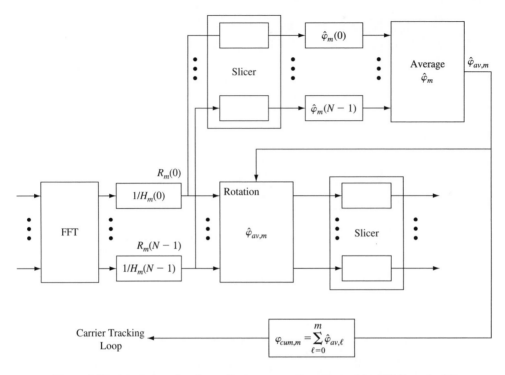

Figure 5.53 A technique for phase offset compensation proposed by [Nikitopoulos01]

5.3.6.8 Peak-to-average Power Ratio and High Power Amplifier Distortion

According to Equation (5.97), the modulated OFDM signal (during the *l*th symbol) before transmission in continuous and sampled form was given before as:

$$s(t) = \frac{1}{\sqrt{T_u}} \cdot \sum_{k=-0}^{N-1} a_k \exp\left(2\pi f_k t\right) \cdot u(t) \tag{5.97}$$

$$s(t_n) = \frac{1}{\sqrt{N}} \cdot \sum_{k=0}^{N-1} a_k \exp\left(j2\pi \frac{kn}{N}\right) = IFFT\{[a_0\, a_1\, a_2 \ldots a_{N-1}]\} \tag{5.98}$$

For relatively large values of the number of subcarriers N, regardless of the modulated data symbols a_k, both the real and imaginary components of $s_l(t)$ have Gaussian distribution. The consequence of that is Rayleigh distribution for the amplitude of the OFDM transmitted symbol $s_l(t)$, whereas power is chi-square distributed with a cumulative distribution function as follows:

$$F(z) = 1 - \exp(-z) \tag{5.167}$$

If we define as peak-to-average power ratio (PAPR) the ratio of the maximum to the average transmitted signal then the following equation can be used:

$$\text{PAPR} = \frac{\max(|s_k|^2)}{\frac{1}{N} \cdot \sum_{k=0}^{N-1} |s_k|^2}, \quad k = 0, 1, \ldots, N-1 \tag{5.168}$$

where s_k represent the kth sample of the transmitted signal.

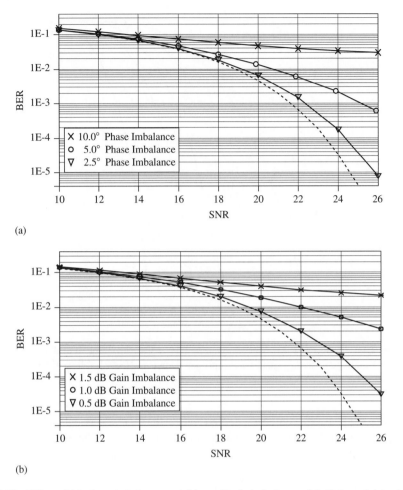

Figure 5.54 Effect of (a) phase imbalance and (b) amplitude imbalance of I–Q demodulator in system performance

With mutually uncorrelated samples s_k, the probability that the PAPR is below z is:

$$P(\text{PAPR} \leq z) = (F(z))^{\alpha N} \tag{5.169}$$

Taking $a = 2.8$, it can be shown by simulations [VanNee98] that the above equation is quite accurate for $N > 64$. This is demonstrated in Figure 5.55, which depicts the CDF of the PAP ratio for $N = 32$, 64, 128, 256 and 1024. From this figure one can see that for CDF $= 10^{-5}$, the difference in PAPR between the formula and simulation is less than 0.2 dB.

In general, PAPR must be reduced in order to ease the demands on the A/D and D/A converters and operate the power amplifiers at higher efficiency. There are several methods that can be used to reduce PAPR. They include clipping, windowing of the signal peak, modification of the operating point of the high power amplifier and peak cancellation. Furthermore, coding is another way to reduce PAPR effectively.

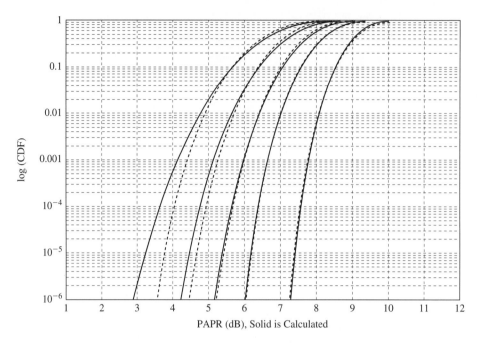

Figure 5.55 CDF of the PAPR for $N = 32, 64, 128, 256, 1024$. Solid lines and dotted lines represent calculated and simulated values respectively (reprinted from [VanNee98], ©1998 IEEE)

By *clipping* the peaks exceeding a specified level in the OFDM signal, PAPR can be reduced drastically. Although this is the simplest technique to control PAPR, it produces distortion which, in turn, introduces self-interference. Furthermore, clipping introduces considerable out-of-band radiation.

Peak windowing entails the multiplication of large signal peaks with a window function like Kaiser, Hamming or other appropriate type. Care must be taken to choose its length such that out-of-band interference is minimized while the BER performance does not deteriorate considerably [vanNee00].

Clipping and peak windowing deliberately introduce nonlinear distortion to the OFDM transmitted signal. However, this nonlinear distortion does not affect BER drastically. For example SNR deteriorates by less than 1 dB for PAP ratio reduced by 4 dB at BER $= 10^{-4}$ for 16-QAM and $N = 64$ [VanNee98]. Despite that, clipping and peak windowing result in considerable increase of out-of-band radiation. By employing back-off at the power amplifier, the spectral mask beyond the transmission band can be confined to acceptable levels. The back-off value is defined as the ratio (in decibels) of the output power of the PA in operation to the output power that the PA can give at the 1 dB compression point. The more out-of-band spectrum distortion is permitted, the lower the effect of PAPR reduction methods when applied jointly with PA back-off. For example applying 6.3 dB backoff on 64 subcarrier OFDM without PAPR reduction has the same effect on out-of band spectrum (-30 dB with respect to in-band power) as applying both peak windowing and back-off of 5.3 dB [vanNee98]. This is illustrated in Figure 5.56.

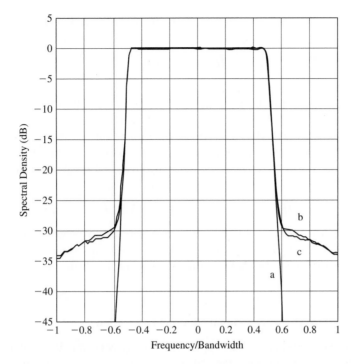

Figure 5.56 (a) Ideal spectrum for 64-subcarrier OFDM; (b) plain OFDM with back-off of 6.3 dB and $p = 0.3$; (c) peak windowing with back-off of 5.3 dB (reprinted from [VanNee98], ©1998 IEEE)

On the other hand linear peak cancellation can be achieved by subtracting a suitable reference function from the OFDM signal like the *sinc* function. This function is shifted in time such that it is subtracted from the signal when a peak is detected. In addition, to avoid spectrum distortion, the bandwidth of the reference function must be close to the signal bandwidth [VanNee].

Coding and selected mapping can be used to improve performance degraded by large PAPR. OFDM symbols with high PAPR can take advantage of coding, since in a coded system the BER performance depends on the power of several neighbouring symbols rather than on each of the symbols separately. Selected sampling is used to transform an OFDM symbol to M different symbols which correspond to exactly the same original symbol. Then, after IFFT, the one having the lowest PAPR is used for transmission. The transformation is done by multiplying the data sequence by a fixed phasor:

$$X_m[k] = X[k] \exp(j\varphi_m), \quad m = 1, 2, \ldots M, \tag{5.170}$$

where φ_m is fixed for a particular m and takes values from 0 to 2π.

5.3.7 Channel Estimation in OFDM Systems

For coherent detection the channel transfer function $H(t, f)$ must be estimated. In discrete form, N complex coefficients $H_{l,k}$ must be estimated corresponding to the kth subcarrier of the lth OFDM symbol. The relation between the estimated and the actual channel coefficients is:

$$\hat{H}_{l,k} = H_{l,k} + n_{l,k}^{H} \tag{5.171}$$

where $n_{l,k}^{H}$ represents additive white noise.

For such an estimation to take place pilot symbols are used in the OFDM time–frequency two-dimensional grid. Such a grid usually has periodical structure, as depicted in Figure 5.57. We assume there are N_{st} OFDM symbols in one time-frame and N_{sf} OFDM subcarriers in one OFDM symbol. Shaded boxes represent pilot symbols located at distances N_t and N_f along the time and frequency axes, respectively. Applying channel estimation on the received data symbols we have:

$$y_{l,k} = \hat{H}_{l,k} \cdot s_{l,k} + n_{l,k}, \quad l = 1,\ldots,N_{st} \quad k = 1,\ldots,N_{sf} \tag{5.172}$$

In applications where the channel changes very slowly in time, one-dimensional channel estimation is used. We will briefly examine two-dimensional channel estimation after we give the most frequently used techniques for one-dimensional channel estimation using pilot symbols.

5.3.7.1 Least Square One-dimensional Channel Estimation

Least squares (LS) is the simplest way to perform channel estimation by dividing the received symbols $\hat{z}_{t,k}$ by the corresponding transmitted pilot symbols $p_{t,k}$ where t represents the set of indices of pilot symbols within OFDM time frames. Thus, the channel transfer function estimates are:

$$\hat{H}_{t,k,LS} = \frac{\hat{z}_{t,k}}{p_{t,k}} = H_{t,k,LS} + \frac{n_{t,k}}{\alpha_{t,k}} = H_{t,k,LS} + n'_{t,k} \tag{5.173}$$

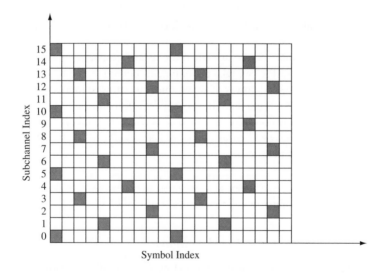

Figure 5.57 Time–frequency grid for OFDM transmission (shaded boxes indicate pilots)

5.3.7.2 Linear Minimum Mean Square Error One-dimensional Channel Estimation

From the above equation by minimizing the error $\Delta H_{t,k}$:

$$\Delta H_{t,k} \equiv \hat{H}_{t,k,\text{LS}} - H_{t,k,\text{LS}} \tag{5.174}$$

in the mean square sense ($\min\{E(|\Delta H_{t,k}|^2)\}$) we obtain the linear minimum mean square error (LMMSE) estimate [Edfors98]:

$$\hat{\mathbf{H}}_{\text{LMMSE}} = \mathbf{R}_{H\hat{H}_{\text{LS}}} \mathbf{R}_{\hat{H}_{\text{LS}}\hat{H}_{\text{LS}}}^{-1} \hat{\mathbf{H}}_{\text{LS}} \tag{5.175}$$

where $\hat{\mathbf{H}}_{LS}$ is the vector representing the LS estimate of channel transfer function at the position of the pilots. $\mathbf{R}_{H\hat{H}_{\text{LS}}}$ is the correlation matrix of the channel with the known pilot symbol and $\mathbf{R}_{\hat{H}_{\text{LS}}\hat{H}_{\text{LS}}}$ is the autocorrelation matrix of the pilot symbols. Its inverse is given by:

$$\mathbf{R}_{\hat{H}_{\text{LS}}\hat{H}_{\text{LS}}}^{-1} = \left(\mathbf{R}_{\text{PP}} + \frac{1}{\text{SNR}}\mathbf{I} \right)^{-1} \tag{5.176}$$

By assuming a known model for the radio channel (for example [Hoeher92]) correlation, the correlation and autocorrelation matrices $\mathbf{R}_{H\hat{H}_{\text{LS}}}$ and $\mathbf{R}_{\text{PP}}$ can be calculated. This results in lower on-line calculation complexity. However, it is necessary to have a satisfactory estimation of SNR. Where using a mistakenly low SNR value (while the actual SNR is high), LMMSE estimation can be worse than LS channel estimation.

5.3.7.3 Two-dimensional Channel Estimation

For two-dimensional (2D) estimation it is necessary to determine the maximum distances N_t, N_f in the time–frequency grid. By assuming that the channel is a stochastic process, in order to have adequate sampling rate the Nyquist criterion is used to give [Fazel97], [Hoeher92]:

$$N_t \leq \frac{1}{2f_{D,\text{max}}T_s} \quad \text{and} \quad N_f \leq \frac{1}{2\frac{1}{T_u}\tau_{\text{max}}} \tag{5.177}$$

where $f_{D,\text{max}}$ represents the maximum Doppler frequency, T_S is the duration of an OFDM symbol, $1/T_u$ is the subcarrier distance and τ_{max} is the maximum delay of the channel impulse response. To get the optimum LMMSE channel estimation, Equation (5.175) is modified to contain both time and frequency indices as follows:

$$\hat{\mathbf{H}}_{t,k,\text{LMMSE}} = \mathbf{R}_{H\hat{H}_{\text{LS}}} \mathbf{R}_{\hat{H}_{\text{LS}}\hat{H}_{\text{LS}}}^{-1} \hat{\mathbf{H}}_{t,k,\text{LS}} \tag{5.178}$$

In the above equation t and k represent pilot indices within the time–frequency grid.

To substantially simplify processing requirements 2D estimation can be implemented by using separate one-dimensional (1D) estimators-filters. Fist all attenuations for a specific sub-carrier for all OFDM symbols (sweep in time) will be determined. After that, all attenuations for a specific symbols for all subcarriers (sweep in frequency) will be obtained. Assuming alignment of pilots in both time and frequency axes, there will be N_{st}/N_f plus N_{sf}/N_t attenuation estimations.

5.4 Spread Spectrum Systems

5.4.1 Introduction and Basic Properties

Spread spectrum is a transmission technique according to which the modulated signal is transformed into a signal of much higher bandwidth before transmission. Two of the most frequently used approaches to generate spread spectrum are direct sequence (DS) and frequency hopping. In both of them the transformation of the modulated signal to a spread spectrum signal is done through a spreading sequence or spreading code (SC), which is a bit sequence independent of the information signal and of much higher rate compared with the information data sequence.

In DS systems the modulated signal is multiplied by the spreading code before upconversion. At the receiver, the downconverted signal is despread by multiplication by the synchronized spreading code before entering the detection stage (see Figure 5.58).

In FH systems the spreading code imposes the carrier frequency to the modulated signal through a frequency synthesizer. At the receiver, after spreading code synchronization, the opposite is done, as illustrated in Figure 5.59.

The basic characteristics distinguishing a spread spectrum signal by a nonspread information signal are (a) the bandwidth of the transmitted signal, which is many times higher than the necessary information bandwidth and(b) the spreading operation, which is achieved by the use of a spreading sequence (or code). In demodulation, the receiver benefits from this spreading transformation by synchronizing to the received spread signal before detection.

Figure 5.60 shows graphically the resulting signal after the spreading operation for DS and FH systems in the time and frequency domains respectively. In DS, multiplication by the spreading code results in a high-frequency signal. In FH the SC dictates the carrier frequency f_C at which the transmitted signal will be centred for a given time period. For example at $t_3, f_C = f_4$.

The main advantages of spread spectrum transmission are associated with the effect of the spreading code on the transmitted signal, which is transformed into a wideband signal at the output of the transmitter. From the spreading and despreading operation stem the principle advantages of spread spectrum systems in the presence of narrowband interference. However, spread spectrum systems exhibit similar improvements in the presence of multipath transmission, intended jammers and in multiple access communications, as will be seen in the following sections.

5.4.1.1 Direct Sequence Transmission

The information signal before spreading at the transmitter is:

$$U(t) = \sum a_n g_T(t - nT_S) \tag{5.179}$$

where α_n is the transmitted symbol sequence and $g_T(t)$ represents the modulation pulse of duration equal to the symbol time T_S.

Subsequently, the spreading code sequence $s_C(t)$ is multiplied by $U(t)$ to give:

$$U(t) \cdot s_C(t) = \sum a_n g_T(t - nT_S) \cdot \sum c_{PN}(n) \cdot p(t - nT_C) \tag{5.180}$$

where $c_{PN}(n)$ is the value (± 1) of the spreading sequence, T_C is the duration of a chip and $p(t)$ represents the unit pulse:

$$p(t) = \begin{cases} 1, & 0 \leq t \leq T_C \\ 0, & \text{elsewhere} \end{cases} \tag{5.181}$$

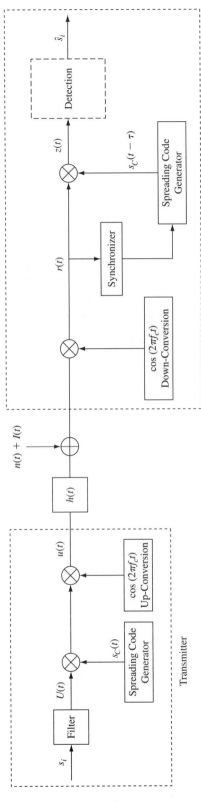

Figure 5.58 Transceiver of DSSS system

343

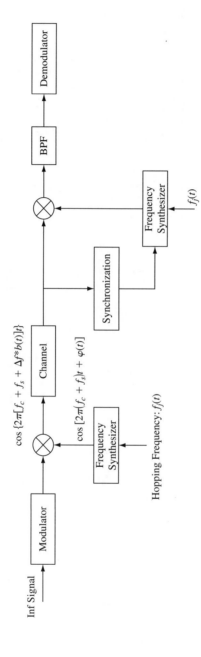

Figure 5.59 Transceiver of a frequency hopping SS system

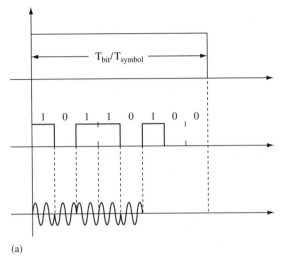

(a)

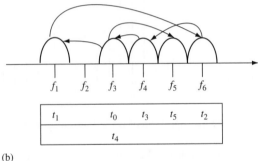

(b)

Figure 5.60 (a) Information and chip sequence pulses for direct sequence SS; (b) frequency hopping pattern occurring at successive time instants $t_0, t_1, \ldots, t_5$ in FHSS system

The transmitted signal at the output of the transmitter is:

$$u(t) = U(t) \cdot s_C(t) \cdot \cos 2\pi f_C t \tag{5.182}$$

In the frequency domain we have a wideband signal produced by the convolution of a narrowband signal (f) with a wideband one $S_C(f)$, as illustrated in Figure 5.61.

Figure 5.62 shows in more detail the spectrum of the spreading sequence. The bandwidth of the main lobe is $2/T_C$ whereas each spectral line has a bandwidth equal to $1/T_b$. If at the input of the receiver, apart from the useful spread signal, there exists a narrowband interfering signal $I(t) \cos(2\pi f_C t)$ as well, after downconversion the signal becomes:

$$r_{BB}(t') = U(t') \cdot s_C(t') + I(t') \tag{5.183}$$

where t' denotes a time index of the received signal shifted by τ corresponding to the delay with respect to that at the transmitter ($t' = t + \tau$).

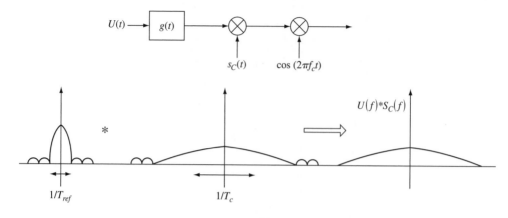

Figure 5.61 Spreading operation in a DSSS system

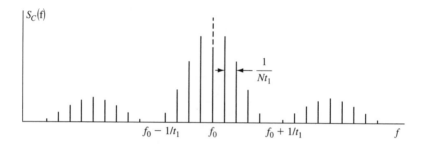

Figure 5.62 Spectrum of DSSS transmitted signal

Multiplication of the above signal by the synchronized (to the useful signal) spreading code gives:

$$z(t) = U(t') \cdot s_C(t') \cdot s_C(t) + I(t') \cdot s_C(t') \qquad (5.184)$$

For perfect synchronization $(t = t')$ the square of the spreading code sequence must be equal to one $[s_C^2(t) = 1]$ in order to recover in the detector the originally transmitted signal $U(t)$.

The first term of the above equation corresponds to the despread narrowband information signal whereas the second product in the frequency domain results in an interference signal spread along the whole spread spectrum bandwidth. By doing that, the interference power P_J is spread along the whole SS bandwidth (S_{BW}) such that the interference spectral density is now $S_J = P_J / S_{BW}$. This can be seen in Figure 5.63. Passing this composite signal through a BPF of bandwidth equal to the information bandwidth I_{BW}, it is easy to see that the total interference power passing through the information bandwidth is equal to

$$P_J(\text{det}) = \frac{P_J \cdot I_{BW}}{S_{BW}} = P_J \cdot \frac{1/T_b}{1/T_C} \qquad (5.185)$$

The above relation states that the interference power is divided by T_b/T_C when it reaches the input of the detector. The ratio T_b/T_C is called the processing gain of the system and represents the reduction of the interference power after passing through the detector. Processing gain stems from the spreading–despreading operation on the useful signal.

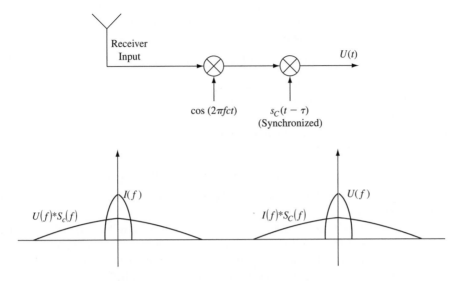

Figure 5.63   Despreading operation in a DSSS system

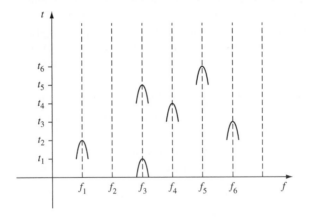

Figure 5.64   Example of hopping sequence (generating $f_3, f_1, f_6, f_4, f_3, f_5$ in successive time instants)

5.4.1.2 Frequency Hopping Transmission Systems

As mentioned before, the FHSS loads the encoded information signal $U(t)$ to a carrier the frequency of which is determined by the spreading code which commands the frequency hopping synthesizer. In this way the transmitted signal remains at a particular carrier frequency for a time interval T_h after which the signal switches to a different carrier frequency. This procedure is illustrated in Figure 5.64, which shows pictorially the hopping process in the time–frequency grid.

Figure 5.59 shows a simple model for a general frequency hopping spread spectrum system. The modulated signal is upconverted using a local oscillator frequency $f_i(t)$, which changes in a pseudorandom fashion through a frequency synthesizer under the command of a PN generator.

Consequently, the carrier frequency of the transmitted signal is 'hopped' among M possible frequencies constituting a hopping pattern:

$$f(t) = f_j, \quad jT_h \leq t \leq (j+1)T_h \tag{5.186}$$

where T_h is the time during which the system stays at frequency f_j. It is also called the dwell time of the frequency hopper-synthesizer.

All frequencies belong to a set $S = \{F_n : 1 \leq n \leq Q\}$ ordered such that $F_n \leq F_{n+1}$ $\forall n$. Q stands for quantity and is the overall number of frequencies that the system can use.

The transmitted signal can be expressed as:

$$s(t) = \sqrt{P} \cos \{2\pi[f_C + f_j + \Delta f \cdot b(t)]t + \phi(t)\} \tag{5.187}$$

where Δf is equal to half the FSK frequency difference between information bits 0 and 1 and $b(t)$ takes the values of ± 1.

At the receiver, the synchronized signal is de-hopped by mixing by $f_i(t)$, which through a hopping synthesizer follows exactly the same hopping pattern as in the transmitter. This guarantees that the signal at the output of the despreader will contain only the information $\Delta f \cdot b(t)$, which can be FSK-demodulated after passing through a BPF.

The main parameters of an FH system are: dwell time T_h being the time during which it remains at the same frequency and its inverse defined as the hopping rate, the number of frequencies Q used for hopping, and the ratio of bit time to dwell time $(T_S/T_h = k)$, which determines whether the FH system is a slow frequency hopping system ($k \leq 1$) or a fast-FH system. In slow FH the system stays at a particular frequency for the duration of several symbols, whereas in fast FH the carrier frequency changes a few times during a symbol.

5.4.2 Direct Sequence Spread Spectrum Transmission and Reception

Figure 5.65 shows the basic demodulation structure of the DSSS receiver, which includes downconversion [multiplication by $\cos(\omega_C t + \varphi)$], synchronization by multiplication by the synchronized spreading code and detection. The last stage consists of a matched filter, which is implemented by an integrator. The received signal before detection is given by:

$$r_b(t) = \{[U(t) \cdot s_C(t) \cdot \cos(\omega_C t)] * h(t)\} \cdot \cos(\omega_C t + \phi) \cdot s_C(t - \tau)$$
$$+ n(t) \cdot \cos(\omega_C t + \phi) \cdot s_C(t - \tau) \tag{5.188}$$

where $n(t)$ is the AWGN noise added by the radio channel while $h(t)$ represents the channel impulse response. In the general case there is a local oscillator phase difference φ and a spreading code misalignment τ.

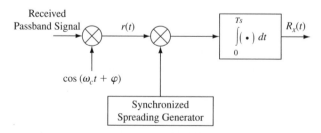

Figure 5.65 Demodulation model of DSSS receiver

5.4.2.1 Performance in the Presence of Noise and Interference

We assume that there is no multipath channel [i.e. $h(t) = A\delta(t)$]. When there is only additive noise, the output of the integrator at the kth integration interval is given by:

$$Rx_{\text{out}}(k) = U_0(k) + n_0(k) \tag{5.189}$$

where

$$U_0(k) = \int_{t_k - T}^{t_k} \{[U(t') \cdot s_C(t') \cdot \cos(\omega_C t')]\} \cdot \cos(\omega_C t' + \varphi) \cdot s_C(t' - \tau) dt' \tag{5.190}$$

and

$$n_0(k) = \int_{t_k - T}^{t_k} n(t') \cdot \cos(\omega_C t' + \varphi) \cdot s_C(t' - \tau) dt' \tag{5.191a}$$

The noise component $n_0(k)$ has a mean value equal to zero and a variance:

$$\overline{n_0^2(k)} = \frac{N_0}{2} \cdot \frac{1}{T_S} \tag{5.191b}$$

where $N_0/2$ is the PSD of $n(t)$.

Considering $U_0(t)$ it can be easily shown [Cooper86] that to maximize the output of the integrator and eliminate interference form neighbouring bits, the integration time must be T_S and the sample time t must be an integer of the bit duration. Hence the integration interval is $(0, T_S)$. In addition it can be shown [Cooper86] that Equation (5.190) reduces to:

$$U_0(k) = A a_n(k) \cdot R_{S_C}(\tau) \cdot \cos\varphi \tag{5.192}$$

where

$$R_{S_C}(\tau) = \frac{1}{N_{S_C}} \int_0^{N_{S_C}} s_C(t) s_C(t + \tau) dt \tag{5.193}$$

Noting that the autocorrelation of $s_C(t)$ takes the maximum value for $\tau = 0$ and that the cosine function has a maximum at $\varphi = 0$, we conclude that for optimized detection the local oscillators and spreading codes of the transmitter and receiver must be properly synchronized.

As a result, the output of the detector for a perfectly synchronized (in code and phase) incoming signal is:

$$Rx_{\text{out}}(k) = A a_n(k) + n_0(k) \tag{5.194}$$

Taking into account narrowband interference $i(t) = I(t) \cdot \cos\omega_C t$ [where $I(t)$ is a narrowband information signal] in a similar way we can obtain [Goldsmith05]:

$$Rx_{\text{out}}(k) = A a_n(k) + n_0(k) + i_0(k) \tag{5.195}$$

where $i_0(k)$ under the assumption of perfect code and phase synchronization is given by

$$i_0(k) = \int_0^{T_S} I(t) \cdot s_C(t) \cdot \cos^2(\omega_C t) dt \tag{5.196}$$

5.4.2.2 Performance in the Presence of Multipath Fading

In this case the received signal before detection becomes:

$$r_b(t) = c_0 U(t) \cdot \cos(\omega_C t) + \left[\sum_{l=1}^{L} c_l U(t - \tau_l) \cdot \cos(\omega_C(t - \tau_l)) \right]$$

$$\cdot \cos(\omega_C t) \cdot s_C(t) + n(t) \cdot \cos(\omega_C t) \cdot s_C(t) \qquad (5.197)$$

where c_l, τ_l represent the amplitude and delay of the lth reflected ray of the multipath profile. In addition, as before, perfect code and phase synchronization is assumed.

After mathematical manipulation the output of the integrator-detector becomes:

$$Rx_{\text{out}}(k) = c_0 \cdot U_0(k) + \int \sum_{l=1}^{L} c_l \cdot R_{S_C}(\tau_l) \cdot \cos(\omega_C \tau_l) + n_0(k) \qquad (5.198)$$

Where delays τ_l are multiples of the symbol time T_S, autocorrelations $R_{S_C}(\tau_l)$ are delta functions $\delta(\tau_l)$ and the second term of the above equation is eliminated. In that case there is no ISI added to the useful signal at the output of the detector. A qualitative explanation is easy to follow since the summation terms represent delayed replicas of the signal $[U(t - \tau_l)]$ which, as explained before, due to the code misalignment [multiplied by $s_C(t)$], will be spread over the SS bandwidth, resulting in the elimination of ISI.

5.4.3 Frequency Hopping SS Transmission and Reception

Below we consider briefly the error performance of a slow FH system in the presence of a partial-band jammer. The evaluation of the performance of FFH is more complicated since jamming changes during a symbol whereas it remains constant for SFH transmission.

Let us consider an M-FSK modulation in which the jammer occupies a bandwith W_J which is a percentage a of the whole transmitted bandwidth $W_{SS}(a = W_J/W_{SS})$. In addition we assume that the hop rate is $R_H = 1$ hop/symbol and that the interferer (jammer) remains in the same frequency band for some time. If E_J is the interference total power and E_b is the energy per bit, the interference power density N'_J within the jamming band and the degraded SNR are respectively:

$$N'_J = \frac{E_J}{W_J} = \frac{E_J}{aW_{SS}} = \frac{N_J}{a}, \quad \text{SNR}_J = \frac{E_b}{N'} = \frac{aE_b}{N_J} \qquad (5.199)$$

From the above it is easily deduced that N_J represents the spectral density of interference when it is spread across the whole FH band. If there were only AWGN, N_J would represent again the same quantity assuming that E_J would be the total noise power.

If we denote by $P_S(N)$ and $P_S(J)$ the probability of symbol error due to AWGN channel and the presence of jamming interference respectively, the overall P_S will be:

$$P_S = p(\text{symbol error|with AWGN only})p(\text{no jamming})$$

$$+ p(\text{symbol error|with jamming only})p(\text{jamming}) = P_S(N)(1 - a) + P_S(J)a \qquad (5.200)$$

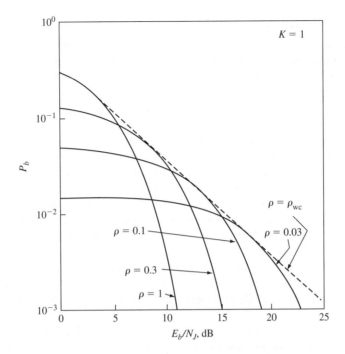

Figure 5.66 Performance of BFSK FHSS system in partial-band interference for various values of a (here denoted as ρ) (reprinted from M. Simon *et al.*, 'Spread Spectrum Communications Handbook', rev. sub. edn, copyright ©1994 by McGraw–Hill)

After some mathematical manipulations it can be shown that the average probability of error is [Simon94]:

$$P_S = aP_S(\text{error} \mid \text{jammer}) = a\frac{1}{M}\sum_{i=2}^{M}(-1)^i \binom{M}{i}\exp\left[\frac{-KE_b}{N'_J}\left(1-\frac{1}{i}\right)\right] \qquad (5.201)$$

where M is the constellation size with $M = 2^K$. In the above formula only the contribution of the jammer-interference term is taken into account while it is assumed that there is no contribution from AWGN. It is easy to see that for binary FSK ($M = 2$, $K = 1$) the average probability of error is:

$$P_{S-\text{BFSK}} = \frac{a}{2}\exp\left[-a\frac{E_b}{2N_J}\right] \qquad (5.202)$$

Figure 5.66 illustrates the performance of FH-BFSK spread spectrum in partial-band interference for several values of a. In small E_b/N_J ratios performance is worse when the whole band is interfered ($a = 1$). On the other hand, when the interference power is gathered in a narrow band ($a < 1$), to reduce BER below a certain value, there must be increased E_b/N_J compared with the case of wider interference ($a \approx 1$).

5.4.4 Spread Spectrum for Multiple Access Applications

In single user spread spectrum applications the correlation properties of the spreading code are used to eliminate narrowband interference and intersymbol interference due to multipath propagation. By exploiting analogous properties of suitable spreading sequences, a spread spectrum receiver can be designed to demodulate properly only one out of many users and reduce the received signal from the rest of the users effectively to the level of noise. To achieve this, each user is assigned a unique spreading sequence, which has very low cross-correlation value with the SCs of the rest of the users. In this way the signals of the other users, since they are not properly despread, are treated as interference of low power level at the detector. According to this scheme, many users can utilize the same spectrum at the same time and still be able to have efficient communication between specific users sharing the same spreading code. Modern wideband cellular communications (CDMA, WCDMA) are based in this principle to have spectrally efficient communication between multiple users. A base station is employed in such a system through which all communication between users takes place.

Figure 5.67 illustrates the arrangement of a cellular multiple access system. There is a base station, which serves as a receiver for all transmissions initiated from mobile terminals. This link is the uplink channel or the reverse link. In addition the BS serves as the transmitter of information directed to mobile terminals. This link is called the downlink channel or the forward link.

To investigate the capabilities of multiple access systems, the downlink and uplink receiver performance must be evaluated. The reception parameters are not the same for the two links and this can be realized if we consider Figure 5.68(a, b). In the downlink channel the BS-transmitter [Figure 5.68(a)] transmits all the signals in a synchronous way. As a result all the receivers at downlink receive the summation of the all the signals aligned in time.

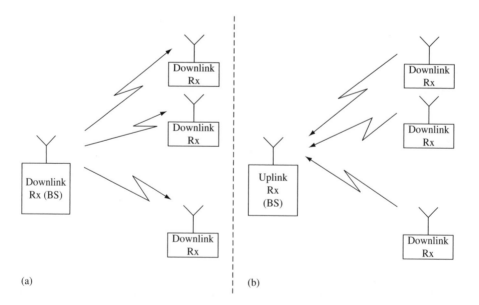

Figure 5.67   (a) Downlink transmission reception; (b) uplink transmission reception

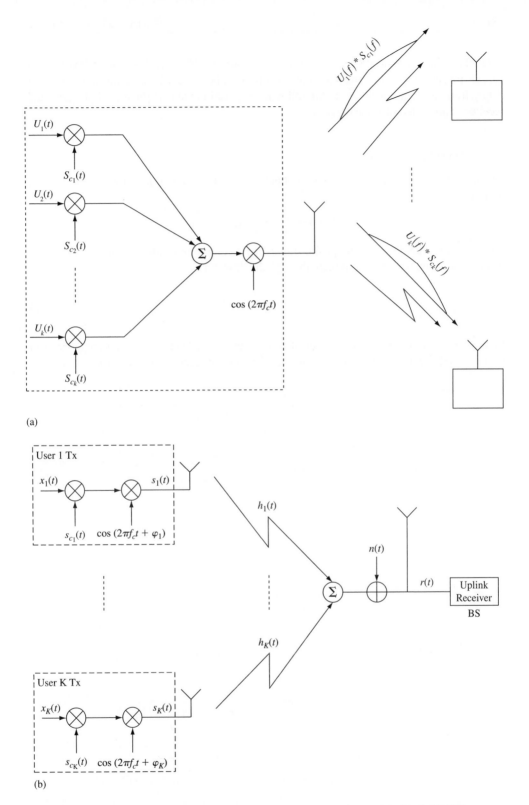

(a)

(b)

Figure 5.68 (a) Transmitter of the downlink DSS system; (b) system model of uplink DSSS system

On the other hand the BS receiver at uplink [Figure 5.68(b)] receives signals from all the various transmitting MTs, which are not synchronized because they are transmitted independently from different locations. In the following subsection the performance of the downlink receiver is presented in some detail.

5.4.4.1 Downlink Reception

Following the mathematical formulation of previous sections the information signal of the kth user is:

$$U_k(t) = \sum a_{nk} g_T(t - nT_S) \tag{5.203}$$

The transmitter BS forms a composite signal $U_D(t)$ consisting of all the information signals from N users and transmits it to them through the downlink:

$$U_D(t) = \sum_{k=1}^{N} U_k(t) \cdot s_{Ck}(t) \tag{5.204}$$

Figure 5.69 shows the receiver model of N different users (MTs) receiving the same composite signal. Taking a specific user (the kth) we briefly present the detector output along

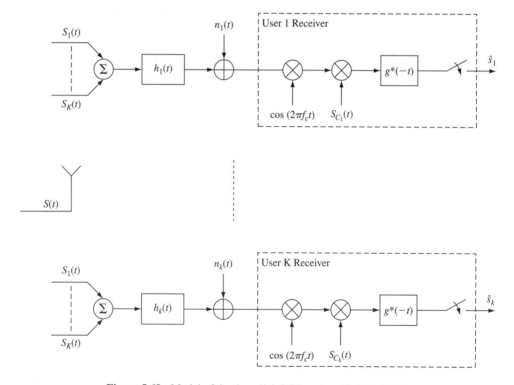

Figure 5.69 Model of the downlink DSS system [Goldsmith05]

with the resulting signal-to-noise-plus-interference ratio. Mathematical manipulations give the following result for the detector output [Goldsmith05]:

$$\hat{a}_k = \int_0^{T_S} \sqrt{P_k} A_k a_{nk} s_{Ck}^2(t) \cos^2(\omega_C t) dt + \int_0^{T_S} \sum_{\substack{i=1 \\ i \neq j}}^{N} \sqrt{P_k} A_k a_{ik} s_{Ci}(t) s_{Ck}(t) \cos^2(\omega_C t) dt$$

$$+ \int_0^{T_S} n(t) s_{Ck}(t) \cos(\omega_C t) dt \approx \sqrt{P_k} A_k a_{nk} + A_k \sum_{\substack{i=1 \\ i \neq j}}^{N} a_{ik} R_{ik}(0) + n_k \qquad (5.205)$$

where A_k represents the channel amplitude response for the kth channel (connecting the BS with the kth user) for a radio channel with no multipath components. R_{ik} is the cross-correlation between the two SCs, $s_{Ci}(t)$ and $s_{Ck}(t)$:

$$R_{ik}(\tau) = \frac{1}{T_S} \int_0^{T_S} s_{Ci}(t) s_{Ck}(t - \tau) dt \qquad (5.206)$$

The second term of the above equation represents the interference to the kth user by the rest of the users:

$$I_{nk} = A_k \sum_{\substack{i=1 \\ i \neq j}}^{N} a_{ik} R_{ik}(0) \qquad (5.207)$$

Comparing Equation (5.205) with Equation (5.194) we realize that the useful term (the first one) is similar to the one derived for single-user DSSS receiver. The same holds for the noise term. The second term (I_{nk}) is the one indicating the impact of the rest of the users to the downlink MT receiver. From the above equation we can see that the cross-corelation at $\tau = 0$ between the two SCs must be as low as possible to reduce interference from other users. Orthogonal codes like the Walsh–Hadamard codes are suitable for the downlink. The selection of spreading codes for multiple access applications is critical in determining the overall system performance.

We will outline briefly the calculation of SINR for this receiver. For the downlink receiver of the kth user SINR is defined as the ratio of the power of the useful signal [first term of Equation (5.205)] to the average power of the $N - 1$ interferers plus noise. This interference plus noise power is:

$$P(I + n) = E\left[\sum a_{ik} R_{ik}(0) + n_k\right]^2 \qquad (5.208)$$

Using the above equation the SINR for the downlink receiver is:

$$\text{SINR} = \frac{P_j}{E\left[\sum a_{ik} R_{ik}(0) + n_k\right]^2} \qquad (5.209)$$

To simplify the analysis we assume that the radio channel gains A_k are all equal to one. In addition the noise term n_k has zero mean and is independent of the signals of all users.

After some mathematical manipulations the noise and interference terms of the above equation become [Cooper86]:

$$E[n_k^2] = \frac{N_0}{2t_{\text{bit}}}, \quad E[a_{ik}R_{ik}(0)]^2 = \frac{1}{2t_{\text{bit}}} \cdot \sum_{\substack{i=1 \\ i \neq k}}^{N} P_i \cdot \int_0^{t_{\text{bit}}} R_S^2(\tau) d\tau \tag{5.210}$$

where R_S is the autocorrelation function of anyone of the N passband information signals $U_D(t)\cos(\omega_C t)$ entering the kth receiver.

An insightful way to associate SINR with processing gain is to use the concept of effective bandwidth B_{eff} of the received signals using the energy density spectrum $S_S(f)$ of an energy signal:

$$B_{\text{eff}} = \frac{\left[\int_0^\infty S_S(f) df\right]^2}{\int_0^\infty S_S^2(f) df} \tag{5.211}$$

It can be easily shown that [Cooper86]:

$$\int_0^\infty R_S^2(\tau) d\tau \cong \int_0^{t_{\text{bit}}} R_S^2(\tau) d\tau \cong \frac{1}{4B_{\text{eff}}} \tag{5.212}$$

From Equations (5.212) and (5.210) we get:

$$2t_{\text{bit}} B_{\text{eff}} = \frac{1}{E[R_{ik}^2(0)]} \tag{5.213}$$

Taking into account Equation (5.212) and the fact that $R_S(\tau)$ occupies a bandwidth close to $1/t_{\text{chip}}$, it turns out that the above product $2t_{\text{bit}} B_{\text{eff}}$ is equal to the processing gain P_G for the downlink of multiuser DSSS system: $P_G = 2t_{\text{bit}} \cdot B_{\text{eff}}$.

As a result the SINR is given as follows:

$$\text{SINR} = \frac{P_G \cdot P_j}{\dfrac{N_0}{2t_{\text{bit}}} \cdot P_G + \sum_{\substack{i=1 \\ i \neq k}}^{N} P_i} \tag{5.214}$$

From the above equation the following conclusions can be deduced:

(1) The transmitted power of all active users except that of the desired signal (P_k) is added to calculated the overall amount of interference for SINR calculation. These components, at the output of the kth detector are treated in the same way as additive noise.
(2) Usually the interference term $\sum_{i \neq k}^{N} P_i$ is much higher than the noise term $(N_0 P_G)/(2T_b)$ and therefore the system SINR can be accurately represented by SIR.

In the case of a multipath channel of the kth user with L components

$$h_{L,k}(t) = \sum_{l=0}^{L} A_{l,k} \delta(t - \tau_{l,k})$$

it can be shown that the interference term in (5.210) will have the form [Goldsmith05]:

$$I_{nk} \approx \sum_{\substack{i=1 \\ i \neq k}}^{N} \sum_{l=1}^{L} A_{l,k} a_i(nT_s - \tau_{l,k}) R_{ik}(\tau_{l,k}) \cos(\omega_C \tau_{l,k}) \tag{5.215}$$

There are two main differences compared with Equation (5.210). I_{nk} has $L \times (N-1)$ components instead of $N-1$. In addition, the code correlation $R_{ik}(\tau_{l,k})$ is taken at $\tau_{l,k} \neq 0$. The latter suggests that there will be significant interference due to the fact that the codes used in MA systems have zero autocorrelation only at $\tau = 0$.

5.4.4.2 Uplink Reception

Figure 5.70 illustrates the BS receiver for demodulation of the transmitted signals from N users (uplink). For this reason, the receiver has N parallel branches each one of which demodulates the corresponding user. This means that the kth branch uses the kth spreading code $s_k(t)$, which for successful reception must be synchronized with the time of arrival of the signal from the kth MT.

Furthermore, we assume there is no multipath during transmission. As there is a difference in the paths followed by the transmitted signals of the different users, we model these time delays by delta functions. Therefore the signal transmitted by the jth user is received by the kth receiver branch through a channel $A_{jk}\delta(t - \tau_j)$.

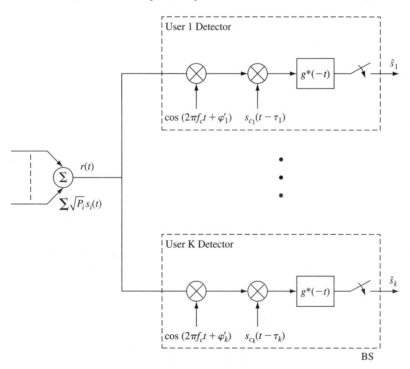

Figure 5.70 Receiver of uplink DSS system (reprinted from A. Goldsmith, 'Wireless Communications', copyright ©2005 by Cambridge University Press)

Taking into account the above, the output of the detector for the receiver of the kth user is [Goldsmith]:

$$\hat{a}_k = \int_0^{T_S} \sqrt{P_k} A_k a_{nk} \{s_{Ck}(t)^* [A_k \delta(t - \tau_k)]\} s_{Ck}(t - \tau_k) \cos^2(\omega_C t + \varphi_k) dt$$

$$+ \int_0^{T_S} \sum_{\substack{i=1 \\ i \neq j}}^{N} \sqrt{P_k} a_{ik} \{s_{Ci}(t)^* [A_{jk} \delta(t - \tau_j)]\} s_{Ck}(t - \tau_k) \cos(\omega_C t + \varphi_k)$$

$$\times \cos(\omega_C t + \varphi_j) dt + n_k \tag{5.216}$$

The three terms appearing in the above equation represent the useful signal, interference from the remaining $N - 1$ users and noise, respectively.

The bracket in the useful term (convolution of the code with the CIR) is equal to $s_{Ck}(t - \tau_k)$, which, with the aid of the synchronization system, is perfectly aligned with the locally generated SC. As a result, the useful term becomes $\sqrt{P_k} A_k a_{nk}$.

The interference I_{nk} (second term) can be easily shown to give [Goldsmith05]:

$$I_{nk} \approx A_{jk} a_{jk} \cos(\Delta \varphi_{ik}) \cdot R_{ik}(\tau_i - \tau_k) \tag{5.217}$$

Comparing this with Equation (5.207) we realize that the most important differences are the channel gain A_{jk}, which can be quite different than A_k and the cross-correlation factor which has an argument $\tau_i - \tau_k$ different from zero.

Finally, considering SINR for the uplink receiver, similar expressions can be obtined [Goldsmith05], [Pursley77]:

$$\text{SINR} = \frac{1}{\frac{N_0}{2E_S} + \frac{K-1}{3P_G}} \tag{5.218}$$

Properties for spreading codes for multiple access DSSS: From the above discussion on downlink and uplink receiver performance we realize that the interference term I_{nk} is proportional to the autocorrelation R_{ik}:

$$I_{nk} \approx R_{ik}(\tau_i - \tau_k) = \frac{1}{T_S} \int_0^{T_S} s_{Ci}(t - \tau_i) s_{Ck}(t - \tau_k) dt \tag{5.219}$$

For synchronized in time interference (like in downlink reception) we have $\tau_k = \tau_i \, \forall i$. Consequently in this case $R_{ik}(0)$ should be as close to zero as possible. For unsynchronized interference $\tau_k \neq \tau_i \, \forall i$. Hence, $R_{ik}(\tau) \to 0$.

As each user in a multiuser system uses a unique SC belonging to the same group of spreading sequences, these SCs must have very low cross-correlation R_{ik}, otherwise the interference terms in both the downlink and the uplink will have high values. Another important property of PN sequences for multiple access applications is that the number of SCs with the same length N must be large in order to accommodate for as many users as possible.

5.4.5 Spreading Sequences for Single-user and Multiple Access DSSS

From the above presentation of DS systems some basic properties of the spreading codes can be deduced. Spreading codes must possess correlation properties close to those of noise while at the same time they must have additional features for ease of implementation and improvement of spectral properties. Apart from autocorrelation, cross-correlation properties and code length are two important characteristics, which have a major impact on the system performance. Furthermore, depending on the application of the DS spread spectrum system different types of codes can be used. In brief, the following basic properties must be fulfilled:

(1) autocorrelation function $R_S(\tau)$ must have a very narrow width (typically equal to the chip time),
(2) spreading code misalignment by more than one chip interval T_C must result in very low correlation $[R_S(\tau) \approx 0$ for $\tau \geq T_C]$; and
(3) for multiple access DS applications cross-correlation between spreading sequences of the same family must have low values.

Most of these requirements are satisfied by pseudo-noise sequences generated easily using shift registers with feedback.

In addition, the following properties are desirable for DS applications:

- *Balance property* – almost equal number of ones and zeros in order to avoid a DC component on the spectrum.
- *Run property* – rare occurrences of consecutive 0s or 1s (called runs). More specifically, long runs must occur very rarely.

Maximal-length sequences (*m*-sequences) satisfy all the above requirements and in theory they can be generated using coding theory. Furthermore, they can be easily implemented using shift registers, as mentioned in Chapter 2. A shift register of length n with feedback can produce an *m*-sequence with period $N = 2^n - 1$. Figure 5.71 shows the structure of the shift register generator (SRG). The coefficients c_i represent switches which when they take the value of zero; they are open (no connection of the output of the respective bit to the feedback path). For $c_i = 1$ the switch is closed. At the rising edge of the clock the contents of the shift register are shifted to the right by one position. The new entry of the shift register (last bit) is produced by the following equation:

$$a_i = \sum_{k=1}^{n} c_k a_{i-k} \qquad (5.220)$$

whereas the generating function takes the form [Dinan98]:

$$G(D) = \sum_{i=0}^{\infty} a_i D^i = \sum_{i=0}^{\infty} \sum_{k=1}^{n} c_k a_{i-k} D^i = \frac{G_0(D)}{F(D)} \qquad (5.221)$$

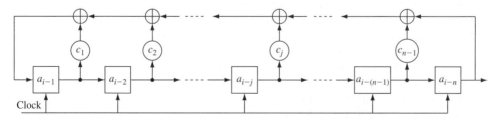

Figure 5.71 Shift register structure for *m*-sequence generation

where D is an operator imposing delay by one unit of time (D^i denotes a delay of i time units). $G_0(D)$ is an initial-condition polynomial and $F(D)$ is the characteristic polynomial of the SRG and is given by:

$$F(D) = 1 + \sum_{k=1}^{n} c_k D^k \tag{5.222}$$

In order for the above sequence a_i to generate mmsequences it is necessary for $F(D)$ to be irreducible [Dinan98]. Spreading sequences generated in this way have a code length (period) $N = 2^n - 1$. In addition they have almost the same number of ones and zeros, and there are very few occurrences of long runs ($1/2^k$ of the runs have length equal to k).

The autocorrelation function over the whole period N is:

$$R_S(\tau) = \frac{1}{N} \sum_{n=0}^{N-1} \sum_{m} a_n a_m R_p[\tau+] \tag{5.223}$$

where R_p is the autocorrelation function of the basic chip waveform $p(t)$

$$R_p(\tau_a) = \frac{1}{T_C} \int_{-\infty}^{\infty} p(t)p(t + \tau_a)dt \tag{5.224}$$

Where the basic chip waveform is a unit rectangular pulse its autocorrelation is:

$$R_p(\tau_a) = \begin{cases} T_C \cdot \left(1 - \dfrac{|\tau_a|}{T_C}\right), & |\tau_a| \leq T_C \\ 0, & \text{otherwise} \end{cases} \tag{5.225}$$

Subsequently the resulting autocorrelation function for unit rectangular pulse is:

$$R_S(\tau) = \begin{cases} 1 - \dfrac{|\tau|(1 + 1/N)}{T_C}, & |\tau| \leq T_C \\ -\dfrac{1}{N}, & |\tau| \geq T_C \end{cases} \tag{5.226}$$

Figure 5.72 shows the above correlation function. Outside the interval $[-T_C, T_C]$ it has a constant value $-1/N$. We must now consider briefly some practical aspects of m-sequence applications to spread spectrum systems. It is of great importance to determine the relation between integration time in the detector and code length. If the code length is equal to the spreading gain ($N = T_S/T_C$) then the code is a short SC and autocorrelation is determined using the complete symbol interval T_S. However, to avoid significant ISI the code length should be many times higher than the spreading gain ($N \gg T_S/T_C$) [Goldsmith05]. These codes are called long SCs. Integrating over a symbol interval T_S and not over the entire period of the code (code length) does not permit us to rely on the above-mentioned properties of m-sequences. Instead we must examine the properties of autocorrelation when correlation is applied over an interval considerably less than the code length.

Regarding application of m-sequences in multiple access communications, two important properties must be satisfied. The first one is associated with cross-correlation which must have low values as mentioned above. However, m-sequences are not characterized by low cross-correlation properties. The other property has to do with the amount of m-sequences

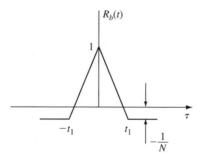

Figure 5.72 Autocorrelation function of pseudonoise sequence

with equal period. In order to use m-sequences effectively in a multiple access system there must exist a large number of SCs with the same length N to accommodate a large number of users. However, this is not true for m-sequences.

From the above discussion two conclusions are drawm: (a) for applications where m-sequences are applied, partial correlation properties must be examined; and (b) although m-sequences can be very effective for single-user spread spectrum communications, they are unsatisfactory for multiuser spread spectrum systems and other spreading sequences (like those Gold, Kasami and Walsh-Hadamard) must be considered.

5.4.5.1 Partial Correlation of m-Sequences

If the integration time in the receiver is shorter than the code length N, then autocorrelation of the SC must be done for a part of the full period NT_C. Denoting Np the number of chip periods over which integration is performed in the receiver we have:

$$R_{Np}(\tau) = \frac{1}{NpT_C} \int_0^{NpT_C} s_C(t)s_C(t+\tau)\mathrm{d}t, \quad Np < N \tag{5.227}$$

After some mathematical manipulations we get [Cooper86]:

$$
\left\{
\begin{array}{l}
\left.
\begin{array}{l}
E[R_{Np}(\tau)] = R_P(\tau) \\[2mm]
\sigma^2[R_{Np}(\tau)] = \dfrac{1}{Np}R_P^2(\tau - T_C)
\end{array}
\right\} 0 \leq \tau \leq T_C \\[6mm]
\left.
\begin{array}{l}
E[R_{Np}(\tau)] = 0 \\[2mm]
\sigma^2[R_{Np}(\tau)] = \dfrac{1}{Np}
\end{array}
\right\} T_C \leq \tau \leq (N-1)T_C
\end{array}
\right. \tag{5.228}
$$

Figure 5.73 illustrates the mean and variance of the partial correlation function $R_{Np}(\tau)$. It is important to observe that for misalignment greater than the chip time, the variance is constant and equal to $1/Np$ which represents the noise of the spreading code. By increasing Np, this noise is reduced.

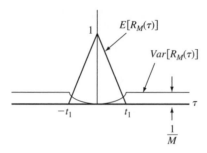

Figure 5.73   The mean and variance of the partial correlation $R_M(\tau)$ of the pseudonoise sequence

5.4.5.2 Spreading Codes for Multiple Access Systems

The main characteristics of spreading sequences for multiuser applications are short lengths, low cross-correlation and adequately large number of codes to accommodate for large number of users.

Gold Sequences

Since Gold codes are generated using m-sequences they have the basic properties which make them pseudorandom, which is when they are balanced and have similar run lengths. However, not all Gold sequences generated in the above-described random fashion have good cross-correlation properties. Following [Dinan98], preferred pairs of Gold sequences must be formulated. These must have uniform and bounded cross-correlation between the various codes among the same set of Gold codes. Preferred pairs are formed by an m-sequence q and its decimation $q' = q \cdot [m]$ which is obtained by using one every m symbols in sequence q. The cross-correlation values of the preferred pair is [Dinan98]:

$$R_{ij}(\tau) = \begin{cases} -t(n)/N \\ -1/N \\ [t(n)-2]/N \end{cases} \quad \text{where} \quad t(n) = \begin{cases} 1 + 2^{(n+1)/2} \ldots n = \text{odd} \\ 1 + 2^{(n+2)/2} \ldots n = \text{even} \end{cases} \qquad (5.229)$$

Subsequently, the group of codes defined by q and q' as follows

$$\{q, q', q + q', q + Dq', q + D^2q', \ldots, q + D^{N-1}q'\} \qquad (5.230)$$

is a set of Gold codes [Dinan98] with uniform and bounded cross-correlation among any pair in the set. D is a unit delay operator. In fact for any chosen pair cross-correlation takes one of the above three values. In addition, their autocorrelation does not take only two values as it was the case for m-sequences.

Kasami Sequences

Kasami sequences are also generated from m-sequences and they are characterized by very low cross-correlation values. They are distinguished in small set Kasami sequences and large set sequences.

   The small set is formed by decimating a sequence q by $2^{n/2} + 1(q' = q[2^{n/2} + 1])$. Subsequently, by adding (modulo-2) the $2^n - 1$ bits of the two sequences q and any of the cyclically shifted q' sequence will produce the small set of Kasami sequences which has $2^{n/2}$ different sequences.

As in the case of Gold sequences they also take three different values [Dinan98]:

$$R_{ij}(\tau) = \begin{cases} -\dfrac{1}{N} \\[2ex] -\dfrac{2^{n/2}+1}{N} \\[2ex] \dfrac{2^{n/2}-1}{N} \end{cases} \tag{5.231}$$

The large set of Kasami sequences has similar properties to the small set, but the number of different sequences is larger (there are $2^n - 1$ sequences). They, like the small set take three different values.

Orthogonal Sequences (Walsh-Hadamard Codes)

In several applications of spread spectrum systems it is necessary to eliminate interference which has the same time signature as the useful signal (i.e. we have synchronized reception). One such system is the downlink receiver in a multiple access system as was briefly noted in Section 5.4.4. Because the BS simultaneously transmits signals for all the users a specific MT receiver must eliminate interference terms of the form Equation (5.207). In order to have $R_{ik}(0) = 0$, orthogonal spreading codes must be used. In other applications, orthogonal codes are used to increase the bandwidth efficiency of the system.

Walsh-Hadamard sequences belong to orthogonal spreading codes. Walsh functions are derived from the rows of Hadamard matrices, which are square matrices generated recursively as follows:

$$H_2 = \begin{bmatrix} 0 & 0 \\ 0 & 1 \end{bmatrix}, \quad H_4 = \begin{bmatrix} 0 & 0 & 0 & 0 \\ 0 & 1 & 0 & 1 \\ 0 & 0 & 1 & 1 \\ 0 & 1 & 1 & 0 \end{bmatrix}, \quad H_{2N} = \begin{bmatrix} H_N & H_N \\ H_N & \overline{H}_N \end{bmatrix} \tag{5.232}$$

In the above equations, N is a power of 2 whereas $\overline{H}_N$ is the 2-complement of matrix H_N. Hence each of the rows of H_{2N} gives a specific spreading sequence of length $2N$ which is orthogonal to all other sequences obtained from the same matrix.

5.4.6 Code Synchronization for Spread Spectrum Systems

In this section synchronization of both DS and FH spread spectrum systems will be briefly examined. We are mostly concerned with synchronizing the code generator at the receiver in time with that of the transmitter.

Figure 5.74 shows the basic block diagram of the synchronizing system for both DS and FH. The principle of operation is to adjust the time delay of the PN generator. The maximum is detected at the output of the integrator. We make two assumptions to facilitate the presentation of such a system. First, we assume that carrier recovery is done by the aid of a different loop. Second we assume that there is only the main component arriving at the receiver (no multipath) with a delay τ_d. Consequently, the transmission channel is modelled by an impulse response $h(t) = \delta(t - \tau_d)$.

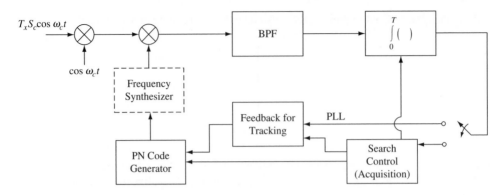

Figure 5.74   Synchronization system for DS/FH receiver (dashed box is used only for FH systems)

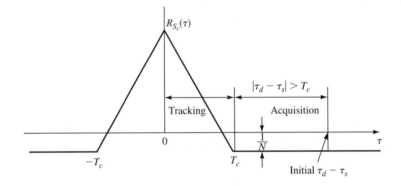

Figure 5.75   Tracking and acquisition regions with respect to correlation $R_{SC}(\tau)$

The signal at the output of the integrator is [Goldsmith05]:

$$r(t) = \frac{1}{T} \int_0^T U(t - \tau_d) s_C(t - \tau_d) \cos^2(\omega_C t) \cdot s_C(t - \tau_S) dt$$

$$\approx \frac{a_k}{2T} \int = \frac{a_k}{2} R_{Sc}(\tau_S - \tau_d) \qquad (5.233)$$

where $s_C(t - \tau_S)$ represents the synchronizing signal generated by the loop.

In more detail, the system synchronizes in two phases, the acquisition and the tracking phase. The synchronization procedure is depicted graphically in terms of the autocorrelation function in Figure 5.75 where the initial code misalignment in time is $\tau_S - \tau_d$. The bold line in arrows shows the trajectory of this misalignment during both acquisition and tracking phases. Looking at the block diagram, during the acquisition phase, the output of the integrator is fed into a sequential search control mechanism which, if its input is below a predetermined threshold, it increments the time index by T_C and in turn it commands the SC generator to produce the code advanced by one chip time. At the same time 'search control' resets the integrator. This procedure is repeated until a maximum is detected. In that case, acquisition is completed and the spreading code phase is within a chip time from the correct phase ($|\tau_d - \tau_S| \leq T_C$). The

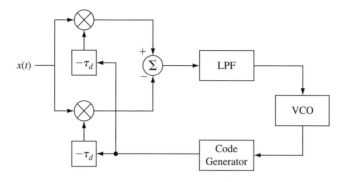

Figure 5.76 The delay-lock loop functional block

system now enters the tracking mode. The switch turns to the upper position and connects a feedback mechanism (usually a modified PLL) to the input of the code generator while at the same time the search control block passes the current value of $\tau_S - \tau_d$ to the feedback circuit. In this way, a fine adjustment is achieved in the code time index until the time index difference becomes zero ($|\tau_d - \tau_S| = 0$). In this case correlation function R_{Sc} takes the maximum value and synchronization is terminated successfully. However, because the radio channel keeps changing dynamically, the tracking mechanism is always on to keep adjusting to new values of the received signal delay τ_d produced by the movement of the radio transmitter or receiver.

Figure 5.76 shows the feedback loop used for the tracking mode which is a delay lock loop (DLL). By delaying and advancing the code phase by τ_δ the autocorrelation function for each branch is displaced accordingly leading to an error signal as shown in the Figure. Detailed analysis of the system is given in [Cooper86], [Glisic97]. Analysis shows that the bandwidth of the loop has to be carefully chosen taking into account the trade-off between the probability of loosing lock and the ability to track fast changes in code misalignment.

Regarding the acquisition phase, care must be taken that its duration is not prohibitively long. To this end, one solution is to have a variable advancement time (for example it can be higher than T_C in the beginning). Another solution is to conduct a parallel search in which there is more than one branch that performs the correlations of the received signal with copies of the spreading code having different delays. Furthermore, the integration time can be chosen such that the acquisition period is shortened.

In an actual system things become more complicated because carrier recovery and spread spectrum synchronization must be performed. In that case synchronization peaks must be found in two dimensions, time and frequency. The later concerns the frequency recovery part of the received signal.

5.4.7 The RAKE Receiver

One method to demodulate signals transmitted over frequency selective channels is to modulate the signal using a broad bandwidth W much higher than the maximum delay spread of the radio channel ($W \gg \max \tau_{\mathrm{rms}}$). In this way, the multipath components that differ in time delay by more than $1/W$ can be resolved. The next step is to design a receiver with parallel paths each of which will demodulate only one of the multipath components while the rest must be eliminated.

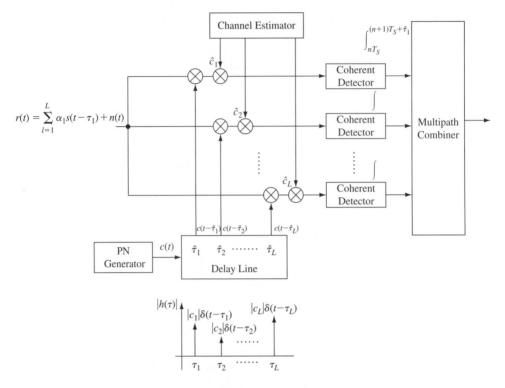

Figure 5.77 RAKE receiver

The latter can be achieved by using time-delayed versions of a PN sequence, each one of which will be correlated to a corresponding branch in the receiver.

Figure 5.77 depicts the structure of a RAKE receiver. The received signal $R_X(t)$ is given as:

$$R_X(t) = \sum_{l=0}^{L} [U(t)s_C(t)] \cdot c_l \delta(t - \tau_l) = \sum_{l=0}^{L} c_l \cdot T_S \left(t - \frac{l}{W} \right) = \sum_{l=0}^{L} c_l \cdot T_S(t - \tau_l) \quad (5.234)$$

where $U(t)$ is the information baring signal, $s_C(t)$ is the spreading signal and $T_S(t)$ is their product representing the transmitted signal.

It is necessary to recall here the form of the received signal after going through a multipath channel. Taking into account the tapped-delay line (TDL) channel model (see Chapter 1) the lowpass impulse response of the channel is:

$$h_L(\tau; t) = \sum_{-\infty}^{\infty} c_l(t)\delta(t - \frac{l}{W}) = \sum_{-\infty}^{\infty} c_l(t)\delta(t - \tau_l) \quad (5.235)$$

Therefore, given a transmitted signal $T_S(t)$ as above, the receiver signal can be expressed as [Proakis01]:

$$r_L(t) = \sum_{l=1}^{L} c_l(t)T_S(t - \tau_l) \quad (5.236)$$

If there is no information transmitted [i.e. we strip information from $T_S(t)$] we have:

$$r_{LS}(t) = \sum_{l=1}^{L} c_l(t) s_C(t - \tau_l) \tag{5.237}$$

In the receiver, the correlation and averaging of the above with a local replica of the spreading waveform $s_C(t)$ gives:

$$\overline{\sum_l c_l \cdot s_C^*(t - \tau_l) \cdot s_C(t - \hat{\tau}_k)} = \sum_l c_l R_{S_c}\left(\tau_l - \frac{k}{W}\right) \tag{5.238}$$

The above equation shows that if there exists a path (in the multipath profile) with delay τ_k, the receiver will find it and correlation of the nonmodulated spreading waveform will give the channel gain factor c_k.

Reconsidering the above, we conclude that the TDL approach suggests that, in a multipath channel at the input of the receiver we have L replicas of the same transmitted signal. If in addition we correlate this composite signal with the delayed versions of the spread spectrum signal, we can demodulate all L components separately (L-order diversity). Furthermore, the channel gains can be estimated by transmitting the spreading sequence with no information on it.

Figure 5.77 shows the operation of the RAKE receiver. We assume there are as many parallel branches as the multipath components (L branches). We also assume that the delays of the multipath components are multiples of the chip duration T_C. By correlating this composite received signal with copies of $s_C(t)$ time shifted by the multipath delays $\hat{\tau}_1, \hat{\tau}_2, \ldots, \hat{\tau}_L$, at a specific branch k we can isolate only the multipath signal component corresponding to delay τ_k. Owing to the corellation properties of PN sequences, all other summation terms except for the kth will be eliminated. This is demonstrated by the equation below, which gives the outcome of the correlation of the composite received signal at the kth branch with a delayed (by τ_k) version of the spreading waveform:

$$R(\text{branch } k) = R_x(t) \cdot s_C(t - \hat{\tau}_k)$$

$$= s_C(t - \hat{\tau}_k) \cdot \sum_{l=0}^{L} c_l[U(t - \tau_l) \cdot s_C(t - \tau_l)] \cdot \cos[\omega_C(t - \tau_l)] \tag{5.239}$$

$$= c_k U(t - \tau_k) \cdot s_C^2(t - \tau_k) + \sum_{l \neq k}^{L} c_l U(t - \tau_l) \cdot R_{S_c}(\tau_l - \tau_k) \cong c_k U(t - \tau_k)$$

Subsequently, the resulting signal at each branch is multiplied by the estimated channel gain and passes through the detector-integrator, as illustrated in Figure 5.77.

Specifically, for branch k we get:

$$\int_{nT_S + \tau_k}^{(n+1)T_S + \tau_k} |c_k|^2 U(t - \tau_k) \tag{5.240}$$

In the next stage, the outputs from all branches are collected and combined using a suitable combiner criterion (MLC, EGC, etc.) in the multipath combiner block which, in turn, produces the detected symbol at the output.

There are a few departures from the ideal processing presented above. One has to do with the fact that autocorrelation $R_{S_C}(\tau_l - \tau_k)$ is not zero, which results in an interference term from the other multipath components. Another shortcoming can arise from the fact that delays τ_j

cannot be exact multiples of the chip duration T_C. In that case analysis is more complicated and interference terms are more powerful in the output of each branch.

To obtain the probability of error, let us assume maximum ratio combining detection is used. In that case, the combiner block in the RAKE receiver performs simple summation of the outputs from each branch. In that case the decision variable is [Glisic97], [Proakis01]:

$$Y = \sum_{l=1}^{L} \sum_{m=1}^{PG} c_l^2 \cdot T_X(t - \tau_m) + \sum_{l=1}^{L} \sum_{m=1}^{PG} n_m c_l \tag{5.241}$$

It can be shown that the probability of error for binary orthogonal signals is [Proakis01]:

$$P_b \approx \binom{2L-1}{L} \prod_{k=1}^{L} \frac{1}{2\bar{\gamma}_k} \tag{5.242}$$

where the assumption holds for $\gamma_k \gg 1$. γ_k represents the average SNR of the kth branch in the receiver:

$$\gamma_k = \frac{\sqrt{P_C}}{N_0} E(c_k^2) \tag{5.243}$$

References

[Alamouti98]: S. Alamouti, 'A simple transmit diversity technique for wireless communications', IEEE J. Selected Areas Commun., vol. 16, October 1998, pp. 1451–1458.

[Chaivaccini04]: E. Chiavaccini, G. Vitetta, 'Maximum likelihood frequency recovery for OFDM signals transmitted over multipath fading channels', IEEE Trans. Commun., vol. 52, February 2004, pp. 244–251.

[Classen94]: F. Classen, H. Meyr, 'Frequency synchronization algorithms for OFDM systems suitable for communication over frequency selective fading channels', IEEE International Conference on Vehicular Technology, 1994.

[Cooper86]: G. Cooper, C. McGillem, 'Modern Communications and Spread Spectrum', McGraw-Hill, New York, 1986.

[Costa02]: E. Costa, S. Pupolin, 'M-QAM-OFDM system performance in the presence of a nonlinear amplifier and phase noise', IEEE Trans. Commun., vol. 50, March 2002, pp. 462–472.

[Diggavi04]: S. Diggavi, N. Al-Dhahir, A. Stamoulis, A. Calderbank, 'Great expectations: the value of spatial diversity in wireless networks', Proc. IEEE, vol. 92, February 2004, pp. 219–270.

[Dinan98]: E. Dinan, B. Jabbari, 'Spreading codes for direct sequence CDMA and wideband CDMA cellular networks', IEEE Commun. Mag., September 1998, pp. 48–54.

[Dlugaszewski02]: Z. Dlugaszewski, 'Comparison of frequency offset estimation methods for OFDM burst transmission in the selective fading channels', IST Mobile Communications Summit, Thessaloniki, 2002.

[Edfors98]: O. Edfors, M. Sandell, J-J van de Beek, S. Wilson, P. Borjesson, 'OFDM channel estimation by singular value decomposition', IEEE Trans. Commun., vol. 46, July 1998, pp. 931–939.

[El-Tanany01]: M. El-Tanany, Y. Wu, L. Hazy, 'OFDM uplink for interactive broadband wireless: analysis and simulation in the presence of carrier, clock and timinh errors', IEEE Trans. Broadcasting, vol. 47, March 2001, pp. 3–19.

[Fazel02]: K. Fazel, S. Kaiser, 'Multi-carrier Spread Spectrum and Related Topics', Springer, Berlin, 2002.

[Gardner79]: F. Gardner, 'Phaselock Techniques', 2nd edn, John Wiley & Sons Inc., New York, 1979.

[Glisic97]: S. Glisic, B. Vucetic, 'Spread Spectrum CDMA Systems for Wireless Communications', Artech House, Norwood, MA, 1997.

[Goldsmith05]: A. Goldsmith, 'Wireless Communications', Cambridge University Press, Cambridge, 2005.

[Hoeher92]: P. Hoeher, 'A statistical discrete–time model for the WSSUS multipath channel', IEEE Trans. Vehicular Technol., vol. 41, November 1992, pp. 461–468.

[Kim01]: Y.-H. Kim, I. Song, S. Yoon, S.-R. Park, 'An efficient frequency offset estimator for OFDM systems and its performance characteristics', IEEE Trans. on Vehicular Technology, vol. 50, September 2001, pp. 1307–1312.

[Leclert83]: A. Leclert, P. Vandamme, 'Universal carrier recovery loop for QASK and PSK signal sets', IEEE Trans. Commun., vol. 31, January 1983, pp. 130–136.

[Li03]: M. Li, W. Zhang, 'A novel method of carrier frequency offset estimation for OFDM systems', IEEE Trans. Consumer Electronics, vol. 49, November 2003, pp. 965–972.

[Lindsey73]: W. Lindsey, M. Simon, 'Telecommunication Systems Engineering', Prentice-Hall, Englewood Cliffs, NJ, 1973.

[Lindsey72]: W. Lindsey, M. Simon, 'Carrier synchronization and detection of polyphase signals', IEEE Trans. Commun., vol. 28, June 1972, pp. 441–454.

[Luise95]: M. Luise, R. Reggiannini, 'Carrier frequency recovery in all-digital modems for burst-mode transmissions', IEEE Trans. Commun., February–April 1995, pp. 1169–1178.

[McNair99]: B. McNair, J. Cimini, N. Sollenberger, 'A robust timing and frequency offset estimation scheme for orthogonal frequency division multiplexing (OFDM) systems', Proceedings of IEEE Vehicular Technology Conference (VTC' 99), vol. 1, May 1999, pp. 690–694.

[Mengali97]: U. Mengali, A. D'Andrea, 'Synchronization Techniques for Digital Receivers', Plenum Press, London, 1997.

[Mochizuki98]: N. Mochizuki, Y. Matsumoto, M. Mizoguchi, T. Onizawa, M. Umehira, 'A high performance frequency and timing synchronization technique for OFDM', Proceedings of IEEE GLOBECOM'98, vol. 6, November 1998, pp. 3443–3448.

[Moose94]: P. Moose, 'A technique for orthogonal frequency division multiplexing frequency offset correction', IEEE Trans. Commun., vol. 42, October 1994, pp. 2908–2914.

[Morelli97]: M. Morelli, A. D'Andrea, U. Mengali, 'Feedforward ML-based timing estimation with PSK signals', IEEE Commun. Lett., vol. 1, no. 3, May 1997, pp. 80–82.

[Nikitopoulos01]: K. Nikitopoulos, A. Polydoros, 'Compensation schemes for phase noise and residual frequency offset in OFDM systems', IEEE Global Telecommunications Conference, GLEBECOM'01, vol. 1, 2001, pp. 330–333.

[Oerder88]: M. Oerder, H. Meyr, 'Digital filter and square timing recovery', IEEE Trans. Commun., vol. 44, May 1988, pp. 605–611.

[Pollet95]: T. Pollet, M. Bladel, M. Moeneclay, 'BER sensitivity of OFDM systems to carrier frequency offset and Wiener phase noise' IEEE Trans. Commun., vol 43, February–April 1995, pp. 191–193.

[Proakis04]: J. Proakis, 'Digital Communications', 5th edn, McGraw Hill, New York, 2004.

[Pursley77]: M. Pursley, 'Performance evaluation for phase-coded spread-spectrum multiple-access communication – part I: system analysis', IEEE Trans. Commun., vol. 25, August 1977, pp. 795–799.

[Rappaport02]: T. Rappaport, 'Wireless Communications, Principle and Practice', 2nd edn, Prentice Hall, Englewood Cliffs, NJ, 2002.

[Rodrigues02]: M. Rodrigues, 'Modeling and Performance Assessment of OFDM communication systems in the presence of non-linearities', Ph.D. Thesis, University College, London, 2002.

[Sandell95]: M. Sandell, JJ Van de Beek, P.O. Borjesson, 'Timing and frequency synchronization in OFDM systems using the cyclic prefix', Proceedings of International Symposium on Synchronization, December 1995, pp. 16–19.

[Schmidl97]: T. Schmidl, D. Cox, 'Robust frequency and timing synchronization for OFDM', IEEE Trans. Commun., vol. 45, December 1997, pp. 1613–1621.

[Shentu03]: J. Shentu, K. Panta, J. Armstrong, 'Effects of phase noise on performance of OFDM systems using an ICI cancellation scheme', IEEE Trans. Broadcasting, vol. 49, June 2003, pp. 221–224.

[Simon94]: M. Simon, J. Omura, R. Scholtz, B. Levitt, Spread Spectrum Communications Handbook, Electronic Edition, McGraw-Hill, New York, 1994.

[Speth99]: M. Speth, S. Fechtel, G. Fock, H. Meyr, 'Optimum receiver design for wireless broad-band systems using OFDM – part I', IEEE Trans. Commun., vol. 47, November 1999, pp. 1668–1677.

[Tarokh98]: V. Tarokh, N, Seshandri, A. Calderbank, 'Space–time codes for high data rate wireless communications: performance criterion and code construction', IEEE Trans. Information Theory, vol. 44, March 1998, pp. 744–765.

[Tomba98]: L. Tomba, 'On the effect of Wiener phase noise in OFDM systems' , IEEE Trans. Commun., vol. 46, May 1998, pp. 580–583.

[Tubbax01]: J. Tubbax, B. Come, L. Van der Perre, L. Deneire, S. Donnay, M. Engels, 'OFDM versus single carrier with cyclic prefix: a system-based comparison', IEEE Vehicular Technology Conference, VTC 2001, vol. 2, Fall 2001, pp. 1115–1119.

[vanNee98]: R. vanNee, A. deWild, 'Reducing the peak-to-average power ratio of OFDM', IEEE Vehicular Technology Conference (VTC'98), 1998, pp. 2072–2076.

[VanNee00]: R. van Nee, R Prasad, 'OFDM for Wireless Multimedia Communications', Artech House, Norwood, MA, 2000.

[Viterbi66]: A. Viterbi, 'Principles of Coherent Communications', New York, McGraw-Hill, 1966.

[Viterbi83]: A. Viterbi, 'Nonlinear estimation of PSK-modulated carrier phase with application to burst digital transmission', IEEE Trans. on Information Theory, vol. 29, July 1983, pp. 543–551.

[Windisch04]: M. Windisch, G. Fettweis, 'Standard-independent I/Q imbalance compensation in OFDM direct-conversion receivers', 9th International OFDM-Workshop (InOWo'04), Dresden, 2004.

[Wu04]: S. Wu, Y. Bar-Ness, 'OFDM systems in the presence of phase noise: consequences and solutions', IEEE Trans. Commun., vol. 52, November 2004, pp. 1988–1996.

[Zou01]: H. Zou, B. McNair, B. Daneshrad, 'An Integrated OFDM Receiver for High-speed Mobile Data Communications', Globecom, 2001.

6

System Design Examples

G. Kalivas, M. Metaxakis, A. Miaoudakis and A. Tzimas[1]

This chapter presents three design examples concerning a DECT digital receiver, a QAM modem for a digital microwave radio system (DMR) and the design of an OFDM digital receiver. The first design emphasizes the detector, carrier recovery and synchronization blocks to obtain a gross bit rate of 1.152 Mb/s using gaussian MSK modulation. The second example presents the design of a demanding receiver modem operating at 61.44 Mb/s including carrier and timing recovery subsystems. Finally, the last case study outlines the design of an OFDM receiver for the Hiperlan/2 standard where emphasis is given to the carrier frequency offset correction subsystem.

The reason for choosing the DECT example is because it includes both system design and digital implementation. The second system was chosen because it also emphasizes the implementation of a demanding high-speed modem where carrier and timing recovery blocks must be considered carefully. Finally, the OFDM receiver was chosen because it focuses on the carrier frequency offset correction, which is one of the most crucial blocks of OFDM systems due to their sensitivity to frequency offset.

6.1 The DECT Receiver

6.1.1 The DECT Standard and Technology

DECT was proposed as a standard by ETSI in the early 1990s [ETSI-standard1], [ETSI-standard2]. It is essentially a wireless access technology with the objective of supporting a wide range of telecommunications applications such as voice (business, residential telephony) and data (ISDN, Fax, modem, etc.). Some of the most representative applications of DECT are:

- cordless terminal mobility for home and business;
- indoor cellular wireless systems for seamless connectivity;

---

[1] 1971–2005.

Digital Radio System Design   Grigorios Kalivas
© 2009 John Wiley & Sons, Ltd

- wireless local loop;
- GSM/DECT internetworking;
- cordless local area networks (CLAN).

The DECT standard operates in the 1880–1900 MHz frequency band using 10 carriers with carrier separation of 1728 MHz. It employees TDMA/TDD access technique and GFSK/GMSK modulation with BT = 0.5.

The DECT TDMA frame is organized in 24 time slots with overall duration of 10 ms, as illustrated in Figure 6.1. As seen in the figure, each time slot consists of 480 bits of which 388 correspond to data, 32 to preamble, 0-4 are z-field bits and the rest represent guard bits.

6.1.2 Modulation and Detection Techniques for DECT

6.1.2.1 Modulation

Gaussian minimum shift keying is by definition an MSK modulated waveform with a gaussian filter preceding the modulation device, as shown in Figure 6.2. MSK is a continuous phase FSK (CP-FSK) with modulation index equal to 0.5 [Proakis02].

As seen in Chapter 5, the GMSK modulated signal is given by:

$$u(t) = \sqrt{\frac{2E_S}{T}} \cdot \cos\left[2\pi f_C t + \theta(t) + \varphi_0\right] \tag{6.1}$$

where

$$\theta(t) = \sum_i m_i \cdot \pi \cdot h \int_{-\infty}^{t-iT} g(\tau)d\tau$$

$$g(t) = h(t) * rect\frac{t}{T} = \frac{1}{2T}\left[Q\left(2\pi BT\frac{t - T/2}{T\sqrt{\ln 2}}\right) - Q\left(2\pi BT \cdot \frac{t + T/2}{T\sqrt{\ln 2}}\right)\right] \tag{6.2}$$

Where $rect(t/T)$ is a rectangular function of value $1/T$ for $|t| < T/2$ (and zero elsewhere), and $Q(t)$ the well known Q-function.

The impulse response $h(t)$ of the gaussian low-pass filter (LPF) is:

$$h(t) = \frac{1}{\sqrt{2\pi}} \cdot \frac{1}{\sigma T} \exp\left(\frac{-t^2}{2\sigma^2 T^2}\right) \tag{6.3}$$

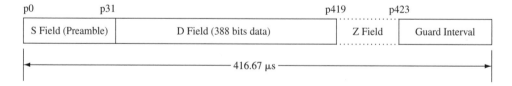

Figure 6.1   The DECT TDMA frame

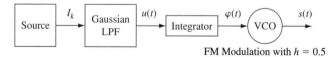

Figure 6.2   GMSK modulator

In addition, $m_i \in \{\pm 1\}$ and represents the modulation information, and $*$ is the convolution operation; for MSK (and GMSK) modulation we have $h = 0.5$.

The frequency separation of the two frequencies in MSK is $\Delta f = f_2 - f_1 = 1/(2T_b)$ and represents the minimum possible frequency separation such that orthogonality of the two sinusoids is preserved for coherent detection.

GMSK is realized by applying a gaussian filter between the information sequence and the waveform modulator. There exist a variety of implementation architectures for GMSK transmission. One of them follows Figure 6.2 whereas another one is depicted in Figure 6.3. In Figure 6.2 a binary bit sequence enters the gaussian LPF, which in turn enters the tuning input of a VCO. Provided that the difference in the two frequencies f_1 and f_2 produced by the VCO is equal to the inverse of twice the bit interval time T_b, we have a GMSK modulated waveform at the output of the VCO.

In Figure 6.3 a quadrature (I–Q) GMSK modulator is depicted. After gaussian filtered data pass through an integrator, they split into two parts implementing the cosine and sine functions of the integrated data. At the output of 'sine' and 'cosine' blocks we have the I and Q data which go through the I–Q modulator part of Figure 6.3 to produce the modulated signal. Because of the limited number of different values of these integrated data, the trigonometric function blocks can be substituted by look-up tables. In this way we have two LUTs, as shown in Figure 6.4.

However, the two quadrature path data are correlated because of the relationship:

$$y(t) = \sin \left\{ \cos^{-1} [x(t)] \right\} \tag{6.4}$$

As a result the two LUTs can be contained in one read-only memory (ROM) called I/Q LUT ROM as in Figure 6.4. Finally, it must be noted that even the functions of gaussian filtering and integration can be accounted for in the modulator LUT, resulting in a more compact structure.

6.1.2.2 Demodulation

Both coherent and noncoherent techniques can be used for GMSK demodulation. In terms of demodulator architecture, two topologies can be used: quadrature detection and discriminator detection, as shown in Figure 6.5(a) and (b), respectively. In all architectures the received signal

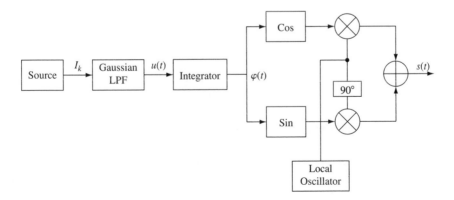

Figure 6.3 Quadrature implementation of GMSK modulator

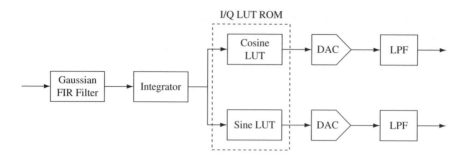

Figure 6.4 Fully digital implementation of GMSK modulator

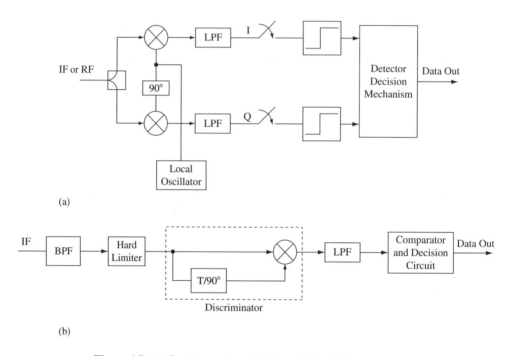

Figure 6.5 (a) Quadrature demodulation; (b) discriminator detection

passes first through a predetection bandpass filter (BPF) to reduce noise power and adjacent channel interference. It can be proved that, in order, to implement coherent GMSK demodulation [Murota81], [DeBuda72], a quadrature receiver is necessary. As a result a coherent GMSK receiver has the general form of Figure 6.5(a).

As noted in Chapter 5, a coherent GMSK detector (in the form of a quadrature receiver) must be augmented by carrier and timing recovery circuits. Murota and Hirade [Murota81] implement carrier recovery using a classical Costas loop. However, carrier and clock recovery systems can be quite complex. Furthermore, as explained in Chapter 5, coherent detection is vulnerable in fast multipath fading environment.

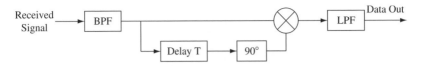

Figure 6.6 Differential detector demodulator

On the other hand since noncoherent detection does not require phase information, simpler receivers can be implemented. As mentioned in Chapter 5, several approaches can be used for noncoherent GMSK detection such as limiter/disciminator, differential detection and sequential decision/feedback detection. The third method is very effective in improving performance under ISI conditions and is based on sequential decision feedback detection [Yongacoglou88].

The limiter/discriminator detector has the structure shown in Figure 6.5(b) and the demodulation principle is based on the fact that the derivative of the phase of the received signal will give the instantaneous frequency which contains the information. The derivation operation is performed by the discriminator which changes the phase of the signal by 90° and delays it by the bit period T. The resulting signal is integrated over a bit interval and it is subsequently sampled at intervals equal to the bit duration.

As presented in detail in Chapter 5, the differential detection demodulator is based on the principle of comparing the phase of the receiver baseband signal with its delayed replica. This is easily done by multiplication, as shown in Figure 6.6. The resulting signal is then lowpass filtered and sampled at multiples of the bit period. Finally, a decision device, which can have the form of a comparator, produces the estimated symbol. As will be shown in the next section, the decision device can have the form of a finite state machine.

6.1.3 A DECT Modem for a Direct Conversion Receiver Architecture

6.1.3.1 Receiver Overview

As discussed in Chapter 3, the direct conversion receiver architecture is considered one of the most promising as it considerably reduces the number of external components and it exhibits lower front-end complexity. This stems from the fact that the received RF signal is translated directly down to baseband, eliminating the need for IF downconverting stages. In Chapter 3 it was also shown that the elimination of IF favours detection schemes based on the existence of in-phase (I) and quadrature (Q) signal components. This, in turn leads to digital processing and implementation in the baseband part of the receiver chain.

In this section, we present the design of the receiver based on the generation of I and Q components. An all-digital implementation approach is investigated with emphasis on low power consumption. A differential detector is employed using a finite state machine (FSM) [Metaxakis98a]. This FSM corresponds to a four-state Trellis operating on symbol rate. Frequency offset is corrected in a feed-forward manner using a phase rotation estimation algorithm operating on the symbol rate. Timing estimation and slot synchronization are performed jointly and implemented by the same functional block. Digital processing for frequency offset correction and timing/slot synchronization operate with oversampling and is based on the correlation between the estimated data sequence and the DECT synchronization preamble.

Figure 6.7 illustrates the overall receiver architecture. The RF signal, through the direct conversion RF front-end, is translated directly to in-phase and quadrature baseband signals. Subsequently there is analogue lowpass filtering to relax the requirements for the digital filters.

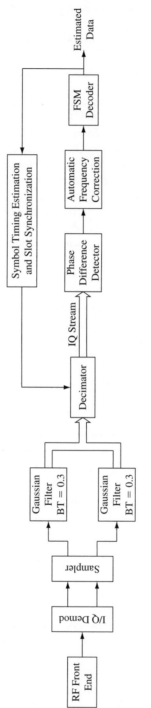

Figure 6.7 Overall direct-conversion receiver architecture

After A/D conversion, the I and Q signals pass through predetection digital lowpass filters and are fed to the baseband detector. Predetection filters are realized as gaussian lowpass FIR filters with BT equal to 0.3. Initially oversampling is used because timing synchronization has not been performed yet. The differential detector uses the oversampled data to calculate the phase difference between two successive samples. At the same time, frequency offset correction and timing estimation are being performed. After timing estimation, the detector operates at symbol rate. The frequency offset estimation algorithm is used to correct the phase difference estimation produced by the differential detector. Now the sample enters the decision device which has the form of a finite state machine, which estimates the transmitted sequence.

6.1.3.2 Differential Detector

Figure 6.8 shows the basic structure of the detector which receives the sampled in-phase (I) and quadrature (Q) components $i(kT)$ and $q(kT)$ which are produced at the output of the two branches of Figure 6.5(a). Mathematical expressions of the quadrature components are as follows:

$$i(kT) = \frac{1}{2}Z_s(t)[\cos(\varphi_s(t) - \varphi_c)] + n_c(t)\Big|_{t=kT} = \frac{1}{2}Z_s(kT)[\cos(\varphi_s(kT) - \varphi_c)] + n'_c(kT)$$

(6.5a)

$$q(kT) = \frac{1}{2}Z_s(t)[\sin(\varphi_s(t) - \varphi_c)] - n_s(t)\Big|_{t=kT} = \frac{1}{2}Z_s(kT)[\sin(\varphi_s(kT) - \varphi_c)] + n'_s(kT)$$

(6.5b)

In the above equations φ_s is the information-bearing phase:

$$\varphi_s(t) = 2\pi f_d \int_{-\infty}^{t} \sum_{n=-\infty}^{\infty} a_n g(u - nT) du$$

(6.6)

where φ_c represents the phase difference in the local oscillator signals of the transmitter and the receiver, whereas n'_c and n'_s represent the in-phase and quadrature noise components after passing through LPFs.

Neglecting noise terms and LO phase difference φ_c we get for $\Delta\psi(KT)$:

$$\Delta\psi(kT) = \tan^{-1}\left(\frac{q(kT)}{i(kT)}\right) - \tan^{-1}\left(\frac{q(kT-T)}{i(kT-T)}\right)$$

$$= \tan^{-1}\left(\frac{\frac{1}{2}Z_s(kT)[\sin(\varphi_s(kT) - \varphi_c)]}{\frac{1}{2}Z_s(kT)[\cos(\varphi_s(kT) - \varphi_c)]}\right) - \tan^{-1}\left(\frac{\frac{1}{2}Z_s(kT-T)[\sin(\varphi_s(kT-T) - \varphi_c)]}{\frac{1}{2}Z_s(kT-T)[\cos(\varphi_s(kT-T) - \varphi_c)]}\right)$$

$$= \tan^{-1}\left[\tan(\varphi_s(kT) - \varphi_c)\right] - \tan^{-1}\left[\tan(\varphi_s(kT-T) - \varphi_c)\right] = \varphi_s(kT) - \varphi_s(kT-T)$$

$$= \Delta\varphi_s(kT)$$

(6.7)

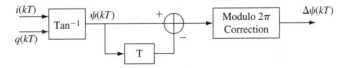

Figure 6.8 Basic structure of differential detector

Table 6.1 q-Values and cumulative phase change calculation as a function of BT

BT	q_{-2}	q_{-1}	q_0	q_1	q_2	q_3	$\Delta\varphi_{min}$	$\Delta\varphi_{min}$ [Yongacoglou88]
0.15	0.0236	0.1972	0.5636	0.5636	0.1972	0.0236	13.98	—
0.2	0.0042	0.1270	0.6542	0.6542	0.1270	0.0042	44.91	1.6
0.25	0.0006	0.0760	0.7088	0.7088	0.0760	0.0006	63.67	29.6
0.3	0.0001	0.0445	0.7408	0.7408	0.0445	0.0001	74.66	51.2
0.4	—	0.0149	0.7705	0.7705	0.0149	—	84.87	80.0
0.5	—	0.0048	0.7806	0.7806	0.0048	—	88.36	97.6
1	—	—	0.7854	0.7854	—	—	90.00	132.8

As expected, without noise terms and local oscillator drifts, the $\tan^{-1}$ operator produces the phase difference between two successive symbols. This phase change $\Delta\varphi_s(kT)$ is estimated from the middle of the previous symbol to the middle of the current symbol. This is shown in the equation below:

$$\Delta\varphi_s(kT) = \varphi_s(kT) - \varphi_s(kT - T) = 2\pi f_d \sum_{n=-\infty}^{\infty} a_n \int_{kT-T}^{kT} g(u - nT)du = \sum_{n=-\infty}^{\infty} a_n q_{k-n} \tag{6.8}$$

where

$$q_{k-n} = 2\pi f_d \int_{kT-T}^{kT} g(u - nT)du = 2\pi f_d \int_{kT-T-nT}^{kT-nT} g(u)du = 2\pi f_d \int_{(k-n-1)T}^{(k-n)T} g(u)du \tag{6.9}$$

Table 6.1 gives calculated values for q_i (in rad) for a variety of values of bandwidth-period product (BT). Along with the values for q_i there is a column giving the phase difference of the closest states $\Delta\varphi_{min}$ for which estimation will result in different symbols. For comparison reasons, an additional column (the last one) is given in Table 6.1 with the corresponding values of the phase change for the 1-bit differential detector by [Yongacoglou88].

As indicated by Equation (6.7), q_0 and q_1 will give the estimate of the phase difference $\Delta\psi(kT)$. The rest of the terms (q_{-2}, q_{-1}, q_2, q_3) represent intersymbol interference. From Table 6.1 we can see that for low BT values some of these terms are comparable with the useful ones. However as BT increases, these terms get very small values. For BT = 0.5 the only significant ISI terms are q_{-1} and q_2.

Keeping only these values the expression of $\Delta\varphi_s(kT)$ becomes:

$$\Delta\varphi_s(kT) = \sum_{n=-\infty}^{\infty} a_n q_{k-n} = a_{k+1} q_{-1} + a_k q_0 + a_{k-1} q_1 + a_{k-2} q_2 \tag{6.10}$$

Table 6.2 gives the values of $\Delta\varphi_s(kT)$ for all possible combinations of information symbols a_i and for BT = 0.5. From the table we observe that $\Delta\varphi_s(kT)$ takes values around three areas: $-90°$, $0°$ and $90°$. As a result all possible combinations are grouped into three different 'states'.

We can now produce a diagram mapping all possible states (R1, R2, R3) on a circle. This is shown in Figure 6.9, where the lines dt_1 and dt_2 represent the boundaries (or decision thresholds) between states 2 and 1 and states 2 and 3, respectively.

Table 6.2 Phase differences and states corresponding to all 4-bit input sequence combinations

Combinations of 4-bit words				State	$\Delta\varphi_s(kT)$ (deg)
a_{k-2}	a_{k-1}	a_k	a_{k+1}		
−1	−1	−1	−1	R3	−90.00
−1	−1	−1	1	R3	−89.45
1	−1	−1	−1	R3	−89.45
1	−1	−1	1	R3	−88.91
−1	1	−1	−1	R2	−0.54
−1	−1	1	−1	R2	−0.54
−1	−1	1	1	R2	0.00
−1	1	−1	1	R2	0.00
1	−1	1	−1	R2	0.00
1	1	−1	−1	R2	0.00
1	−1	1	1	R2	0.54
1	1	−1	1	R2	0.54
−1	1	1	−1	R1	88.91
−1	1	1	1	R1	89.45
1	1	1	−1	R1	89.45
1	1	1	1	R1	90.00

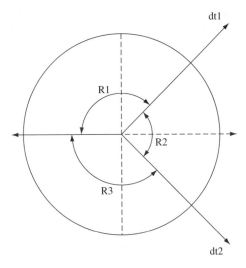

Figure 6.9 Diagram of possible phase states and boundaries

From the above we have:

(1) The differences between two successive samples (phase transitions) are given by $\pm\alpha\pi h + \text{ISI}$ where $\alpha \in \{0, 1\}$.
(2) Since we have three states, a hard decision device can map the phase difference in a 2-bit word according to Table 6.3.
(3) Using the decision on the state (or phase transition), and the previous detected symbol, we can make a decision on the current received symbol (as in Table 6.3).

Table 6.3 Decision mechanism as a function of phase transition and previous symbol

Phase transition	Condition	Encoded condition	Previous symbol	Current symbol
Positive	$\Delta\varphi > \pi h/2$	0 1	1	1
No phase transition	$|\Delta\varphi| < \pi h/2$	0 0	1	0
			0	1
Negative	$\Delta\varphi < -\pi h/2$	1 0	0	0

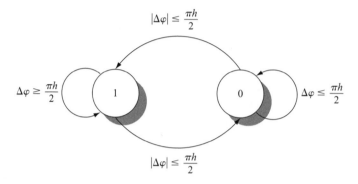

Figure 6.10 Finite state machine corresponding to detection procedure (reprinted from N. Zervas *et al.*, 'Low-power design of a digital baseband receiver for direct conversion DECT systems', IEEE Trans. Circuits and Systems II, vol. 48, © 2001 IEEE)

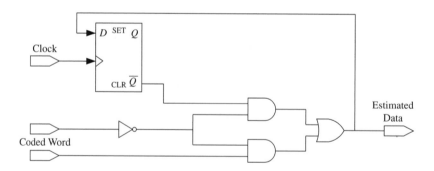

Figure 6.11 Low-complexity implementation of detection scheme

Encoding the above procedure we can represent it graphically by a two-state process using the state diagram of Figure 6.10 [Metaxakis98a] where the 1 and 0 inside the circles represent the decisions taken.

A low-complexity implementation of this detection scheme is illustrated in Figure 6.11, which has inputs from a clock and the encoded 2-bit word. In addition to handling the problem created by the fact that the $\tan^{-1}$ function is only defined in the $[-\pi/2, \pi/2]$ region, a 'modulo-π correction' block can be used which, depending on the sign of the I and Q components, it maps the phase value onto a $[-\pi, \pi]$ region. For example, if $i(kT)$ is negative and $q(kT)$ is positive, on the result of $\tan^{-1}[\bullet]$ we add π rads because the angle of $\psi(kT)$ is located in the region $(\pi/2, \pi]$.

6.1.3.3 Frequency Correction Subsystem

Local oscillator frequencies of the transmitter and the receiver ($f_{\text{TxLO}}, f_{\text{RxLO}}$ respectively) usually exhibit a small frequency difference, which has to be corrected at the receiver before detection. The phase rotation resulting from this frequency difference is:

$$\varphi_r(t) = 2\pi \cdot (f_{\text{TxLO}}(t) - f_{\text{RxLO}}(t)) \cdot t \qquad (6.11)$$

Automatic frequency correction (AFC) in the proposed system is achieved in a feed-forward manner such that no external feedback loops are necessary. As a result, AFC does not affect the system PLL and receiver integration becomes easier.

The algorithm used in this implementation resembles the one used by [Chuang91], but in this implementation one sample per symbol is used whereas Chuang uses oversampling $\times 16$. The objective is to estimate $\Delta\varphi(t)$, which represents the phase rotation between two successive symbols, which is:

$$\Delta\varphi(t) = 2\pi \cdot \Delta f(t) \cdot T \pm a(t) \cdot \pi \cdot h \qquad (6.12)$$

where $a(t)$ stands for the transmitted symbol taking the value of 0 or 1 and $\Delta f(t)$ represents the difference between the transmitter and receiver local oscillator frequencies.

As $h = 0.5$ for GMSK modulation, the dependence of $\Delta\varphi(t)$ is eliminated by multiplying it by 4. In measuring phase offset, angles must be mapped such that there is a correspondence between quadrant, phase value and the absolute phase offset. For example, 300° of phase offset is not the same as −60°, although using polar coordinates the two points coincide. What we just described is known as the 'wrap-around' problem and is created when we have large phase offsets. To avoid this kind of problems it is necessary to transform the values of phase offsets into rectangular coordinates [Chuang91].

Therefore, by quadrupling the estimated phase offset and analysing it in in-phase and quadratic components, we get the phase estimate:

$$\Delta\hat{\varphi}(kT) = \frac{1}{4} \tan^{-1} \frac{\sum_{m=1}^{k} \sin\left(4 \cdot \Delta\varphi(t - mT)\right)}{\sum_{m=1}^{k} \cos\left(4 \cdot \Delta\varphi(t - mT)\right)} \qquad (6.13)$$

In both numerator and denominator we perform averaging over a length k, which practically can be equal to the TDMA slot length. In this way, estimation noise due to additive noise and ISI can be considerably reduced.

Figure 6.12 shows the block diagram of the AFC subsystem. As indicated in this figure, a last block called 'lead-lag decision' is necessary to determine whether the LO of the receiver leads or lags that of the transmitter.

6.1.3.4 Symbol Timing and Slot Synchronization

Symbol timing and slot synchronization are performed in the receiver using the same basic principles. In addition, both these functions are performed jointly and implemented using the same functional block.

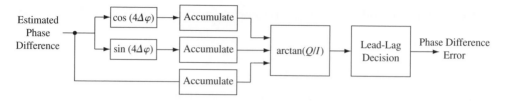

Figure 6.12 Block diagram of automatic frequency correction system

Symbol timing estimation chooses the best (in terms of minimizing the BER) sampling instant in each symbol. The system uses adaptive sampling in the sense that, for each data burst, the sampling time is chosen differently.

As mentioned in the beginning of this chapter, each DECT data slot starts with a synchronization field consisting of 32 bits, namely AAAAE98A (in hexadecimal format). According to the standard, this field must be used for clock and packet synchronization.

In the implementation outlined in this section we use the correlation C_i between the 32 bits of the preamble and the demodulated data corresponding to the same part of the DECT slot which seems to have the best performance [Molisch97], [Benvenuto97].

During the synchronization period the received preprocessed signal is oversampled (with oversampling ratio N) at a rate of N/T. In this way N time-multiplexed sequences are created:

$$S(i) = s\left(i\frac{T}{N}\right), s\left(T + i\frac{T}{N}\right), \ldots, s\left(nT + i\frac{T}{N}\right), \ldots, \quad 0 \le i \le N - 1 \qquad (6.14)$$

The samples

$$s_{n,i} = s\left(nT + i\frac{T}{N}\right)$$

denote the ith sample of the nth symbol.

The structure of the synchronizer (timing estimation) is illustrated in Figure 6.13 where some implementation details are also shown. The oversampled preamble sequence $S(i)$ passes through the differential detector which produces the sequence $D(i)$ of the estimated oversampled preamble:

$$D(i) = d\left(i\frac{T}{N}\right), d\left(T + i\frac{T}{N}\right), \ldots, d\left(nT + i\frac{T}{N}\right), \ldots, \quad 0 \le i \le N - 1 \qquad (6.15)$$

$D(i)$ is first demultiplexed in N parallel streams $D(1), D(2), \ldots, D(N)$, each of which represents the estimated sampled preamble at a particular time instant during the symbol period. For example, $D(i)$ represents the sequence of the ith samples of all 32 bits of the preamble.

Next, the demultiplexed sequences $D(i)$ are fed into N shift registers of 32 bits. Three different comparisons are performed between the data b_n ($0 \le n \le 31$) of each shift register and the known DECT preamble p_n.

The first comparison deals with the last 16 bits of the shift register, which are compared with the synchronization word E98A. If the resulting number of errors is less than a predetermined threshold, the system enters a 'synchronization confirmation' state. According to this, a second comparison is performed between the last 4 bits of the shift register and those of the preamble (representing A in hexadecimal format). If the number of errors is again below

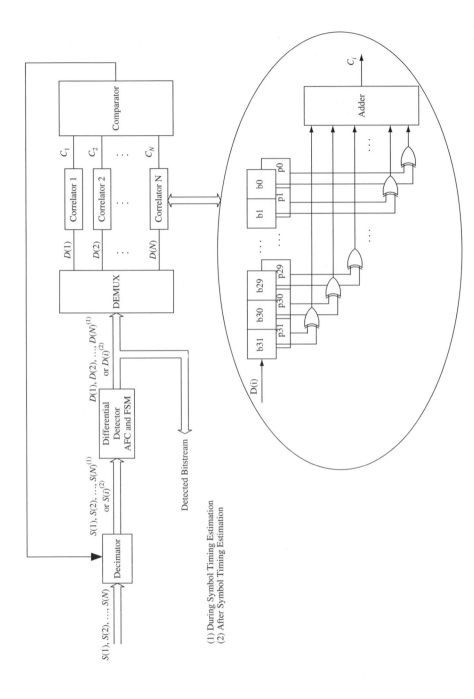

Figure 6.13 Slot and symbol timing synchronizer with implementation details [reprinted from E. Metaxakis, A. Tzimas, G. Kalivas, 'A low complexity baseband receiver for direct conversion DECT-based portable communication systems', IEEE Int. Conference on Universal Personal Communications (ICUPC) 1998, pp. 45–49, © 1998 IEEE]

a second predetermined threshold, the receiver has achieved slot synchronization. When the receiver has achieved synchronization and it has also confirmed it (through the second comparison), it proceeds to select the best sampling instant based on the overall comparison between the data of the shift register and the 32 bits p_n of the known preamble which, as indicated by the figure, takes place continuously as the shift registers are filled. However, the results of this comparison are valid only after the system has passed successfully the previous two comparisons.

From this third comparison, the optimum sampling instant (for the current burst) is determined and is the one that minimizes the expression:

$$C_i = \sum_{n=0}^{31} p_n \otimes b_{n,i}, \quad 0 \leq n \leq N - 1 \tag{6.16}$$

where $\otimes$ denotes XOR operation between the preamble bits p_n and the contents of the shift registers ($b_{n,i}$ represents the nth symbol of the ith shift register). Minimizing the value C_i produced by Equation (6.16) minimizes the BER in the preamble. This is equivalent to maximization of the correlation between the preamble sequence and the estimated data sequences. This leads to description of the functional block for slot and symbol timing synchronization as a correlator.

As shown in Figure 6.13, after the optimum sampling time is determined through the above procedure, decimation is used prior to the overall receiver to feed the following stages with just one sample per symbol. In this way power requirements are significantly reduced.

To have a practical, low-complexity system the oversampling ratio must be kept as small as possible, which is translated to a finite resolution of the symbol timing algorithm. This leads to a residual timing error the maximum value of which is (after decimation) $\varepsilon_{r,\max} = 1/2N$. We will examine the impact of finite resolution in the following section.

Concluding, it must be noted that the presented synchronization algorithm operates in a feedforward manner, eliminating the need for analogue circuitry to control the timing instants. In addition, the acquisition time of the presented system is constant and independent of the initial timing error. This is very important for packet-based communication with relatively short data bursts like DECT.

6.1.3.5 Simulation Results of Digital Receiver

In this section we examine the performance of the digital receiver as depicted in Figure 6.7 in AWGN and co-channel interference (CCI). We also examine the effect of residual timing error, and the effectiveness of the previously proposed frequency offset correction and synchronization estimation algorithms. We examine this performance considering only digital predetection filters. In the next section we will see that composite analogue/digital filters are necessary to adequately suppress CCI.

Initially, the impact of the predetection filters (gaussian) was examined on the probability of error for the DECT detector in AWGN. Figure 6.14 shows the BER performance as a function of the $B \cdot T$ product for fixed SNR. One can see that the best performance is achieved for $B \cdot T \cong 0.3$. This is in good agreement with [Murota81], where the gaussian IF filter with $B \cdot T \cong 0.63$ exhibits the best performance. Hence, the equivalent baseband filter would have half the bandwidth, as is actually the case for this detector. For the rest of the performance results the receiver baseband predetection filters are set to the optimum value of $B \cdot T \cong 0.3$.

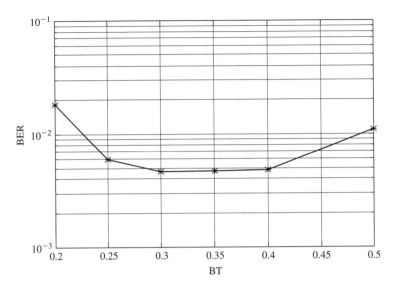

Figure 6.14 BER as a function of the pre-detection product BT [reprinted from E. Metaxakis, A. Tzimas, G. Kalivas, 'A low complexity baseband receiver for direct conversion DECT-based portable communication systems', IEEE Int. Conference on Universal Personal Communications (ICUPC) 1998, pp. 45–49, © 1998 IEEE]

Figure 6.15 gives the BER performance of the differential detector under AWGN when there exists perfect synchronization and there is no frequency offset. Figure 6.16 demonstrates the performance of the receiver when frequency offset is introduced in the system. The results are given in terms of the normalized (with respect to bit period) percentage frequency offset for constant signal-to-noise ratio, SNR $= 12$ dB, and under perfect synchronization. The top curve shows the deterioration when there is no automatic frequency offset correction (no AFC), whereas the bottom curve shows the performance when the AFC algorithm described in Section 6.1.3.1 was used. One can see that frequency offsets up to $\Delta f \cdot T = \pm 12\%$ can be corrected, which corresponds to an absolute correction range of ± 140 kHz for DECT. The systematic error increase with $\Delta f \cdot T$, while AFC is operating (bottom curve), is due to the fact that, as frequency offset increases, the signal energy being cut-off from the tight filters becomes larger [Mehlan93]. On the other hand, it must be noted that the BER performance deteriorates as the SNR becomes lower. Figure 6.17 shows the performance of AFC when SNR is very low (6 dB).

As mentioned in the previous section, the value of the oversampling ratio (N) corresponds to a finite resolution of the timing estimation algorithm, which in turn produces an irreducible residual error, the maximum value of which is $\varepsilon_{r,\max} = 1/2N$. This effect of the finite resolution of the timing algorithm is illustrated in Figure 6.18, where the performance of the receiver is given for timing errors of $T/8$ ($N = 4$) and $T/16$ ($N = 8$). At BER equal to 10^{-4} the system performance with $N = 4$ is degraded by 0.7 dB whereas for $N = 8$ there is no significant degradation.

Figure 6.19 illustrates the impact of CCI on the receiver performance. One can see that, while an SNR of 13 dB, without CCI, will give BER $= 10^{-3}$, a CCI interferer with power ratio of $P_{\text{useful}}/P_{\text{CCI}} = 12$ dB will increase the BER by 20 at 2×10^{-2}. The large increase in BER indicates that using gaussian base-band filters of $B \cdot T \cong 0.3$ does not provide satisfactory performance. In the next section we examine briefly base-band filtering for the DECT receiver.

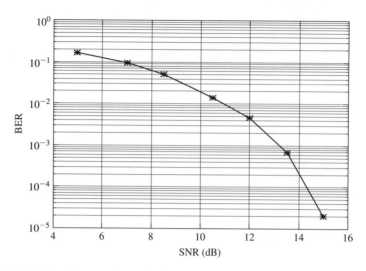

Figure 6.15 Probability of error of the GMSK detector in AWGN [reprinted from E. Metaxakis, A. Tzimas, G. Kalivas, 'A low complexity baseband receiver for direct conversion DECT-based portable communication systems', IEEE Int. Conference on Universal Personal Communications (ICUPC) 1998, pp. 45–49, © 1998 IEEE]

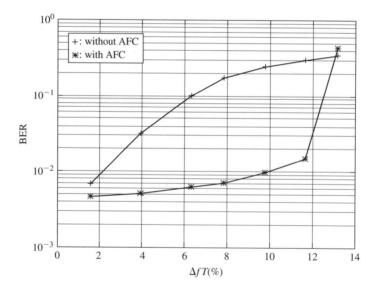

Figure 6.16   Performance of the detector as a function of frequency offset (SNR = 12 dB)

6.1.3.6 Overall (Analogue/Digital) Filter Design

The main function of base-band filtering is to pass the desired channel without any attenuation and distortion, whereas at the same time it must suppress adjacent channel interference produced by neighbouring (in terms of frequency) users. Specifications for the above parameters are given by the standard. In addition to the above, in direct conversion receivers, filters should

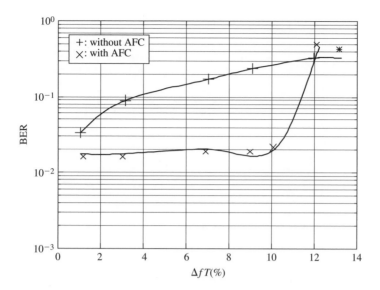

Figure 6.17 Performance of the detector as a function of frequency offset (SNR = 6 dB)

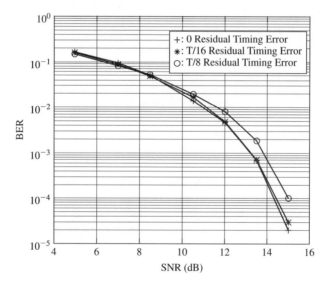

Figure 6.18 Performance with residual timing error (no frequency offset) [reprinted from E. Metaxakis, A. Tzimas, G. Kalivas, 'A low complexity baseband receiver for direct conversion DECT-based portable communication systems', IEEE Int. Conference on Universal Personal Communications (ICUPC) 1998, pp. 45–49, © 1998 IEEE]

adequately suppress higher-order products created by mixing the desired signal with the local oscillator signal. These products are located at twice the local oscillator frequency.

Filtering in the DECT modem is achieved by a combination of analogue and digital filters. To meet the DECT standard requirements, fourth-order Butterworth analogue filters with

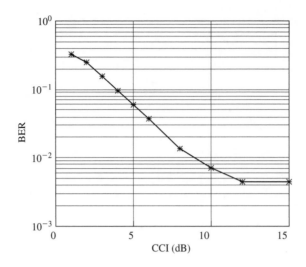

Figure 6.19 Performance with co-channel interference (zero timing and frequency offsets)

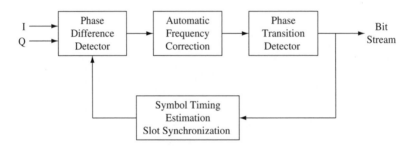

Figure 6.20 The functional diagram of the DECT receiver

cut-off frequency $f_C = 0.8 \times R$ ($R = 1.152$ Mb/s is the DECT gross bit rate) are used followed by finite impulse response filters of 13 taps. FIR is the preferred structure because it is easier to obtain some necessary characteristics like highly accurate response and linear phase response. The receiver FIR filters use an interpolation factor of 4 and have a cut-off frequency $f_C = 0.6 \times R = 691$ MHz. In addition, a Kaiser window is used with $\beta = 3.5$. These filters exhibit satisfactory BER performance in the DECT receiver.

6.1.3.7 Digital Receiver Implementation

System Overview

The baseband receiver block diagram is depicted in Figure 6.7. From this figure we can see that the heart of the baseband receiver is the blocks following the LPFs. In Figure 6.20 the functional structure of this section is illustrated, which consists of four blocks: the phase difference decoder (PDD), the automatic frequency control, the phase difference transition mapper/FSM decoder (PDTM) and the symbol timing estimation and synchronization (STES).

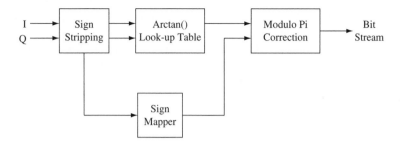

Figure 6.21 Functional diagram of the phase difference decoder

In this section we outline the implementation of the DECT receiver as a digital system. More details of the implementation can be found in [Perakis99] and [Zervas01].

The received signal passes through the RF receiver and I/Q demodulator and is split into in-phase (I) and quadrature (Q) paths, which are first sampled and filtered by predetection lowpass filters (implemented as FIR gaussian filters with a $BT = 0.3$). Quantization produces 6-bit I–Q streams, each pair of which is received in each clock cycle.

In the beginning, there is no timing synchronization and oversampling $N = 4$ is used in the receiver. The data streams at the output of the filters enter the PDD block, which estimates the phase difference of two consecutive samples. Subsequently, this phase difference is corrected by the AFC block and is then decoded by the PDTM. Before phase demodulation takes place, slot synchronization and symbol timing recovery are achieved using the DECT preamble and the symbol timing recovery block in the receiver. In the subsections below we give some detail of the implementation of the receiver blocks described above.

Phase Difference Decoder
Figure 6.21 illustrates the functional diagram of the PDD. The main block of the PDD is a look-up table used to calculate angles based on the I and Q components, i.e. calculate $\theta = \tan^{-1}(Q/I)$. By computing the difference between the angles of two successive symbols, $\Delta\varphi = \theta(t) - \theta(t-1)$ we can assign $\Delta\varphi$ values to detected symbols using the decision mechanism for the suggested detection scheme in Table 6.3.

Taking advantage of some properties of the trigonometric functions, the computation of $\tan^{-1}[\bullet]$ and $\Delta\varphi$ functions can be simplified greatly using the following procedure. The $\tan^{-1}[\bullet]$ function takes values only in the field $(-\pi/2, \pi/2)$, but removing the sign information from the I and Q values (done by the 'sign-stripping' block), maps the $\tan^{-1}[\bullet]$ value only to the first quadrant $(0, \pi/2)$. Hence 64 angles belonging only to the first quadrant $(0, \pi/4)$ need to be stored in the LUT.

To achieve further reduction of the LUT size, only the arctan values corresponding to $Q < I$ are stored (in the range $[0, \pi/4)$. To obtain values I in the range $(\pi/4, \pi/2]$ the values of Q and I are exchanged using the trigonometric identity $\tan^{-1}(Q/I) = \pi/2 - \tan^{-1}(I/Q)$. As only angles in the range $[0, \pi/4)$ are stored, the final LUT size is:

$$\left(\frac{(2^5 \times 2^5)}{2} - 2^5\right) \times 5 = 2400 \text{ bits}$$

Furthermore, the LUT can be accessed as a one-dimensional array and can be easily implemented because it only needs additions, shifting and the easy to see the squaring function $[\bullet]^2$.

Finally to map the resulting values of the function $\tan^{-1}[\bullet]$ from the domain of $(-\pi/2, \pi/2)$ to the entire domain of values that the calculated angle $\theta(t)$ can take [in the $(-\pi, \pi)$ range] the 'sign mapper' block has to be inserted. Nevertheless, the additional hardware needed for this remapping to $(-\pi, \pi)$ field, the exchange of values between I and Q and the addressing scheme for the LUT is insignificant compared with the initial LUT hardware, which would be 13.5 times larger.

Automatic Frequency Correction

The estimated phase difference from the PDD block is the input for the AFC block, which is depicted in Figure 6.12. For convenience we give below Equation (6.13), which represents the selected frequency correction algorithm:

$$\Delta\hat{\varphi}(kT) = \frac{1}{4}\tan^{-1}\frac{\sum_{m=1}^{k}\sin(4\cdot\Delta\varphi(t-mT))}{\sum_{m=1}^{k}\cos(4\cdot\Delta\varphi(t-mT))} \tag{6.13}$$

From Figure 6.12, which illustrates the implementation of the algorithm, we can see that the main processing blocks are $\sin(4\theta)$, $\cos(4\theta)$, accumulators which implement the summation operation and $\arctan(Q/I)$, which gives the inverse tangent function of the ratio of the summations. Figure 6.22 gives the implementation approach for the AFC block. The angle θ enters the block from PDD and is multiplied by 4 (by shifting to the left by two bits) to get (4θ) arguments. To obtain sine and cosine values for this argument look-up tables are again used. In this way the need for time-consuming calculations is eliminated and substituted by assigning arguments to values from the LUT, which is much simpler. The 8-bit input to the AFC block comes from the phase difference decoder block and is reduced to 6 bits by performing left-shift by two bits to obtain the 4θ values. Subsequently the quadrant information is also removed (corresponding to the two most significant bits, MSBs) and as a result the input streams are represented by 5-bit words which are accessed using 4-bit addresses to index the LUTs. In turn, LUTs store only values of sinuses because cosines can be produced from sine values using the trigonometric identities:

$$\cos\left(\frac{\pi}{2}-x\right) = \sin x \text{ and } \sin\left(\frac{\pi}{2}-x\right) = \cos x$$

There is no need to perform calculations to determine $\cos x$ but only to assign to it the value of $\sin(\pi/2-x)$. We have two LUTs (one for I and one for Q) with size equal to 16×5 bits each.

A distinguishing functional characteristic in this block is the summation of several values of $\sin(4\theta)$ and $\cos(4\theta)$ saved in the accumulators. As a result, these summations increase in significant bits. To produce a constant number of bits needed by the next functional block $\tan^{-1}(x)$, it is necessary to locate the position of the MSB of these sums. This then can be used as the 'select' input of a multiplexer, which will provide the 6 bits (5 bits + sign bit) input of the $\tan^{-1}(x)$ function block. This in turn is divided by 4 (2 bit shifting) and added to the estimated phase (through the delay element) in order to correct the frequency offset.

Phase Difference Transition Mapper

The output of the AFC block gives the corrected phase difference and is used as the input of a 'phase difference to binary number' transformation block. This is illustrated in the left-hand part of Figure 6.23. This transformation represents the decision mechanism of Table 6.3, where the value of the phase difference entering the block is compared with two bounds ($-\pi/4$ and $\pi/4$) and determines one of three transitions: positive, zero or negative transition. Subsequently the

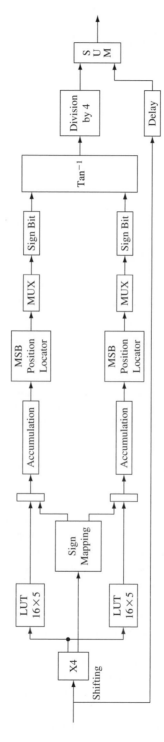

Figure 6.22 Implementation approach of the AFC subsystem

391

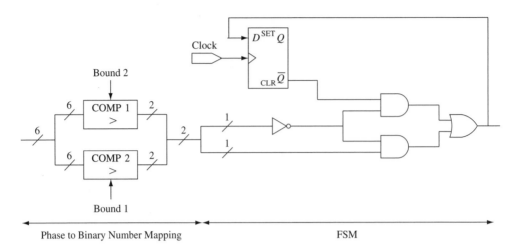

Figure 6.23 Implementation approach of the phase difference transition mapper

resulting transition is used as input of the finite state machine to translate this to a 1-bit decision. The digital implementation of the FSM is depicted on the right-hand side of Figure 6.23.

Symbol Timing Estimation

As mentioned in Section 6.1.3(d) above, three comparisons must take place in this block in order to achieve slot synchronization and optimum sampling estimation. Oversampling by 4 is performed during the synchronization period and there are 4 bit streams of size 32 bits. The result is a 128 bit stream appearing as a time-multiplexed signal representing four difference 32-bit sequences from which one must be chosen as the optimum.

Therefore, from this 128-bit register, the first time, the following bits are chosen as some of the inputs of an exclusive-OR (XOR) gate:

$$D_n(1) = d_n(0), d_n(4), d_n(8), d_n(12), \ldots d_n(127) \tag{6.17}$$

where n represents the time index for the input data and numbers in parentheses represent the bit position within the 128-bit shift register. This is depicted at the top of Figure 6.24, which illustrates the digital implementation of the overall symbol timing estimation block. The second input of the XOR is the given by the standard preamble sequence AAAAE98A.

The first comparison compares the last 16 bits of the preamble word (E98A) to the respective input sequence [last 16 bits of sequence (6.17)]. For the 16-bit comparison 5 bits are needed to express the result. This comparison takes place in the 'COMP 1' block in Figure 6.24. One input to the comparator comes from the XOR operation after passing through weight counter. The second input of the comparator is the chosen threshold 'thr1'. The second comparison involves the last 4 bits of the shift register sequence D_n and those of the preamble and takes place in the 'COMP 2' block.

When both the above comparisons are successful, slot synchronization has been achieved and the third comparison involves four comparisons of sequences $D_n(1), D_n(2), D_n(3), D_n(4)$ with the 32-bit preamble AAAAE98A. This third comparison is used to achieve timing synchronization and, as stated in Equation (6.16), from the four sequences $D_n(i), i = 1 \ldots 4$, the one minimizing C_i is chosen. This is performed in the 'Comp Min{ }' block of Figure 6.24.

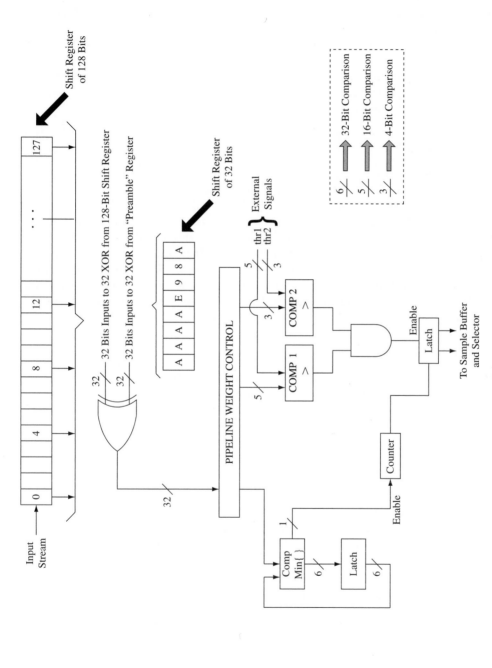

Figure 6.24 Digital implementation of of overall symbol timing estimation subsystem (reprinted from N. Zervas *et al.*, 'Low-power design of a digital baseband receiver for direct conversion DECT systems', IEEE Trans. Circuits and Systems II, vol. 48, © 2001 IEEE)

6.2 QAM Receiver for 61 Mb/s Digital Microwave Radio Link

6.2.1 System Description

Modern digital microwave links are designed to transfer huge amounts of data and for this reason they operate at very high bit rates using multilevel modulations with high spectral efficiencies. In addition, they operate at frequencies in the order of few GHz with line-of-sight configurations.

In this section we will present in detail the design of a 64-QAM modem with a gross bit rate of 61.44 Mb/s for a digital microwave radio application implemented as a part of a sub-SDH system operating at 15 GHz.

6.2.1.1 Transmitter

Figure 6.25 shows a block diagram of the transmitter part of the modem. The input to the modulator comes from a framer unit, where FEC (forward error correction) coding also takes place. The framer delivers a bit stream at 61.44 Mb/s along with a synchronized clock signal to the mapper, which in turn produces I and Q signals for the 64-QAM constellation. Each QAM symbol consists of 6 bits. The first two bits correspond to the quadrant whereas the last 4 bits are allocated using Gray coding for best demodulation. Filtering is necessary to reduce ISI and confine the transmit spectral mask to the required specifications. For this purpose, the I and Q streams are fed to FIR filters which implement raised cosine function in order to minimize the ISI. Digital to analogue converters followed by lowpass filters are used to transform the I and Q streams into analogue signals for feeding the I–Q modulator. This circuit is responsible for transforming the two quadrature components into a QAM modulated signal, which is then ready for upconversion and transmission.

Receiver

After the microwave front-end, the receiver consists of an analogue IF stage and a digital demodulator/detector. Figure 6.26 shows a functional block diagram of the downconverter and digital demodulator of the 64-QAM receiver [Metaxakis98b].

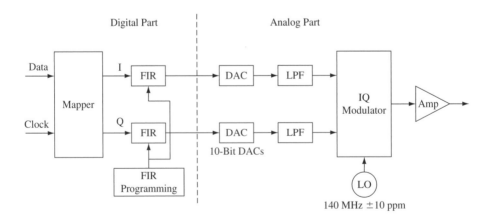

Figure 6.25   Functional block diagram of the modulator/upconverter

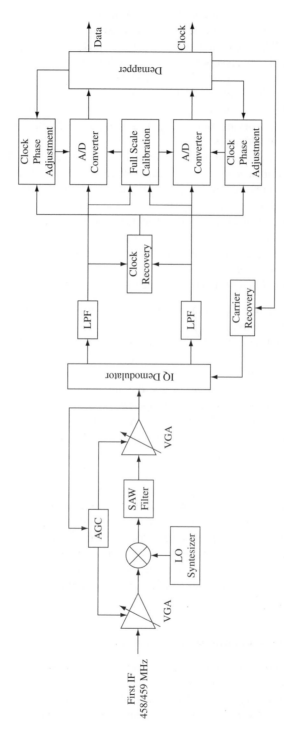

Figure 6.26 Functional block diagram of the downconverter and digital receiver

395

The analogue IF stage (which constitutes the second downconverter) receives the modulated signal at 458 MHz from the first downconverter and passes it through a mixer to further downconvert it to 140 MHz. Following the mixer, system filtering takes place using a SAW bandpass filter and then the received signal goes through a variable gain stage incorporating an automatic gain control (AGC) loop. This is done because the received signal at the first IF has a dynamic range of 60 dB. Applying AGC will bring the signal to the centre of the dynamic range of the I–Q demodulator. Overall, the second downconverter converts the first IF received signal (458/459 MHz) with varying input power from -67 to -8 dBm, to a constant average power signal (at 0 dBm) at 140 MHz.

The IF signal, after downconversion, passes through a quadrature demodulator for coherent detection. To minimize imbalances in the two channels and the effect of feedthrough signals, the gain and phase errors must be lower than 0.5 dB and 1.5° [DMR90]. The I/Q demodulator is followed by lowpass filters used mostly to remove the high-frequency products resulting from demodulation (since channel selection and adjacent channel rejection were performed at the IF stage). However, to minimize distortion, LPFs should exhibit low group delay (in the order of 20 ns) within the bandwidth of the baseband received signals.

At this point (at the output of the LPFs) the incoming signal passes to the digital modem which, after digitizing the I and Q signal components, performs the following operations:

(1) timing recovery to recover the optimum timing instant at which the analogue signal must be sampled at the A/D converter;
(2) carrier recovery (CR), which is used to estimate and correct the phase of the received signal such that it can be demodulated in a coherent fashion – when CR is achieved, the received signal follows the transmitted in frequency and phase;
(3) synchronizer, which is used to align in time the digital samples of I and Q channels in order to enter simultaneously into the demapper; and
(4) demapper, which transforms the I–Q composite received points in the cartesian coordinates into specific constellation points representing one of the 64 symbols.

6.2.2 Transmitter Design

Considering again Figure 6.25, the first half on the left of the dotted line represents the digital part whereas the last half includes D/A converters, analogue LPFs and the balanced modulator, altogether constituting the analogue part of the transmitter. As we will see in greater detail, the digital section of the transmitter is implemented in FPGA. To satisfy the specification for the level of spurious signals at the output of the transmitter, the spurious signal-free dynamic range of the D/A converter and the carrier and image rejection of the I–Q balanced modulator must be chosen appropriately.

6.2.2.1 Mapper and Finite Impulse Response Filters

The mapper uses 6-bit blocks and transforms them to two 3-bit I and Q corresponding to the coordinates of the constellation point. Apart from data, it has an input clock as well. Each FIR produces outputs of 10 bits such that the succeeding D/A converter can guarantee quantization noise lower than -50 dBc. Oversampling by 2 or 4 must be used to remove the image response of digital filtering.

6.2.2.2 D/A Converters and LPFs

To ensure conversion rates beyond 50 Msps (which takes into account oversampling) fast settling D/A converters must be employed. In addition, the current-to-voltage amplifier-converters must be fast enough to match the above rate.

The analogue LPFs are used to remove remaining spurious signals created by the digital section and the D/A converter. Their selectivity must be such as to satisfy the transmission spectrum mask specification at the output of the I–Q balanced modulator. The requirements for the LPFs are shown in Table 6.4.

6.2.2.3 I–Q Modulator and Synthesized LO (Upconversion Subsystem)

I–Q Modulator

The I–Q modulator exhibits a similar form to that described in Section 6.1 and depicted in Figure 6.3. It consists of two balanced mixers driven by an LO signal through a 90° phase shifter to produce quadrature information-bearing signals at their outputs. The resulting signals at the two branches are in turn driven into an in-phase combiner to produce the composite high-frequency analogue signal at 140 MHz. To satisfy the transmission spectrum mask specifications the I–Q modulator must fulfil the requirements in Table 6.5.

Synthesized LO

The synthesized LO must produce a stable output frequency of 140 MHz with low phase noise. A programmable PLL can be used to implement the local oscillator. To satisfy the above general requirements it is important to use a very clear reference frequency and a VCO with low phase noise capabilities. It is also desirable to utilize a PLL including programmable dividers at both phase detector inputs such that the frequency of operation of the loop can be selected to optimize in terms of noise. Since this LO does not implement the channelization of the overall system, the frequency of operation of the PLL can be chosen as low as possible to minimize the resulting overall phase noise.

6.2.3 Receiver Design

Figure 6.27 shows a detailed block diagram of the receiver modem, where one can see the functional partition of its blocks and operations [Kalivas97]. Only the front-end of this receiver

Table 6.4 Lowpass filter requirements

Rejection (dB)	Frequency (MHz)
−3	10.2
−70	40

Table 6.5 Requirements for I–Q modulator

Sideband rejection	35 dBc @ 140 MHz
Carrier rejection	37 dBc @ 140 MHz
DC offset voltage	0.12 mV @ 140 MHz

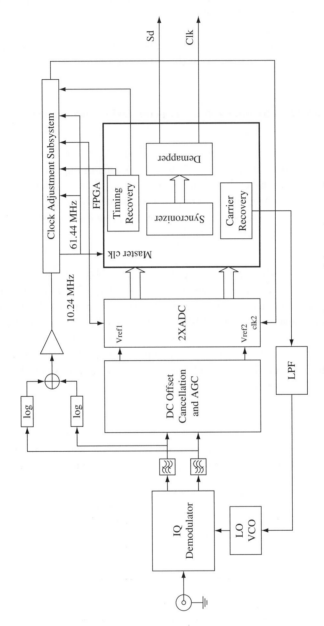

Figure 6.27 Detailed block diagram of the receiver modem

modem is analogue and consists of the *I–Q* demodulator, a VCO and an LPF, all of them comprising part of the carrier recovery loop. In addition, two more LPFs are connected at the *I* and *Q* outputs of the demodulator to deliver cleaner signals to the digital receiver subsystem.

The blocks between the LPFs and the A/D converters constitute a DC-offset cancellation subsystem which uses comparators and buffers to compensate for a DC component included in the *I* and *Q* received data. On the top line of the figure there exists the analogue part of the timing recovery subsystem responsible for adjusting the phase of the sampling clock. This block receives input from the timing recovery algorithm. Following the A/D converters there is an FPGA in which the most critical timing and carrier recovery algorithms are implemented. Furthermore, the FPGA implements the synchronizer and the demapper part of the digital receiver. The synchronizer is necessary for the synchronization of the *I* and *Q* streams before they are both fed to the demapper which in turn, based on the values of *I* and *Q*, makes decisions about the QAM symbols being transmitted.

6.2.3.1 DC Offset Cancellation and AGC (DCOC/AGC)

Figure 6.28 is a functional diagram of the DCOC/AGC subsystem. Owing to hardware imbalances and active signals (useful and interfering) at the input of the antenna, DC and low-frequency components may appear at the output of the *I–Q* demodulator, creating considerable distortion. In addition, depending on the distance between relay stations and weather conditions, received signals have variable power. More specifically, this system calibrates the input to the digital part of the receiver such that the 64 constellation points appear symmetrical at the centre of the Cartesian coordinate system and utilizing all the dynamic range of the A/D converters. This means that any possible systematic DC offset (which would be added to all constellation points) is accounted for by the DCOC circuit. It guarantees that the inputs A_{in1} and A_{in2} to the A/D converters take values in the 0–2 V range. In addition, when there is high attenuation, the constellation diagram seems compressed to a rectangle of a small size. The AGC part of this subsystem gives the appropriate gain to the system such that the size of the rectangle in the constellation formation takes the optimum value. This means that the mapping of the transmitted

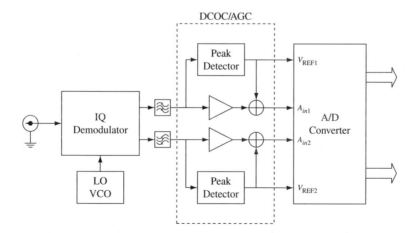

Figure 6.28   Functional diagram of the DCOC/AGC subsystem (in dashed-line box)

symbols uses all the dynamic range provided by the A/D converters. Hence the DCOC/AGC system functions as a full-scale A/D converter calibration system in both DC offsets and gain.

6.2.3.2 A/D Converters

The A/D conversion subsystem contains two matched A/D converters to convert similarly the two streams I and Q. Since we have a 64-QAM modulation, the 6 bits per symbol correspond to a symbol rate of 10 MHz approximately, which is the frequency of the analogue signal at the input of the A/D converters. The converters use two clock inputs produced by the timing recovery system which consists of two separate units, one for I and one for Q. Furthermore, the converters use two separate voltage reference inputs (V_{REF1}, V_{REF2}) produced by the DCOC/AGC subsystem as described above. The A/D provides two 8-bit outputs with excellent dynamic performance, exhibiting an effective number of bits equal to 7.4 bits when the input signal has a frequency of 10.3 MHz.

6.2.3.3 Timing Recovery System

Since digital microwave radio applications involve continuous transmission, usually there is no known pattern in the form of a preamble attached between successive data transmissions. Therefore timing recovery will be based on data-aided techniques, which do not involve a specific preamble pattern. In addition, owing to the high transmission rate of the application, it is not possible to use algorithms based on oversampling techniques. A robust, blind technique operating at the baud rate was presented by Mueller and Muller [Mueller76]. This technique uses timing functions which are linear combination of a few terms of the impulse response. Based on the symmetry error of the sampled IR that is produced by samples located on both sides of the principle factor h_0, it can be shown that a simple timing estimation function with low variance has the following form:

$$z_k = f_k(a_k, x_k) = x_k \cdot a_{k-1} - x_{k-1} \cdot a_k \qquad (6.18)$$

where x_k represent the input sampled symbol (at time k) to the estimator, whereas a_k are the MLSE decoded symbols.

The above function gives satisfactory results in binary PAM, exhibits low jitter and is implemented in a very simple way. In our application, the signal at I and Q channels have the form of PAM-8 modulated data.

It was found that the following function helps the system to converge faster:

$$F_k = 1 - \frac{x_k}{a_k} \qquad (6.19)$$

Furthermore, a combination of the above-mentioned functions was found to provide satisfactory speed of convergence and stability:

$$D = C_1 \cdot f_k(a_k, x_k) + C_2 \cdot F_k(a_k, x_k) \qquad (6.20)$$

However, when the initial timing instance falls outside the eye opening, convergence time may be very long. A ramp function r_k can help the loop to reach the correct timing instant faster. This function must have a slope that directly depends on the distance from the maximum eye opening such that, when the first sampling instant is away from the eye opening, it can converge faster than in other cases.

So r_k must include the absolute difference $x_k - a_k$, a derivative and a threshold *thr* and can be expressed as follows:

$$r_k = \sum |x_k - a_k| \cdot s \qquad (6.21)$$

where s is given below:

$$s = \begin{cases} 1, & \text{for } \dfrac{d\left(g_1 r_k + g_2 \bar{z}_k\right)}{dt} \geq thr \\[4mm] 0, & \text{for } \dfrac{d\left(g_1 r_k + g_2 \bar{z}_k\right)}{dt} \leq thr \end{cases} \qquad (6.22)$$

In the above formula, *thr* is a measure properly set to ensure faster convergence whereas g_1, g_2 are appropriate gain factors.

Figure 6.29 illustrates the basic timing recovery algorithm [Equation (6.18)] as presented above. Figure 6.30 shows the complete TR subsystem and the algorithm integration in it. The top box corresponds to the timing recovery algorithm as presented, above which is used to tune the phase of the pulse output of a PLL such that it creates a sampling pulse at the correct timing instance. This pulse is subsequently used as the sampling clock of the A/D converter. Apart from the input from the TR algorithm, the sampling pulse generating PLL needs an input corresponding to the symbol clock of the digital system (equal to 10.24 MHz). In Section 6.2.5 we will see how the complete TR system is implemented.

It must also be mentioned that signals I and Q are independently sampled and processed by the TR algorithm and subsystem in order to reduce implementation losses due to circuit impairments, which may introduce phase delays between the I and Q channels. As a result, two separate TR systems are implemented (for I and Q) which operate in parallel but independently.

6.2.3.4 Carrier Recovery System

The carrier recovery system can consist of a feedback loop to track the phase of the incoming signal in the form of a classical PLL or a Costas loop. However, the phase detector must acquire some special properties in order to make locking easier in a system where phase is not uniformly distributed (like in QAM-modulated transmission). In addition, the servo-loop must have unique and stable solutions. Figure 6.31 presents the general structure of such a system. A robust phase detector for our application is the Leclert–Vandamme approach [Leclert83], according to which the error function is expressed as follows:

$$\varepsilon(\varphi) = \text{sgn}(x_I - a_I) \cdot \text{sgn}(x_Q) - \text{sgn}(x_Q - a_Q) \cdot \text{sgn}(x_I) \qquad (6.23)$$

where x_I, x_Q, as before, represent the input symbol (sampled at time k) to the estimator, and a_I, a_Q are the MLSE decoded symbols. For reasons of simplicity the time index k representing the sampling instant is omitted.

The phase detector, along with the rest of the loop components is depicted in Figure 6.32. As it is a digital loop, A/D conversion is necessary in front of the detector inputs. The sequence of error increments $\varepsilon(\varphi)$ is lowpass filtered and fed to the VCO for frequency and phase locking.

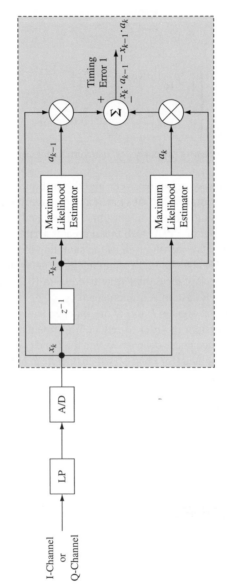

Figure 6.29 Timing recovery algorithm

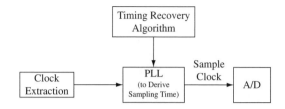

Figure 6.30 Functional diagram of complete TR subsystem

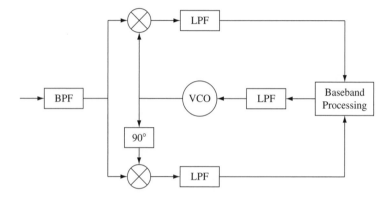

Figure 6.31 Functional diagram of carrier recovery subsystem

Some important properties of this digital phase comparator are as follows [Leclert83]:

(1) It uses all 64-constellation points for locking which implies that the detector is not sensitive to data distribution. This is because, due to the nature of the error function, it can be used in any modulation exhibiting symmetry in all four quadrants.
(2) The theoretically calculated phase detector characteristic for QAM modulation reveals that there is no false lock point.
(3) It is suitable for low-complexity implementation.

However, in DMR applications the frequency offset between the transmitter and the receiver can take large values. On the other hand, if the acquisition range of the CR loop is increased, the phase jitter inherent in the recovery process will also be increased. For this reason, a frequency sweeping subsystem is used to aid the loop by bringing the initial frequency offset within the limits of the capture range of the Costas loop. A detailed presentation of the overall CR system including the frequency sweeping subsystem will be given in Section 6.2.5.

6.2.4 Simulation Results

Simulations were carried out to evaluate the operation of the most critical blocks of the receiver, the TR and the CR systems.

6.2.4.1 Timing Recovery System

The TR system was simulated with x_k and a_k representing the received symbol and the corresponding MLSE decoded one at sampling instant k, respectively. Initially, there is perfect

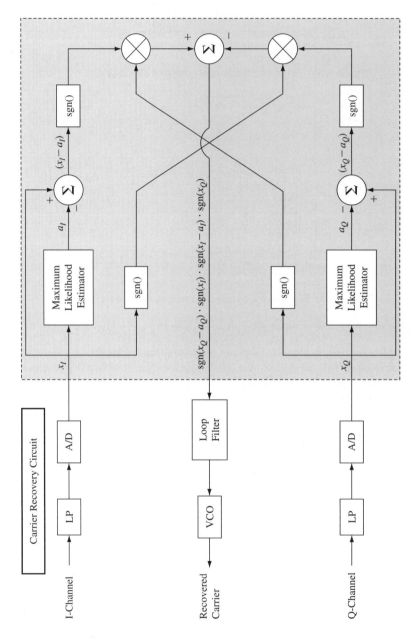

Figure 6.32 Carrier recovery algorithm (in shaded box)

404

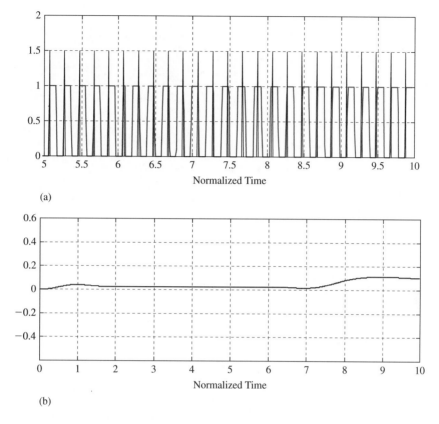

Figure 6.33   The convergence of timing recovery system

synchronization, and after some time a random offset at the sampling clock is introduced. Figure 6.33 shows the result taken after a clock offset is introduced at $t=6.1$ (normalized time). In Figure 6.33(a) the pulse stream represents the clock periodic signal whereas the spike corresponds to the sampling instant. One can see the different position of the spike from $t=6.1$, when the offset is applied, until $t=8.5$ when the sampling instant has again converged to the right value (at the rising edge of the clock). The same result can be verified by Figure 6.33(b), where one can see the dynamic response of the system. There we can observe that, although the offset is applied at $t=6.1$, the loop starts responding at $t=7.2$ and sampling time reaches it steady-state correct value at $t=8.5$ when the line becomes again horizontal.

6.2.4.2 Carrier Recovery System

In a similar manner, a carrier offset is applied on the receiver LO (with respect to the transmitter carrier frequency). Figure 6.34 demonstrates the resulting eye diagram and the constellation mapping when the CR algorithm described above is employed. It can be seen that eye closure is very slight, representing a very low degradation. More specifically, the resulting (from simulations) rms value of the phase jitter due to the application of the algorithm is found to be equal to 0.57° corresponding to an SNR degradation of approximately 0.5 dB for 64-QAM

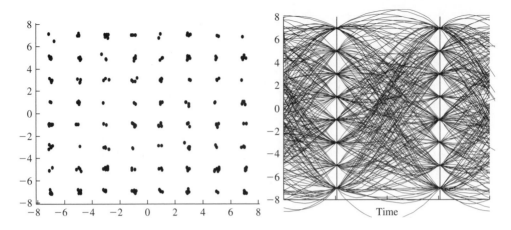

Figure 6.34 Constellation and eye diagram showing the effect of the CR system

modulation [DMR90]. This figure represents the implementation loss of the CR system on the overall receiver.

6.2.5 Digital Modem Implementation

Analysis and design of QAM modems has matured over more than 20 years in the system design and algorithmic development aspects. However, it is only during the 1990s that implementation efficiency and complexity were looked at more systematically. Until the mid 1980s QAM modems were implemented using mostly Ics and discrete components and for this reason they were bulky and power-consuming. When QAM modulation started appearing in almost every wireless communication application, and with the advent of integrated technology. QAM modems can achieve a high degree of integration. Therefore we see that techniques based on LUTs and direct mapping [Welborn00] are very efficient, mainly in the transmitter side, and can help put everything on a single FPGA. In addition, ASIC design using recent CMOS integration [Tan98], [D'Luna99] technologies has produced highly integrated modems with low power consumption.

We follow an intermediate approach in the implementation described in the following section: FPGAs are used to implement the digital transceiver. However, analogue ICs are used for the IF part connecting the radio subsystem and the digital modem. In this way there is the flexibility of evaluating in practice various algorithms on the digital part and choose the best performing ones. Furthermore, the chosen algorithms have low complexity and are suitable for ASIC integration, as will be seen.

6.2.5.1 Transmitter Implementation

Figure 6.35 shows a functional diagram of the digital part of the transmitter which, as mentioned above, implements the 64-QAM mapper and the digital filtering. Serial data enter the system at a rate of 61.44 Mb/s and are converted to parallel 6-bit worlds. This is done using a serial-to-parallel block. The 6-bit words correspond to 64-QAM symbols and are transformed to two

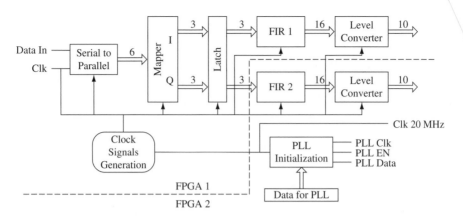

Figure 6.35 Detailed diagram of the digital part of the transmitter implemented in FPGA

worlds of 3 bits each which represent the coordinates (I and Q) of the constellation point. This transformation takes place in the 'mapper' block, which uses logic equations to produce the 64 values of the I and Q components for a particular constellation point. The 'mapper' block uses a 10.24 MHz clock corresponding to the symbol rate. The 3-bit I and Q streams are clocked at 20.48 MHz (twice the symbol rate) through a latch and feed the two FIRs at this rate. This is done in order to operate the FIRs with oversampling ratio equal to 2 such that aliasing distortion can be easily removed because it is centred at $3f_S$. Each FIR implements the equation:

$$y_n = \sum_{i=0}^{M-1} x_{n-i} \cdot b_i \tag{6.24}$$

where x_i is the input at time moment i and b_i is the ith filter coefficient. M is the overall number of coefficients. In our implementation $M = 17$ and internal FIR accuracy $= 16$ bits [Chorevas96]. Following the FIRs there are level converters to transform the two-complement 16-bit words produced by the filters to a 10-bit word which in turn feeds the D/A converters. In addition, as noted in the figure, there is a 'clock signal generator' block, which uses the 61.44 input clock to generate all other clocks (10.24 and 20.48 MHz). Finally, there is a 'PLL initialization' block, which is used to create the necessary signals for programming the PLL.

The 10-bit outputs from the FIRs are used as inputs at the two D/A converters. Figure 6.36 illustrates the D/A conversion subsystem which, apart from the DACs, contains Op-amps to transform the current output of the D/A converters to voltage and precision low-offset voltage Op-amps such that the digital to analogue conversion is accurate and is not affected by offset voltages.

The produced analogue signals in the I and Q paths go through lowpass filtering to achieve the necessary rejection in order to satisfy the spectrum mask. Apart from the ampliturde response, the filters must not exceed a group delay distortion of 20 ns within the passband. This is necessary in order to pass the signal pulses with minimum distortion in time. Commercial passive filters are used and the measured (through a Network analyser) amplitude and group delay responses are given in Figure 6.37(a) and (b) respectively. In the first figure, the dotted

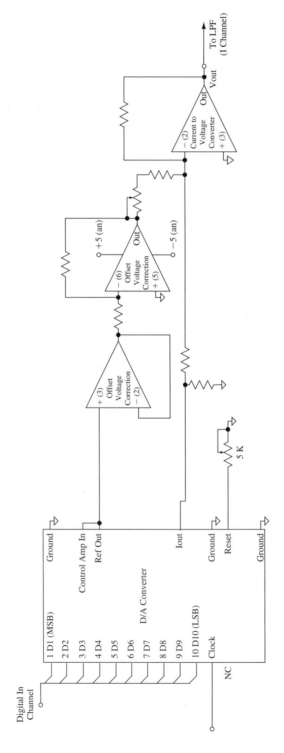

Figure 6.36 D/A conversion subsystem

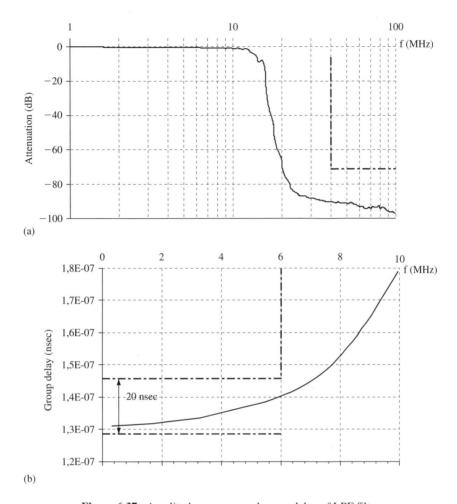

Figure 6.37 Amplitude response and group delay of LPF filters

line represents the system requirements in terms or rejection. From these figures it can be seen that the filters satisfy both the attenuation and group delay specifications.

Subsequently, the two analogue I and Q signals enter the I/Q modulator/upconverter, which driven by the LO will give a composite output signal at 140 MHz. An additional feature of the I–Q modulator is a tunable phase shifter. To satisfy the specifications of Section 6.2.2 in sideband rejection, carrier rejection and maximum allowable DC offset the chosen circuit has the ability to tune the phase difference between the two branches of the feeding LO signals in order to achieve the 90° phase difference.

The LO consists of a programmable synthesizer IC and external voltage controlled oscillator (VCO) along with the necessary LPF at the output of the phase detector. Figure 6.38 illustrates a detailed diagram of the synthesizer–LO system. The main structural elements are the programmable PLL IC, an active filter connected at the output of the phase detector (PD) of the PLL and an integrated VCO (located at the bottom of the figure). The PLL IC, apart from the

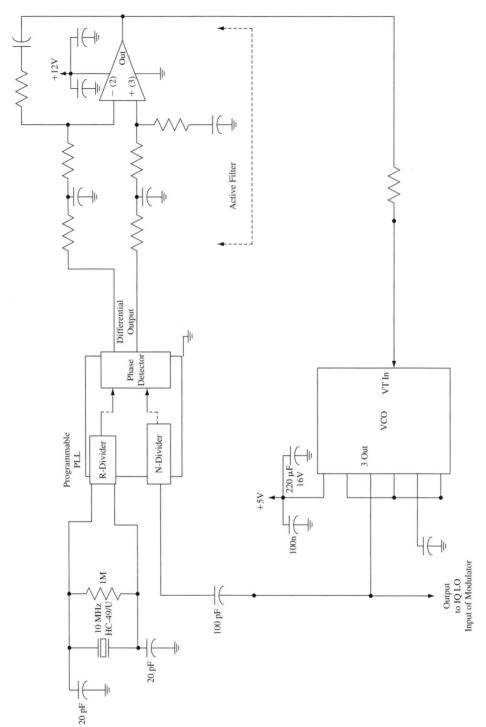

Figure 6.38 LO–synthesizer subsystem

PD, contains two programmable dividers, R and N. R is used to divide the reference frequency and N is used to divide the VCO output frequency before it is applied at the second input of the PD. One can see that, in order to have an active filter with response close to a perfect integrator, the differential output of the PD is used to drive the filter-amplifier at the top right side of the figure. Finally, it must be noted that the PLL is programmed through the transmitted FPGA using the 'PLL initialization' block as described in the beginning of this section.

6.2.5.2 Receiver Implementation

Figure 6.39 shows the prototype of the implemented receiver where the details of all subsystems are now illustrated. As explained in the design section, there is an analogue front-end consisting of the I–Q demodulator, LPFs and the DC offset cancellation/AGC block. In addition, a VCO and an LPF in front of it is the analogue part of the carrier recovery subsystem (bottom part of the figure) whereas there is an analogue section for the timing recovery subsystem, which is responsible for generating the initial clock and will be discussed in detail later. In addition, the timing recovery system includes three phase-locked loops to interface the analogue part with the algorithmic part of the system and produce the clocks and the sampling pulses for proper timing.

Hence, the overall implemented receiver is a hybrid system constituting from an analogue front-end, digital ICs mostly used as interfaces (A/D converters for digital conversion and PLLs for the TR subsystem) and a digital programmable part (in the form of an FPGA), which implements all digital functions of the receiver. Below, we will describe briefly the implementation of the DCOC/AGC and TR and CR blocks, which deserve some special attention.

6.2.5.3 DC Offset Cancellation and AGC

Implementation details of this subsystem are depicted in Figure 6.40. The main element of this subsystem is a high-frequency comparator, which at every time instant compares the amplitude of the input signal (I or Q) to the voltage across a capacitor. If the voltage across the capacitor is lower than the signal amplitude (applied at the other input of the comparator), the comparator activates a tri-state buffer, which acts as a charge pump and increases the capacitor voltage. In this way, the voltage across the capacitor becomes equal to the peak value of the input signal (I or Q). The second part of the circuit consists of an analogue adder used to add to the instantaneous value of the signal the peak value from the capacitor. In this way the DC value of the signal is corrected. The resulting corrected signal gets amplified and is applied to the A/D converter, while its peak value (from the capacitor) is used as the reference voltage of the A/D converter.

6.2.5.4 Timing Recovery System

Figure 6.41 shows the implementation of the TR system. The first block uses a nonlinearity (diode rectifiers) along with buffers and filtering to produce a spectral line at the symbol rate 10.24 MHz. The recovered symbol clock is then used as an input to a PLL (denoted as PLL1) to produce the master clock of 61. 44 MHz which is used to clock the FPGA . This can be seen in both Figures 6.41 and 6.39, where it can be seen that PLL2 and PLL3 can use the 10.24 MHz clock and the output from the timing recovery algorithm as inputs to produce the recovered sampling clock, which will be used as the clock input to sample the two signal streams (I and Q)

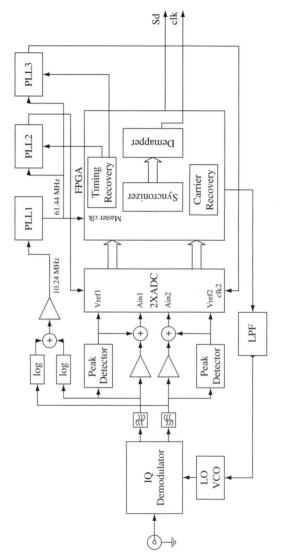

Figure 6.39 The overall implemented receiver

412

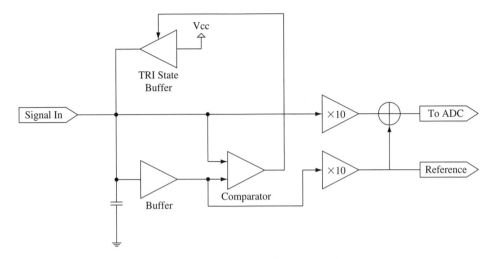

Figure 6.40   DC offset cancellation and AGC subsystem

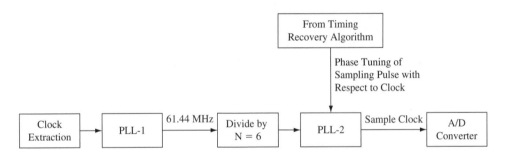

Figure 6.41   Implementation of TR subsystem

at the A/D converters. The TR algorithm control input to PLL2 and PLL3 is produced by a circuit shown in Figure 6.29, which is implemented in the FPGA.

6.2.5.5 Carrier Recovery System

The implemented carrier recovery system is illustrated in Figure 6.42 and can be divided into two subsystems, one at the left of point O and one at the right of it as indicated by the dashed line. The subsystem at the left of point O is the implementation of the CR algorithm as it was presented in Section 6.2.3.3 above. The inputs XQ7 and XI4 of the first XOR gate implement the functions sgn x_Q and sgn$(x_I - a_I)$ of the first term of the phase detector function as given in Section 6.2.3.3. The same is true for the inputs XI7 and XQ4, which implement the second term of the PD equation containing the terms sgn x_I and sgn$(x_Q - a_Q)$ respectively. The products are implemented by the XOR gates the outputs of which enter a look-up-table, which is used to transform the composite Pd function into a 1-bit output. The LUT output can take one of the three states. A 'high', which is used to produce a positive control increment, a 'low', which is used to give a negative control increment, and a Z-state which does not produce any change

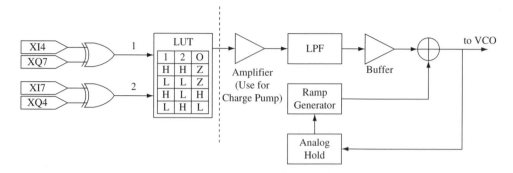

Figure 6.42 Implementation of CR subsystem

at the output. To put it more concisely, the subsystem at the left of point O implements the Leclert–Vandamme digital phase detector with a three-state FSM using only XOR gates and a LUT. This subsystem is implemented into the FPGA.

On the right side of point O we have two paths producing analogue voltages, which are added to obtain the overall control voltage of the VCO. The upper part (connected to point O) consists of an amplifier, a lowpass filter and a buffer which constitute the standard components needed to interface a digital PD to the analogue control voltage of the VCO in a carrier recovery PLL. The lower part is a pull-in system creating a sawtooth analogue voltage. This helps the VCO frequency to get close to the transmitter carrier frequency such that the resulting frequency difference is within the PLL locking range. At this point, the ramp 'freezes' (by using an analogue comparator) and the CR algorithm is activated to lock the PLL at the desired frequency and phase.

6.2.5.6 Digital Part of the Receiver Implemented on FPGA

Figure 6.43 shows a detailed diagram of the logic units built in the FPGA and their interconnection. The FPGA accepts as input the digitized versions of I and Q channels from the two A/D converters. Furthermore, it accepts a master-clock input produced by a PLL driven by the analogue part of the timing recovery subsystem. The receiver FPGA produces five output signals, as can be seen in Figure 6.39. Two of them correspond to the outputs of the two separate timing recovery algorithms. We must emphasize that the two channels I and Q, implement separately the timing recovery operation. Consequently, these two signals represent the timing error in channels I and Q respectively and drive the timing synchronization PLLs which produce the clocks for the I and Q channels. These clocks are used as 'clock' inputs for the I and Q A/D converters. A third signal comes out from the carrier recovery block of the FPGA and drives the analogue part of the CR subsystem as it has been discussed in Section 6.2.5.3. Finally, the demodulated serial data and a clock signal are the last two output signals of the FPGA.

Below we present briefly the operation of the four functional blocks of the receiver FPGA:

(1) *TR block* – there are two TR blocks (I and Q) as explained above. As seen in the figure, each one consists of two ROMs, two latches and an adder. The first ROM takes as input the first term of Equation (6.20), which is repeated below for convenience:

$$D = C_1 \cdot a_k \cdot x_{k-1} - C_1 \cdot a_{k-1} \cdot x_k + C_2 \cdot \left(1 - \frac{x_k}{a_k}\right) \qquad (6.20)$$

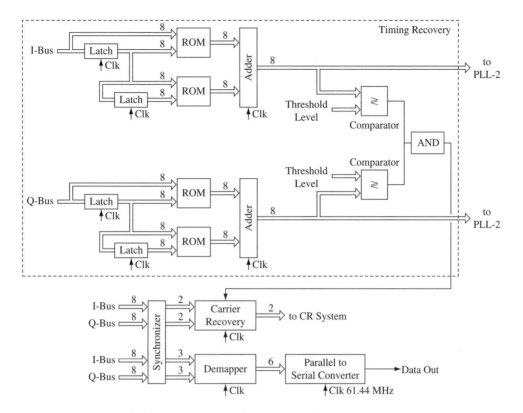

Figure 6.43   Detailed diagram of the digital part of the receiver implemented in FPGA

(2) Using x_{k-1} and a function implemented inside the ROM, it then produces in its output the first of the three terms of function D above. In a similar way, the second ROM implements both the second and third term of Equation (6.20) and produces at its output the respective terms. Subsecquently the results of the two ROMs are added to produce D, which will be used to drive the TR subsystem PLLs (PLL2 and PLL3) in order to correct the timing instant at which the A/D should sample the analogue data.

(3) *CR block* – as mentioned in the previous section, the CR block is implemented using two XOR gates and a look-up-table. The output resembles a three-state FSM which can give a logic high level (H), a logic low level (L) and a high impedance state (Z). This three-state digital signal drives through an LPF the VCO of the CR PLL to correct for the carrier phase.

(4) *Synchronizer* – this block is used to synchronize the I and Q streams before they enter the demapper. It contains two FIFOs where the data from the I and Q A/D converters are stored. The input of each FIFO is clocked by the sampling clock of each channel (I and Q) whereas the outputs are clocked by the common symbol clock (which runs at 1/6 the sampling clock speed).

(5) *De-mapper* – in the de-mapper the inputs are the I and Q streams (3-bit words) coming from the synchronizer. This block addresses a LUT which transforms 3-bit words to one of the eight possible levels the I and Q signals can assume. Then, depending on the combination, it decides which of the 64 constellation points has been transmitted.

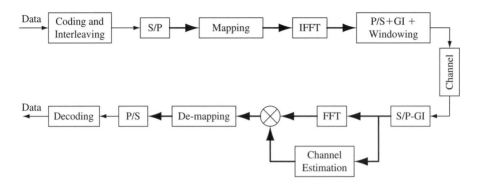

Figure 6.44 General block diagram of OFDM transceiver

6.3 OFDM Transceiver System Design

6.3.1 Introduction

As mentioned in Chapter 5, OFDM transmission techniques can be used in broadband applications because the division of the transmission bandwidth in subchannels with flat frequency response provides effective and computationally simple ways to combat dispersive environments and achieve good performance in high transmission rates.

In this section we outline the design of an OFDM modem for Hiperlan/2-802.11a type of WLANs. ETSI and IEEE have developed Hiperlan/2 (H/2) and IEEE 802.11a respectively to address low-mobility, high-bandwidth applications such as WLANs. For this purpose, they specified physical layers (PhyL) with high degree of similarity in the standards [H/2-standard1], [802.11a-standard]. The main difference between the two standards lies in the MAC layer mechanisms. Consequently both standards can support OFDM transmission with seven modes of operation corresponding to different transmission rates, modulation and coding. In addition, they both use 64 subcarriers, the same channel bandwidth (20 MHz) and the same value for the cyclic extension.

Both of the above-mentioned standards target low-mobility indoor WLAN applications to achieve high-speed, high-performance transmission within 20 MHz of information bandwidth. The block diagram in Figure 6.44 illustrates the structure of the digital transceiver as indicated by the standards. The standard defines forward error correction, interleaving and puncturing for coding adjustment. Then QAM mapping and pilot insertion blocks deliver the data to a serial-to-parallel converter which feeds the data to the IFFT block. Through the IFFT and a parallel-to-serial transformation the data are transformed to the time-domain. Subsequently a cyclic extension is added in front of the symbol to guard the data against multipath and, after windowing, the OFDM symbol is upconverted for radio transmission. The opposite processing takes place at the receiver. In addition, the receiver must include blocks for equalization and time- and frequency-synchronization. Table 6.6 shows the main transmission parameters for both standards:

In general, the number of subcarriers in an OFDM system is chosen such that the complex attenuation in every subchannel can be considered to have a constant value throughout the bandwidth of the subchannel. The duration of an OFDM symbol is chosen such that the attenuation of each subchannel can be considered constant throughout (at least) one OFDM symbol.

Table 6.6 Basic characteristics of OFDM transmission for Hiperlan/2

Parameter	Value
Sampling rate $f_S = 1/T$	20 Msamples/s
Duration of useful data symbol T_u	64/20 MHz = 3.2 μs
Duration of cyclic extension T_g	16/20 MHz = 800 ns (8/20 MHz = 400 ns)
Overall symbol duration $T_S = T_g + T_u$	80/20 MHz = 4 μs
Number of subcarriers containing data	48
Number of subcarriers containing pilots	4
Overall number of subcarriers	64
Frequency distance between the subcarriers $1/T_u$	312.5 kHz
Total bandwidth	20 MHz
Used bandwidth	312.5 kHz × (48 + 4) = 16.250 MHz

In the case of indoor high-speed wireless LAN applications the number of subcarriers is chosen to be 64 whereas the band occupied by each subcarrier spans 312.5 kHz. A guard interval equal to 800 ns is added to the symbol as a shield to protect the symbol from ISI due to multipath fading. For this reason its duration must be equal to the time of the last arriving reflected ray at the receiver and should be close to the maximum delay of the channel. The guard interval is chosen to be a cyclic extension of the symbol (and not filled with zero samples) for two reasons: to keep the transmitted spectrum without distortion and to ease the requirements for precision in timing synchronization.

Forty-eight out of 64 channels are used for transmission of data. There are four pilot channels with fixed values to use for channel estimation and synchronization, and the remaining 12 subcarriers loaded with zeros for reasons of spectral shaping at the edges of the transmitted spectrum.

We will outline the design of an H/2 direct conversion transceiver and will elaborate on the most critical blocks of the digital modem. In addition, we will consider the main aspects of the radio front-end subsystem, which affect the overall system performance.

Although, the IFFT–FFT operations are the heart of the OFDM transmission technique, they constitute relatively standard building blocks, which can be optimized in different ways to increase speed of operation. However, channel estimation and synchronization (time and frequency) subsystems can be implemented in different ways depending on accuracy, speed, complexity and performance criteria, and therefore they characterize the identity of the digital modem. For this reason, we will concentrate in the design of these processing blocks for the application at hand.

As the designed system address H/2 we will briefly present some technical characteristics of the physical layer that can be exploited in order to give efficient low-complexity modem design.

6.3.1.1 Hiperlan/2 Physical Layer and Physical Bursts

To support the various functionalities as proposed by the standard, H/2 defines logic channels, which are transmitted through physical bursts (PhyB). Five different types of physical bursts are defined in H/2 standard. They are: *the broadcast PhyB*, which is used to transmit information

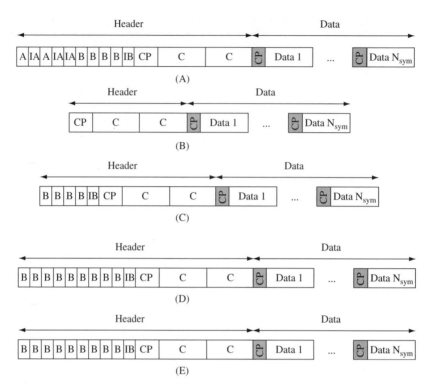

Figure 6.45   Hiperlan/2 physical bursts

from the access point (AP) to all mobile terminals (MT); *the downlink PyhB*, which is used to transmit information from AP to one MT; *the short uplink PhyB* and the *long uplink PhyB*, which are used to transmit information from an MT to AP; and *the direct link PhyB*, which is used in special cases for direct conection between two MTs without the intervention of an AP. Their structures are depicted in Figure 6.45. As seen from their structures, all physical bursts contain a preamble and a data field. The data field consists of a sequence of long transport channel (LCH) or short transport channel (SCH) data constituting protocol data units (PDUs).

The physical layer defines seven modes of operation, each of which combine a specific modulation and coding schemes. This results in different transmission speeds, as in Table 6.7.

6.3.2 Channel Estimation in Hiperlan/2

As explained in the previous chapter, the estimated complex response of the radio channel $\hat{H}_{l,k}$ for the lth OFDM symbol at subcarrier k takes the general form $\hat{H}_{l,k} = H_{l,k} + n^H_{l,k}$, where the noise term $n^H_{l,k}$ has the characteristics of white noise. Regarding channel statistics, we consider a wide-sense stationary uncorrelated scattering channel [Hoher92] as described in a previous chapter, for which autocorrelation functions $R_f(\Delta f)$ and $R_t(\Delta t)$ can be easily calculated.

As presented in Chapter 5, the LS estimation is the simplest way to perform channel estimation and is based on the transmission of training symbols $a_{l,k}$ (lth symbol, kth subcarrier)

Table 6.7 Modes of operation for the Physical layer of Hiperlan/2

Mode	Modulation	Code rate	Gross bit rate	PHY bit rate	Bytes / OFDM symbol
1	BPSK	1/2	12 Mb/s	6 Mb/s	3.0
2	BPSK	3/4	12 Mb/s	9 Mb/s	4.5
3	QPSK	1/2	24 Mb/s	12 Mb/s	6.0
4	QPSK	3/4	24 Mb/s	18 Mb/s	9.0
5	16-QAM	9/16	48 Mb/s	27 Mb/s	13.5
6	16-QAM	3/4	48 Mb/s	36 Mb/s	18.0
7	64-QAM	3/4	72 Mb/s	54 Mb/s	27.0

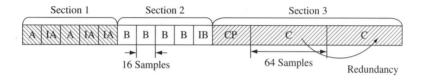

Figure 6.46 Header of the broadcast physical burst

containing known complex data for all subcarriers in the beginning of each physical packet. In this way, the channel response is obtained by dividing the detected data symbols $\hat{z}_{l,k}$ by the known training symbols $a_{l,k}$ as shown below:

$$\hat{H}_{l,k,LS} = \frac{\hat{z}_{l,k}}{a_{l,k}} = H_{l,k} + \frac{n_{l,k}}{a_{l,k}} = H_{l,k} + n'_{l,k} \tag{6.25}$$

The resulting $\hat{H}_{l,k,LS}$ is used for equalization of all symbols of the current physical packet because, as noted before, we consider that the channel response is constant over the duration of a physical packet. From Figure 6.45 we can see the structure of the broadcast physical burst in $H/2$. Its header consists of three sections and is depicted in Figure 6.46. Section 1 contains pilots at positions $\pm 2, \pm 6, \pm 10, \pm 14, \pm 18$ and ± 22 whereas section 2 also contains pilots in different locations ($\pm 4, \pm 8, \pm 12, \pm 16, \pm 20$ and ± 24). Section 3 consists of two symbols transmitting energy in all positions except those at the edges of the band (-32 to -27, 27 to 31) and at DC.

Channel estimation can be performed using section 3 of the header in two ways:

(1) *Averaged-LS.* Averaged least square estimation uses the values $a_{l1,k}, a_{l2,k}$ of the two symbols $l1$ and $l2$ of section 3 of the header in order to estimate an average channel response:

$$\hat{H}_{k,LS} = \frac{\hat{H}_{l1,k,LS} + \hat{H}_{l2,k,LS}}{2} \tag{6.26}$$

Because noise in uncorrelated with zero mean, noise power in the ALS estimation is reduced by 3 dB compared with the LS estimation using only one symbol. Figure 6.47 gives BER simulated results for 64 QAM without coding for Hiperlan/2. For comparison purposes, it includes ideal estimation, LS estimation (no averaging) and ALS estimation (averaged LS). Substantial improvement can be seen comparing simple LS (no averaging)

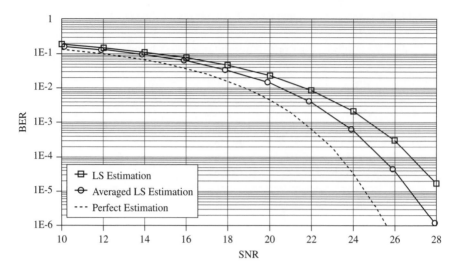

Figure 6.47 BER performance for mode-7 in H/2 for different channel estimation

and ALS (averaging using two pilot symbols) estimation. Complexity is only marginally increased by an addition per subcarrier coefficient.

(2) *LMMSE estimation.* On the other hand, further improvement can be achieved if channel statistics is taken into consideration and LMMSE channel estimation can be used according to the following equation [Edfors98]:

$$\hat{\mathbf{H}}_{\text{LMMSE}} = \mathbf{W} \cdot \hat{\mathbf{H}}_{k,\text{LS}} = \mathbf{R}_{HH} \cdot \left(\mathbf{R}_{HH} + \frac{\beta}{\text{SNR}}\mathbf{I} \right)^{-1} \cdot \hat{\mathbf{H}}_{k,\text{LS}} \tag{6.27}$$

The OFDM symbol index l has been omitted for simplicity. $\mathbf{W}$ is usually called the LMMSE weight matrix. $\mathbf{I}$ represents the identity matrix, $\mathbf{R}_{HH}$ is the channel autocorrelation matrix and $\beta = 1$ in case all pilots have the same power. The first two terms in Equation (6.27) can be calculated off-line and only $\hat{\mathbf{H}}_{k,LS}$ is estimated in real time.

According to the H/2 specifications the mobile speed can reach 3 m/s, corresponding to a maximum Doppler frequency equal to $f_{D,\max} = 57$ Hz for a carrier frequency of 5.7 GHz. Using the mathematical expressions for autocorrelation as presented in Chapter 5, it can be easily deduced that, due to the low Doppler, the autocorrelation function varies very slowly with time. After some mathematical manipulations we get for the autocorrelation matrix:

$$R_{HH} = [E\{h_{k,l}h^*_{k',1'}\}] = r_t(l-1)R_{hh}$$

$$= r_t(l-1) \begin{bmatrix} r_f(0) & r_f^*(1) & \cdots & r_f^*(25) & r_f^*(27) & r_f^*(28) & \cdots & r_f^*(52) \\ r_f(1) & r_f(0) & \cdots & r_f^*(24) & r_f(26) & r_f^*(27) & \cdots & r_f^*(51) \\ \cdots & \cdots & \cdots & \cdots & \cdots & \cdots & \cdots & \cdots \\ r_f(25) & r_f(24) & \cdots & r_f(0) & r_f(2) & r_f(3) & \cdots & r_f(27) \\ r_f^*(27) & r_f^*(26) & \cdots & r_f^*(2) & r_f(0) & r_f^*(1) & \cdots & r_f^*(25) \\ r_f^*(28) & r_f^*(27) & \cdots & r_f^*(3) & r_f(1) & r_f(0) & \cdots & r_f^*(24) \\ \cdots & \cdots & \cdots & \cdots & \cdots & \cdots & \cdots & \cdots \\ r_f^*(52) & r_f^*(51) & \cdots & r_f^*(27) & r_f(25) & r_f(24) & \cdots & r_f(0) \end{bmatrix} \tag{6.28a}$$

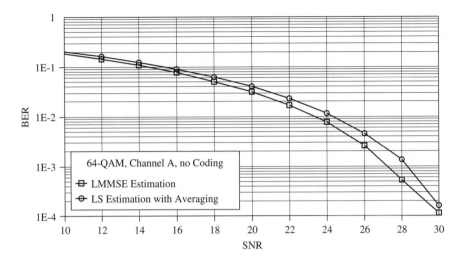

Figure 6.48 BER performance of H/2 for 64-QAM, ETSI-BRAN channel A

where $r_f(k)$ is given by the following equation:

$$r_f(k) = \frac{1 - e^{-\tau_{max}\left(\frac{1}{\tau_{rms}} + j2\pi\frac{k}{T_u}\right)}}{\left(1 - e^{-\frac{\tau_{max}}{\tau_{rms}}}\right)\left(1 + j2\pi\frac{k}{T_u} \cdot \tau_{rms}\right)} \tag{6.28b}$$

Figure 6.48 presents the BER performance for 64-QAM, channel BRAN-A with no coding for both averaged-LS and LMMSE. It is seen that the resulting improvement from using LMMSE is below 1 dB. However, complexity is increased considerably. A 52×1 vector representing the LS estimation $\hat{H}_{k,LS}$ is multiplied by the 52×52 weight matrix involving 2704 complex multiplications and 2652 complex additions per estimation.

6.3.2.1 Singular Value Decomposition

As presented in Chapter 5, the LMMSE channel estimation can be expressed as:

$$\hat{\mathbf{H}}_{LMMSE} = \mathbf{U}\mathbf{\Delta}\mathbf{U}^H\hat{\mathbf{H}}_{LS} \tag{6.29}$$

where $\mathbf{U}$ is an ortho-normal matrix containing the eigenvectors of $\mathbf{R}_{HH}$ and $\mathbf{\Delta}$ is a diagonal matrix containing the eigenvalues $\lambda_0 \geq \lambda_1 \geq \lambda_2 \geq \cdots \geq \lambda_{N-1} \geq 0$ along its diagonal.

$$\mathbf{\Delta} = \begin{vmatrix} \dfrac{\lambda_0}{\lambda_0 + \frac{\beta}{SNR}} & 0 & \cdots & 0 \\[3ex] 0 & \dfrac{\lambda_1}{\lambda_1 + \frac{\beta}{SNR}} & \cdots & 0 \\[3ex] \vdots & \vdots & \ddots & \vdots \\[3ex] 0 & 0 & \cdots & \dfrac{\lambda_{N-1}}{\lambda_{N-1} + \frac{\beta}{SNR}} \end{vmatrix} \tag{6.30}$$

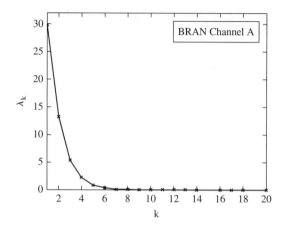

Figure 6.49   Eigenvalues for ETSI-BRAN channel A

As a result, after some manipulation a reduced-order p approximation $\hat{\mathbf{H}}_{p,\mathrm{LMMSE}}$ of $\mathbf{H}_{\mathrm{LMMSE}}$ can be estimated as follows:

$$\hat{\mathbf{H}}_{p,\mathrm{LMMSE}} = \mathbf{U} \begin{vmatrix} \boldsymbol{\Delta}_p & 0 \\ 0 & 0 \end{vmatrix} \mathbf{U}^H \hat{\mathbf{H}}_{\mathrm{LS}} \tag{6.31}$$

where $\boldsymbol{\Delta}_p$ represents the upper-left part of matrix $\boldsymbol{\Delta}$ with elements $\delta_0, \delta_1, \ldots, \delta_{p-1}$ along its diagonal.

If K represents the number of estimated coefficients and p represents the reduced order of estimation, then channel estimation using the SVD method requires $2pK$ complex multiplications and $2p(K-1)$ complex additions.

In our applications $K = 52$. To determine the reduced order p, we calculate and plot the eigenvalues for the case of ETSI-BRAN channel A. Figure 6.49 indicates that in this case the reduced order p is equal to 10, since eigenvalues are very small beyond this point [Miaoudakis04].

With SVD the resulting complexity for our applications is $2 \times 10 \times 52$ complex multiplications and $2 \times 10 \times 51$ complex additions. However, even such a considerable reduction is not satisfactory to substitute the robust and simple averaged-LS with SVD LMMSE. This is due, as mentioned before, to the marginal (below 1 dB) performance improvement achieved when LMMSE is used instead of average LS.

Finally, we briefly examine whether two-dimensional channel estimation techniques are necessary or, equivalently, how fast the autocorrelation function changes. As expected, due to a radio channel which varies with time, the autocorrelation function also varies with time. However, for H/2 the variation of autocorrelation function is very slow, not only with respect to the duration of an OFDM symbol, but even when compared with the length of a physical packet. Figure 6.50 gives the BER performance as a function of the length of a physical packet. Results for ETSI-BRAN channels A and E, 64-QAM modulation and Doppler frequency equal to 52 Hz are given. Channel estimation was performed only once in the beginning of the packet. It is observed that the performance does not deteriorate even for long packets, a fact that verifies the very slow change of autocorrelation with time.

Based on the above observations and results, we decided, for Hiperlan2, to use averaged-LS for channel estimation.

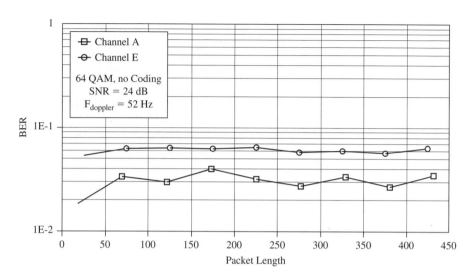

Figure 6.50   H/2 Physical layer performance as a function of the packet length

6.3.3 Timing Recovery

Timing synchronization techniques in OFDM modems can use either the cyclic extension of the OFDM symbol or pilots. As presented in Chapter 5, techniques based on cyclic extension calculate the autocorrelation of the received signal over the length of the cyclic extension. Therefore, to obtain satisfactory results the cyclic extension should have a considerable length N_g. This is not true for H/2 or 802.11a standards where $N_g = 16$.

For this reason we prefer to synchronize the receiver with the transmitter in time using pilots. We use two consecutive similar sections in the structure of the preamble [Schmidl97]. These are the two C blocks in section 3 of the preamble. In front of the two Cs there is a CP interval such that the overall length of this structure is $2D_S + D_g$, where D_S and D_g represent the duration of the C blocks and of the CP respectively.

If we denote as s_k the transmitted data symbols (after IDFT), the noiseless receiver data sample is:

$$\tilde{r}_k = r_{0k} \cdot e^{+j2\pi \Delta f_{\text{CFO}} kT} + n_k \tag{6.32}$$

where Δf_{CFO} denotes the carrier frequency offset between the transmitter an the receiver oscillators as presented in detail before; n_k represents AWGN with variance $\sigma_n^2 = N_0/T$, where N_0 is the power spectral density of the white noise.

An asymptotically optimum frame synchronization metric in white noise is [Muller-Weinfuntner98]:

$$M_k = \sum_{a=0}^{D_S-1} |r_{k+D_S+a}|^2 + \sum_{a=0}^{D_S-1} |r_{k+a}|^2 - 2|S_k| \tag{6.33}$$

where S_k represents the complex correlation:

$$S_k = \sum_{a=0}^{D_S-1} r_{k+a}^* \cdot r_{k+D_S+a} \tag{6.34}$$

The frame position can then be estimated using the following $\hat{k}$ for which a minimum is obtained:

$$\hat{k} = \arg\min_{k} (M_k) \tag{6.35}$$

Figure 6.51 demonstrates the peaks of the autocorrelation metric using the above technique on section 3 of the preamble. This is for BRAN channel A and SNR $= 10$ dB. The correlation produces clear peaks and the receiver is synchronized satisfactorily to the transmitter.

6.3.4 Frequency Offset Correction

Carrier frequency offset represents the difference in frequency between the local oscillators of the transmitter and the receiver. If we have a frequency difference equal to ΔF_C, the general equation describing the receiver signal is given by:

$$r(t_n) = \exp\left(j2\pi\Delta F_c n'T\right)\sum_i h_i(n'T) \cdot s(n'T - \tau_i) + n(n'T) \tag{6.36}$$

where $h_i(n'T)$ represents the impulse response at time instants $n'T$ where $n' = nT - N_g - lN_S$. In addition, the OFDM symbols have length $T_S = T_u + T_g$. T_u is the length of the data part of the OFDM symbol and T_g represents the guard interval length. Similarly, for numbers of samples we have $N_S = N_u + N_g$.

Finally $s(n'T - \tau_i)$ represents the sampled transmitted signal snapshots while τ_i is the delay of the ith tap of the multipath channel. Although there are various approaches to CFO estimation and correction, the ones performed in two steps (coarse and fine correction) can be the most effective. This is due to the fact that the coarse correction stage can be designed for broad range of operation while allowing the fine CFO correction mechanism to reduce considerably the mean square error of the estimated frequency at steady state.

If $\Delta\varphi_k$ and φ_k represent the phase increment from one OFDM symbol to the next and the subcarrier local frequency offset, we have the relations:

$$\frac{\Delta\varphi_k}{2\pi T_S} = \frac{\varphi_k}{T_u} = \Delta F_C \tag{6.37}$$

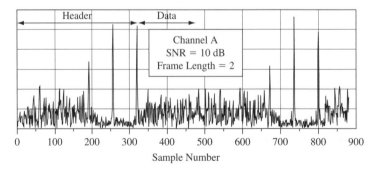

Figure 6.51 Matched filter output for H/2 ETSI-BRAN channel A

As presented in Chapter 5, successive OFDM symbols containing only pilots can be used to obtain a maximum likelihood estimation of the CFO given by [Moose94]:

$$\Delta\hat{F}_C = \frac{1}{2\pi T_u} \tan^{-1}\left[\frac{\sum_{k=-\frac{N}{2}}^{\frac{N}{2}+1} \text{Im}\left\{P_{l,k}P^*_{l-1,k}\right\}}{\sum_{k=-\frac{N}{2}}^{\frac{N}{2}+1} \text{Re}\left\{P_{l,k}P^*_{l-1,k}\right\}}\right] \tag{6.38}$$

Pilots located within data symbols can also be used in the same way to estimate CFO. According to the H/2 standard there are four pilots transmitted along with data symbols. Their values are produced by a pseudorandom noise sequence [H/2-standard1] such that the transmitted reference signal is randomized. This randomness has to be compensated for before ΔF_C estimation.

If we define as $P_{l,k}$ the transmitted pilots of OFDM symbol l at subcarrier k, estimation of CFO, ΔF_C is given as:

$$\Delta\hat{F}_C = \frac{1}{2\pi T_s} \tan^{-1}\left\{\frac{\sum_k \text{Im}\left\{P_{l+1,k}P^*_{l,k}\right\}}{\sum_k \text{Re}\left\{P_{l+1,k}P^*_{l,k}\right\}}\right\}, \quad k = -21, -7, +7, +21 \text{ f} \tag{6.39}$$

In Figure 6.52 the CFO estimation vs the actual CFO is shown for AWGN and SNR = 15 dB. From this we can see that the range of operation is $[-\Delta F_{SC}/2, \Delta F_{SC}/2]$, where ΔF_{SC} represents the subcarrier frequency spacing (for H/2 it is equal to 156.25 kHz). This is in agreement with the properties of this estimator as presented in the literature [Moose94]. In addition, there is a deterioration of the estimation at the edges of the operation region. This is due to the small number of pilots (four), resulting in degradation of the estimation in presence of AWGN. To increase the estimator immunity to noise, averaging must be performed.

Figure 6.53 shows the performance of the estimator in presence of AWGN with SNR = 15 dB while averaging of nine estimations (using 10 OFDM symbols) is applied [Miaoudakis02a]. From that figure the improvement resulting from averaging is evident. As mentioned before,

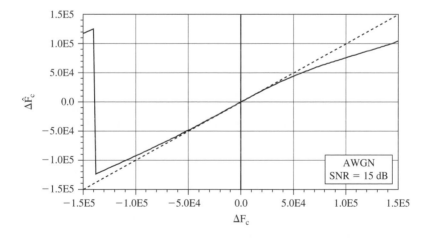

Figure 6.52 CFO estimation using Equation (6.38) for H/2 in AWGN

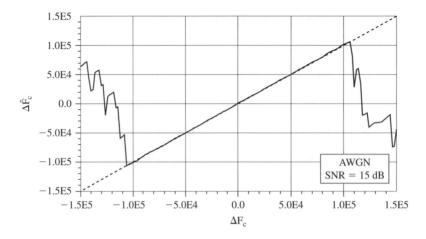

Figure 6.53 CFO estimation using Equation (6.38) and estimation averaging ($N = 10$)

pilots exhibit randomness in phase. This must be removed before estimation of ΔF_C takes place. A simple way to achieve it is to modify Equation (6.39) as follows:

$$\Delta \hat{F}_C = \frac{1}{2\pi T_S} \cdot \sum_k \left[\arg \left(P_{l+1,k} \cdot P_{i,k}^* \right) - \left\langle \arg \left(P_{i+1,k} \cdot P_{i,k}^* \right) > \frac{\pi}{2} \right\rangle \cdot \pi \right] \qquad (6.40)$$

In the above equation arg ($\bullet$) denotes argument and $\langle \bullet \rangle$ represent a logic operation returning 1 for true and 0 for false.

The phase of the product $P_{l+1,k} \cdot P_{l,k}^*$ is equal to $\Delta \varphi_k \pm \pi$ as a result of the random way pilots are generated. $\Delta \varphi_k$ represents the corresponding phase change due to CFO and from Equation (6.37) it is equal to:

$$\Delta \varphi_k = \Delta F_C \cdot 2\pi T_S \qquad (6.41)$$

If we assume that

$$\Delta \varphi_k \in \left[-\frac{\pi}{2}, \frac{\pi}{2} \right]$$

the unknown random fluctuation of $\pm \pi$ in the phase of the above product is cancelled out by subtracting the term $\langle \arg\{P_{l+1,k} \cdot P_{l,k}^*\} > \pi/2 \rangle \cdot \pi$. However, by doing this, the operating range of the estimator is reduced. This is demonstrated by the simulation results in Figure 6.54, which indicate that the region of operation has reduced to approximately 45 kHz.

From the above discussion it is obvious that CFO estimation cannot be based exclusively on pilots. The main reason is that the region of operation even with averaging is reduced to $\pm \Delta F_{SC}$. This is not satisfactory due to the frequency stability requirements imposed by the H/2 standard, which imply that CFO can take higher values. In addition, the small number of pilots demands the application of averaging in order to have satisfactory performance in AWGN.

In our application overall CFO estimation and correction are performed in two steps: first a coarse CFO correction takes place using the header of the H/2 broadcast physical burst, and after that fine CFO cancellation is performed using the four pilots transmitted along with data.

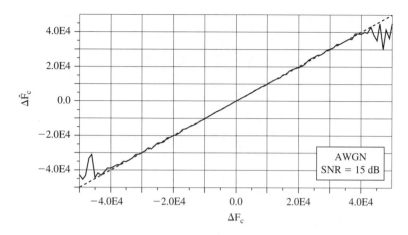

Figure 6.54 CFO estimation using Equation (6.40) and estimation averaging ($N = 10$) [reprinted from A. Miaoudakis *et al.*, 'Carrier frequency offset estimation and correction for Hiperlan/2 WLANs', Int. Conference on Computers and Communications (ISCC02), © 2002 IEEE]

6.3.4.1 Coarse CFO Correction

As noted in Section 6.3.2, there are five types of physical bursts through which all data transmission take place in H/2. Figure 6.46 depicts the structure of the header of the broadcast physical burst in H/2 as defined by the standard consisting of three sections. We note that section 2 (shaded) consists of five short OFDM symbols (B, B, B, B, IB) of specific values. Each of the first four symbols (B) constitutes a regular OFDM symbol consisting of 12 loaded subcarriers ±4, ±8, ±12, ±16, ±20 and ±24). This allows a 16-point DFT to be performed on each short symbol if we add four zero points, one at DC and the rest at the edge of the spectrum [Miaoudakis02b]. Figure 6.55 illustrates the sampled spectrum of the transmitted pilots along with the corresponding phases. The phase change between two successive samples of every nonzero subcarrier at the output of the DFT is proportional to the CFO and, as mentioned above, is expressed by:

$$\Delta\varphi = \Delta F_C \cdot 2\pi \cdot \frac{T_S}{4} \tag{6.42}$$

This approach exhibits three major advantages:

(1) The phase drift between two short symbols, due to the CFO, is four times lower than the corresponding phase drift between two successive OFDM data symbols. This is indicated by the above equation.
(2) The subcarrier spacing at each of the four short symbols is four times higher than the corresponding of a 64-point DFT performed on data symbols.
(3) The received sampled spectrum consists only of constant pilots, which are affected by the CFO in the same way.

The first item is important because the estimation of the CFO is always based on sampling the resultant phase drift in time. If this phase drift between two successive samples is higher than

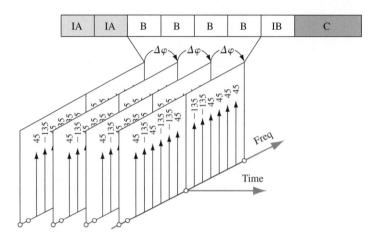

Figure 6.55 Sampled spectrum of section 2 of H/2 preamble [reprinted from A. Miaoudakis *et al.*, 'Carrier frequency offset estimation and correction for Hiperlan/2 WLANs', Int. Conference on Computers and Communications (ISCC02), © 2002 IEEE]

$\pm\pi$ the estimation is not correct. The increased value of subcarrier spacing is important because the limit of width of the operation region also increases. Finally, when the value of CFO is large, the resulting ICI disturbance of adjacent subcarriers is discarded as it is highly correlated. This is the reason why employing constant pilots improve estimator performance. From the above discussion it can be concluded that the expected operating range of the preamble-based estimation is extended by 4 times.

Denoting as b_k, $0 \leq k \leq 79$ the samples of section 2 of the header of the received signal and as $B_{m,n}$, $0 \leq n \leq 79$, $0 \leq m \leq 3$ the output of the DFT-16 for the m short preamble B, we have the relation:

$$B_{m,n} = \sum_{k=0}^{15} b_{16m+k} \cdot e^{-j2\pi n \frac{k}{16} T_S}, \quad 0 \leq m \leq 3 \tag{6.43}$$

Then, similar to Equation (6.39) (ML ΔF_C estimation [Moose94]) we have:

$$\Delta \hat{F}_C = \frac{1}{8\pi T_S} \cdot \frac{1}{3} \cdot \sum_{m=0}^{2} \tan^{-1} \left[\frac{\sum_{n=0}^{15} \text{Im} \left(B_{m+1,n} \cdot B_{m,n}^* \right)}{\sum_{n=0}^{15} \text{Re} \left(B_{m+1,n} \cdot B_{m,n}^* \right)} \right] \tag{6.44}$$

From the equation above, it is seen that the estimate is actually the mean of three successive CFO estimations. Figure 6.56 gives the performance of this estimator where we observe that the edges of estimation are ± 625 kHz. We must also note that, because the header subcarriers are constant, this estimator operates in linear fashion even for large CFO values. Indeed, when we impose ± 4 MHz CFO, operation is linear and is only limited by the phase-wrapping out of the interval $[-\pi/2, +\pi/2]$.

Because we use section 2 of the header (containing only pilots) and DFT is a linear transformation, an equivalent way to estimate CFO is given by:

$$\Delta \hat{F}_C = \frac{1}{8\pi T_S} \cdot \frac{1}{3} \cdot \sum_{m=0}^{2} \tan^{-1} \sum_{i=0}^{15} \left[b_{16m+i} \cdot b_{16(m+1)+i}^* \right] \tag{6.45}$$

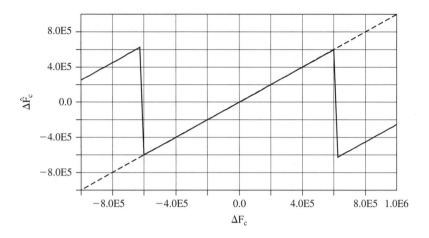

Figure 6.56   CFO estimation using Equation (6.44)

It is straightforward to modify the inner summation of Equation (6.45) as follows:

$$\Delta \hat{F}_C = \frac{1}{4\pi T_S} \tan^{-1} \left(\sum_{i=0}^{31} b_i \cdot b_{32+i}^* \right) \tag{6.46}$$

The root-mean-square-error (RMSE) for the two estimators is calculated from simulations as follows:

$$\text{RMSE} = \frac{1}{M} \sqrt{\sum_{i=0}^{M-1} \left(\frac{\Delta F_C - \Delta \hat{F}_{Ci}}{\Delta F_C} \right)^2} \tag{6.47}$$

where $\Delta \hat{F}_{Ci}$ represents the ith CFO estimation.

Simulation results are used to compute the CFO estimation. Figure 6.57 shows RMSE results for both estimators [Equations (6.44) and (6.45)]. It is demonstrated there that both estimators produce similar results (as expected).

As expected, the region of operation is half of the previous but this is not a problem as it satisfies the H/2 specifications. In addition, calculation of RMSE from simulated results gives somewhat higher values (as expected), but the computational complexity is significantly reduced.

Figure 6.58 shows results for the RMSE for Equations (6.45) and (6.46), where we can observe the small amount of deterioration. However, since we use a second stage for tracking (fine CFO correction), this degradation is not important.

An improvement could be achieved if a weight function matrix W_n is used in the above equation, which will compensate for the channel multipath effect. In that case CFO estimation takes the form:

$$\Delta \hat{F}_C = \frac{1}{8\pi T_S} \cdot \frac{1}{3} \cdot \sum_{m=0}^{2} \tan^{-1} \left[\mathbf{B}_n^H \cdot \mathbf{W}_{n\cdot} \cdot \mathbf{B}_{n+1} \right] \tag{6.48}$$

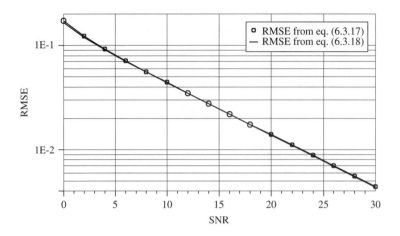

Figure 6.57   RMSE of CFO estimation from Equations (6.44) and (6.45)

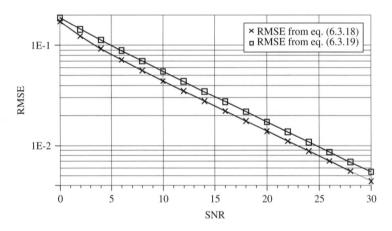

Figure 6.58   RMSE of CFO estimation from Equations (6.45) and (6.46) [reprinted from A. Miaoudakis *et al.*, 'Carrier frequency offset estimation and correction for Hiperlan/2 WLANs', Int. Conference on Computers and Communications (ISCC02), © 2002 IEEE]

where $[\bullet]^{H}$ represents Hermitian transpose, $\mathbf{W}_n$ is a 16×16 gain diagonal matrix and $\mathbf{B}_n$ is the short preamble column vector of size 16, as shown below.

$$
\mathbf{W}_n = \begin{bmatrix} w(0+16n) & 0 & 0 & \cdots & 0 \\ 0 & w(1+16n) & 0 & \cdots & 0 \\ 0 & 0 & w(2+16n) & \cdots & 0 \\ \vdots & \vdots & \vdots & \ddots & \vdots \\ 0 & 0 & 0 & \cdots & w(15+16n) \end{bmatrix}, \quad \mathbf{B}_n = \begin{bmatrix} b_{0+16n} \\ b_{1+16n} \\ b_{2+16n} \\ \vdots \\ b_{15+16n} \end{bmatrix}
$$
$$(6.49)$$

Figure 6.59 shows RMSE values vs CFO from simulated results for BRAN channel A with and without weight function for SNR values 10, 20 and 30 dB. For comparison purposes the RMSE when we have only white gaussian noise is also illustrated.

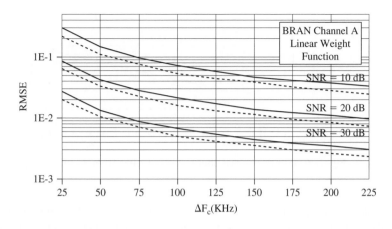

Figure 6.59   RMSE of approximate CFO estimation using linear weight function

It can be seen that only when SNR is large (30 dB) is there substantial improvement in RMSE. However, since a second stage is used for CFO fine correction, it is not important to obtain the lowest possible RMSE at the expense of increased complexity.

6.3.4.2 Fine CFO Correction (Tracking)

We have already indicated at the beginning of this section how pilots, transmitted within data, can be used for CFO estimation. In this section we demonstrate how fine CFO estimation and tracking can take place in the time domain. Because of the relationship between phase and frequency, a constant CFO will produce a linear excursion of the respective phase as follows:

$$\varphi_n = M \cdot n + A \tag{6.50}$$

where M is proportional to ΔF_C and A represents the initial phase difference between transmitted and received signal, which can be due to channel distortion and the random time of arrival of the received signal. A could be eliminated if we had ideal equalization but this is not usually the case.

Figure 6.60 shows an example of the correction mechanism. It contains a header and 10 OFDM symbols (indicated by D). Each symbol starts with a CE part (shaded region). Each symbol also contains four pilots, which are used to calculate an average (of four) for the phase distortion φ_n. The top part of the figure shows the linear relationship between φ_n and n. The circles represent the local estimated of φ_n for each OFDM symbol.

Therefore, the estimated phase (circles) can be used with interpolation to calculate the phase shift at any point in time. A modification of Equation (6.40) gives the estimated sequence of phase distortion Φ_k:

$$\Phi_k = \sum_{i=1}^{k} \sum_{n=0}^{3} \left[\arg\left(P_{i+1,n} \cdot P_{i,n}^*\right) - \left\langle \arg\left(P_{i+1,n} \cdot P_{i,n}^*\right) > \frac{\pi}{2} \right\rangle \cdot \pi \right] \tag{6.51}$$

where $P_{i,n}$ is the nth pilot ($n = 0$–3) of the kth OFDM symbol.

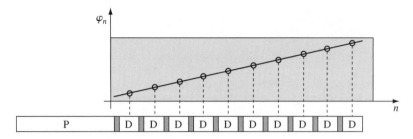

Figure 6.60 Linear phase shift and its sampling from the pilots

Sequence Φ_k constitutes an undersampling of the sequence φ_n. This sampling is actually represented by the circles in the above figure. Since φ_n is a linear function of time, interpolation can be used to to estimate it using Φ_k values as calculated from Equation (6.51). As a result we have the following equations for the estimation of quantities M and A:

$$\hat{M} = \frac{\begin{vmatrix} T_S \sum i\Phi_i & T_S \sum i \\ \sum \Phi_i & I \end{vmatrix}}{\begin{vmatrix} T_S^2 \sum i^2 & T_S \sum i\Phi_i \\ T_S \sum i & \sum i \end{vmatrix}} \quad \text{and} \quad \hat{A} = \frac{\begin{vmatrix} T_S^2 \sum i^2 & T_S \sum i\Phi_i \\ T_S \sum i & \sum \Phi_i \end{vmatrix}}{\begin{vmatrix} T_S^2 \sum i^2 & T_S \sum i\Phi_i \\ T_S \sum i & \sum i \end{vmatrix}} \tag{6.52}$$

Having estimated M and A, elimination of fine CFO is achieved by multiplying the received signal by $\exp[j(\hat{M}n + \hat{A})]$. CFO elimination is done in the packet level, so it is necessary to delay the received signal until linear interpolation is finished.

The above method is evaluated by calculating the RMS value of error vector magnitude. EVM is defined as the magnitude of the error vector on the constellation diagram. It is used because it is independent of the constellation scheme of the transmission. It is associated with SNR by the following equation:

$$\text{EVM} + \text{SNR} = 10 \log_{10} \left(\frac{N_u}{N} \right) \tag{6.53}$$

In this way, by calculating EVM from simulation results we can determine the equivalent SNR loss due to system nonidealities and algorithmic shortcomings. Figures 6.61 and 6.62 show the RMS EVM (in dB) as a function of CFO for packet lengths 20 and 10 with SNR as a parameter.

We note that for a relatively high SNR (SNR > 20 dB) the algorithm performs satisfactorily for CFOs up to 50 kHz. Beyond that limit, the induced ICI distorts phase estimation and the fine CFO algorithm fails. Furthermore, in Figure 6.63 it is demonstrated that with a packet length equal to 2, the algorithm fails completely due to insufficient data to perform averaging. However, no such small packet exists in Hiperlan/2.

Finally, Figure 6.64 gives the SNR loss as a function of CFO. It can be seen there that the loss is below 1 dB for ΔF_C lower than 45 kHz, whereas it is reduced to 0.5 dB when ΔF_C is lower than 20 kHz. Similarly, Figure 6.65 shows SNR loss vs CFO for packet length equal to 20. In this case the loss remains below 0.5 dB for ΔF_C lower than 45 kHz.

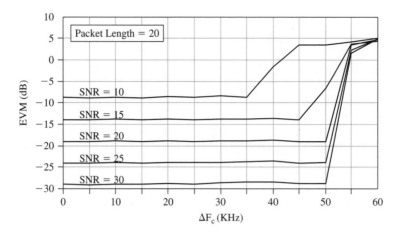

Figure 6.61 RMS of EVM due to CFO cancellation in H/2 for packet length = 20 [reprinted from A. Miaoudakis, A. Koukourgiannis, G. Kalivas, 'An all-digital feed-forward CFO cancellation scheme for Hiperlan/2 in multipath environment', IEEE Portable Indoor, Mobile and Radio Conference (PIMRC'02), Lisbon, September 2002, © 2002 IEEE]

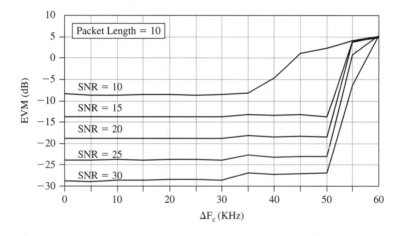

Figure 6.62 RMS of EVM due to CFO cancellation in H/2 for packet length = 10

6.3.4.3 Overall CFO Correction Subsystem

Phase locked loops are vulnerable to noise. In addition, in multipath fading they lose track and occasionally they get out of lock. PLL may have some advantages when concerning continuous transmission. However, for packet transmission like our application of WLAN, initial PLL acquisition is required in each frame, which is a waste of resources. For this reason, feed-forward techniques like the one we present exhibit superior performance.

The overall feed-forward CFO cancellation system is depicted in Figure 6.66. The first part constitutes the coarse CFO cancellation mechanism as presented in Section 6.3.4.1 whereas the

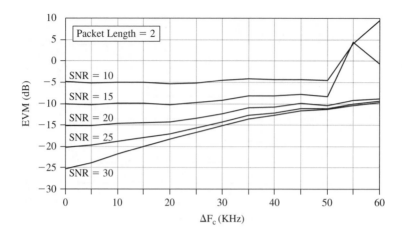

Figure 6.63 RMS of EVM due to CFO cancellation in H/2 for packet length $=2$ [reprinted from A. Miaoudakis *et al.*, 'An all-digital feed-forward CFO cancellation scheme for Hiperlan/2 in multipath environment', IEEE Portable Indoor, Mobile and Radio Conference (PIMRC'02), Lisbon, © 2002 IEEE]

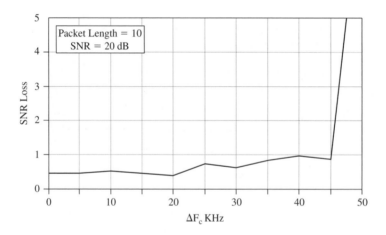

Figure 6.64 Impact of CFO cancellation mechanism to SNR loss in H/2 (packet length $= 10$)

second part receives a partially CFO-corrected signal and eliminates the remaining frequency offset (fine CFO cancellation), as described in detail in Section 6.3.4.2.

The coarse CFO mechanism is based on the preamble of the physical burst and consists of two parts: the estimation part and the correction part. The analysis of the technique according to which coarse CFO estimation takes place was given in 6.3.4.1. Although the analysis and simulation took into account broadcast physical burst, the same apply for other types of bursts as well. At the data link control level, ETSI defines [H/2-standard1] as a MAC frame of duration 2 ms. At least one estimation can be performed every 2 ms because, according to the standard, every MAC frame contains at least one broadcast physical burst.

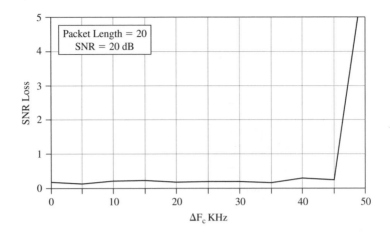

Figure 6.65 Impact of CFO cancellation mechanism to SNR loss in H/2 (packet length = 20)

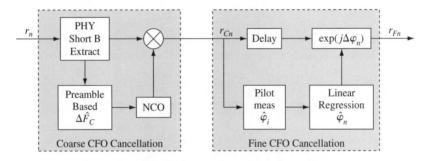

Figure 6.66 Overall CFO cancellation architecture

After CFO has been coarsely estimated it can be corrected using a numerically controlled oscillator, the output of which is digitally multiplied to the received burst in order to remove the coarse frequency offset. The output of the multiplier is applied to the fine CFO cancellation block, which is based to the Moose [Moose94] algorithm using the four pilots transmitted along with data symbols (see Section 6.3.3.2). In this case, the algorithm does not produce a CFO estimate but an estimate for the phase rotation (time domain) $\hat{\varphi}_n$, which according to Equation (6.37) is equal to $\hat{\varphi}_n = \angle \exp(j2\pi\Delta f_C nT)$ between n OFDM symbols. This correction is implemented by complex multiplication of the incoming coarsely corrected samples r_{Cn} by $\exp(-j\varphi_n)$.

6.3.5 Implementation and Simulation

In Figure 6.67 a block diagram of the overall system is illustrated. After downconversion the received data r_n enters the timing synchronization and CFO correction blocks. At the output, the cyclic prefix is removed and after serial to parallel conversion they enter the FFT. The two symbols of section 3 at the output of the FFT are used to perform averaged-LS channel estimation. The channel response coefficients are fed to the input of the FFT to equalize the

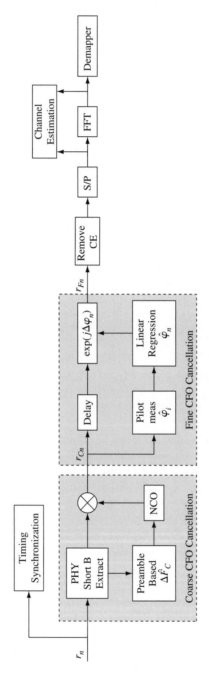

Figure 6.67 Overall OFDM baseband receiver for H/2

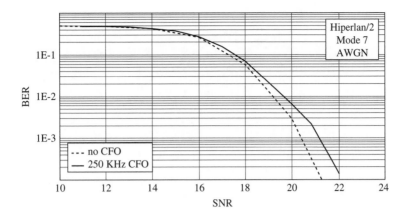

Figure 6.68 BER performance with and without CFO in AWGN [reprinted from A. Miaoudakis *et al.*, 'An all-digital feed-forward CFO cancellation scheme for Hiperlan/2 in multipath environment', IEEE Portable Indoor, Mobile and Radio Conference (PIMRC'02), Lisbon, © 2002 IEEE]

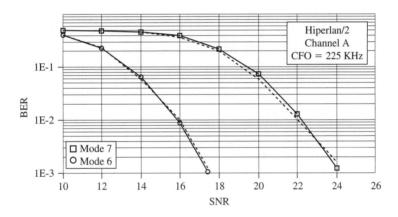

Figure 6.69 BER performance with and without CFO in ETSI-BRAN channel A [reprinted from A. Miaoudakis *et al.*, 'An all-digital feed-forward CFO cancellation scheme for Hiperlan/2 in multipath environment', IEEE Portable Indoor, Mobile and Radio Conference (PIMRC'02), Lisbon, © 2002 IEEE]

OFDM data symbols. At the output of the FFT block constellation data symbols are received which are demodulated through the demapper.

In Figure 6.68 the BER performance of the overall system for mode 7 in AWGN channel is presented for two cases. The dotted line gives the system performance when there is no CFO affecting it, whereas the solid line gives the BER performance when a CFO of 250 kHz is applied to the system. In the second case, the overall CFO correction subsystem is activated. We can see that the degradation is lower than 1 dB in the 10^{-3} BER range.

In Figure 6.69 the BER performance is illustrated for channel BRAN-A. In this case, the applied CFO is equal to 225 kHz. It can be seen that the CFO algorithm operates very satisfactorily, introducing almost no degradation. Figure 6.70 gives BER for the case of BRAN-E channel where the degradation is in the order of 1.5 dB.

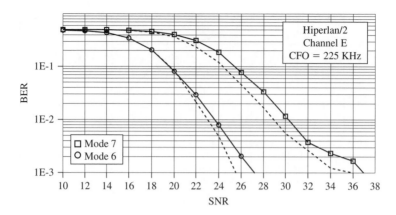

Figure 6.70 BER performance with and without CFO in ETSI-BRAN channel E [reprinted from A. Miaoudakis *et al.*, 'An all-digital feed-forward CFO cancellation scheme for Hiperlan/2 in multipath environment', IEEE Portable Indoor, Mobile and Radio Conference (PIMRC'02), Lisbon, © 2002 IEEE]

The resulting degradation in multipath channel transmission is smaller than expected. This can be explained by the CFO subsystem acting as an extra equalization mechanism, reducing the phase offset produced by the multipath channel only.

References

[802.11a-standard]: IEEE Std 802.11a-1999, 'Wireless LAN medium access control (MAC) and physical layer (PHY) specifications: high speed physical layer in the 5GHz band', 1999.

[Benvenuto97]: N. Benvenuto, A. Sallum, L. Tomba, 'Performance of digital DECT radio links based on semianalytical methods', IEEE J. Selected Areas Commun., vol. 15, May 1997, pp. 667–676.

[Chorevas96]: A. Chorevas, D. Reisis, E. Metaxakis, 'Implementation of a pulse shaping digital FIR filter for a 64-QAM modem', IEEE Int. Conference on Electronics, Circuits and Systems (ICECS), 1996.

[Chuang91]: J.C.I. Chuang, N.R. Sollenberger, 'Burst coherent demodulation with combined symbol timing, frequency offset estimation and diversity selection', IEEE Trans. Commun., vol. 39, July 1991, pp. 1157–1164.

[D'Luna99]: L. D'Luna, L. Tan, D. Mueller, J. Laskowski, K. Cameron, J.-Y. Lee, D. Gee, J. Monroe, H. Law, J. Chang, M. Wakayama, T. Kwan, C.-H. Lin, A. Buchwald, T. Kaylani, F Lu, T. Spieker, R. Hawley, H. Samueli, 'A single-chip universal cable set-top box/modem transceiver', IEEE J. Solid State Circuits, vol. 34, November 1999, pp. 1647–1660.

[DeBuda72]: R. de Buda, 'Coherent demodulation of frequency shift keying with low deviation ratio', IEEE Trans. Commun., vol. COM-20, June 1972, pp. 466–470.

[DMR90]: 'Digital Microwave Radio', Collected papers, IEEE Press, New York, 1990.

[Edfors98]: O. Edfors, M. Sandell, J.-J. van de Beek, S. Wilson, P. Borjesson, 'OFDM channel estimation by singular value decomposition', IEEE Trans. Commun., vol. 46, July 1998, pp. 931–939.

[ETSI-standard1]: ETS 300 175-1: 'Radio equipment and systems (RES); digital European cordless telecommunications (DECT) common interface part 1: overview'.

[ETSI-standard2]: ETS 300 175-2: 'Radio equipment and systems (RES); digital European cordless telecommunications (DECT) common interface part 2: physical layer'.

[H/2-standard1]: ETSI Technical Specification, 'Broadband radio access networks (BRANs); Hiperlan type 2; physical layer', Document ETSI TS 101 475 v1.1.1, April 2000.

[H/2-standard2]: ETSI Technical Specification, 'Broadband radio access networks (BRANs); Hiperlan type 2; data link control (DLC) layer; part 1: basic data transport funtions', Document ETSI TS 101 761-1, v1.1.1, April 2000.

[Hoher92]: P. Hoher, 'A statistical discrete-time model for the WSSUS multipath channel', IEEE Trans. Vehicle Technology, vol. 41, November 1992, pp. 461–468.

[Kalivas97]: G. Kalivas, E. Metaxakis, A. Tzimas, G. Koutsogiannopoulos, A. Kiranas, J. Papananos, D. Reisis, A. Chorevas, G. Korinthios, 'Design and implementation of IF 64-QAM modem', Deliverable D2.2.2, Microelectronic Products for Telecommunications Industry, 1997.

[Leclert83]: A. Leclert, P. Vandamme, 'Universal carrier recovery loop for QASK and PSK signal sets', IEEE Trans. Commun., vol. COM-31, January 1983, pp. 130–136.

[Mehlan93]: R. Mehlan, Y.-E. Chen, H. Meyr, 'A fully digital feedforward MSK demodulator with joint frequency offset and symbol timing estimation for burst mode mobile radio', IEEE Trans. Vehicle Technol., vol. 42, November 1993, pp. 434–444.

[Metaxakis98a]: E. Metaxakis, A. Tzimas, G. Kalivas, 'A low complexity baseband receiver for direct conversion DECT-based portable communication systems', IEEE Int. Conference on Universal Personal Communications (ICUPC) 1998, pp. 45–49.

[Metaxakis98b]: E. Metaxakis, A. Tzimas, G. Kalivas, 'A 64-QAM IF modem for digital microwave radio links', International Conference on Telecommunications (ICT), Chalkidiki, 1998.

[Miaoudakis02a]: A. Miaoudakis, A. Koukourgianis, G. Kalivas, 'Carrier frequency offset estimation and correction for Hiperlan/2 WLANs', Int. Conference on Computers and Communications (ISCC02), 2002.

[Miaoudakis02b]: A. Miaoudakis, A. Koukourgiannis, G. Kalivas, 'An all-digital feed-forward CFO cancellation scheme for Hiperlan/2 in multipath environment', IEEE Portable Indoor, Mobile and Radio Conference (PIMRC'02), Lisbon, September 2002.

[Miaoudakis04]: A. Miaoudakis, 'Methods for carrier frequency offset correction and study of non -ideal RF receiver in wireless OFDM systems', Ph.D. Dissertation, University of Patras, 2004.

[Molisch97]: A.F. Molisch, L.B. Lopez, M. Paier, J. Fuhl, E. Bonek, 'Error floor of unequalized wireless personal communications systems with MSK modulation and training-sequence-based adaptive sampling', IEEE Trans. Commun., vol. 45, May 1997, pp. 554–562.

[Moose94]: P. Moose, 'A technique for orthogonal frequency division multiplexing frequency offset correction', IEEE Trans. Commun., vol. 42, October 1994, pp. 908–914.

[Mueller76]: K. Mueller, M. Muller, 'Timing recovery in digital synchronous data receivers', IEEE Trans. Commun., vol. COM-24, May 1976, pp. 516–531.

[Muller-Weinfuntner98]: S. Muller-Weinfuntner, J. Rossler, J. Huber, 'Analysis of a frame-and frequency synchronizer for (bursty) OFDM', Proc. 7th Communication Theory Mini-Conference (CTMC at GLOBECOM'98), November 1998, Sydney, pp. 201–206.

[Murota81]: K. Murota, K. Hirade, 'GMSK modulation for digital mobile radio telephony', IEEE Trans. Commun., vol. 29, July 1981, pp. 1044–1050.

[Perakis99]: M. Perakis, A. Tzimas, E. Metaxakis, D. Soudris, G. Kalivas, C. Katis, C. Der, C. Goutis, A. Thanailakis, T. Stouraitis, 'The VLSI implementation of a baseband receiver for DECT-based portable applications', IEEE Int. Symposium on Circuits and Systems (ISCAS), 1999.

[Proakis02]: J. Proakis, M. Salehi, 'Communication Systems Engineering', 2nd edn, Prentice Hall, Englewood Cliffs, NJ, 2002.

[Schmidl97]: T. Schmidl, D. Cox, 'Robust frequency and timing synchronization for OFDM', IEEE Trans. Commun., vol. 45, December 1997, pp. 1613–1621.

[Tan98]: L. Tan, J. Putnam, F. Lu, L D'Luna, D. Mueller, K. Kindsfater, K. Cameron, R. Joshi, R. Hawley, H. Samueli, 'A 70 Mb/s variable-rate 1024-QAM cable receiver IC with integrated 10-b ADC and FEC decoder', IEEE J. Solid State Circuits, vol. 33, December 1998, pp. 2205–2218.

[Welborn00]: M. Welborn, 'Flexible signal processing algorithms for wireless communications', Ph.D. dissertation, MIT, June 2000.

[Yongacoglou88]: A. Yongacoglu, D. Makrakis, K. Feher, 'Differential detection of GMSK using decision feedback', IEEE Trans. Commun., vol. 36, June 1988, pp. 641–649.

[Zervas01]: N. Zervas, M. Perakis, D. Soudris, E. Metaxakis, A. Tzimas, G. Kalivas, C. Goutis, 'Low-power design of a digital baseband receiver for direct conversion DECT systems', IEEE Trans. Circuits and Systems II, vol. 48, December 2001, pp. 1121–1131.

Index